Supercritical Carbon Dioxide

in Polymer Reaction Engineering

Edited by
Maartje F. Kemmere and Thierry Meyer

WILEY-
VCH

WILEY-VCH Verlag GmbH & Co. KGaA

chem

0142454137

Chemistry Library

Editors:

Dr. ir. Maartje F. Kemmere
Process Development Group
Department of Chemical Engineering
and Chemistry
Eindhoven University of Technology
PO Box 513
5600 MB Eindhoven
The Netherlands

MER Dr. Thierry Meyer
Swiss Federal Institute of Technology
Institute of Chemical Science & Engineering
EPFL, ISIC-GPM
Station 6
1015 Lausanne
Switzerland

Library of Congress Card No.: applied for

British Library Cataloguing-in-Publication Data:
A catalogue record for this book is available from the British Library.

Die Deutsche Bibliothek – CIP Cataloguing-in-Publication Data: A catalogue record for this publication is available from Die Deutsche Bibliothek

© 2005 WILEY-VCH Verlag GmbH & Co. KGaA, Weinheim, Germany

Composition   K+V Fotosatz GmbH, Beerfelden
Printing   betz-druck GmbH, Darmstadt
Bookbinding   Litges & Dopf Buchbinderei GmbH, Heppenheim

Printed in the Federal Republic of Germany

Printed on acid-free and chlorine-free paper

ISBN-13: 978-3-527-31092-0
ISBN-10: 3-527-31092-4

Foreword

Supercritical fluid technology encompasses a very broad field, which includes various reaction, separation, and material formation processes that utilize a fluid at a temperature greater than its critical temperature and a pressure greater than its critical pressure. Supercritical fluids generally are compressed gases, which combine properties of gases and liquids in a chemically interesting manner. Supercritical fluids have physicochemical properties in between a liquid and a gas. They can have a liquid-like density and no surface tension while interacting with solid surfaces. They can have gas-like low viscosity and high diffusivity and, like a liquid, can easily dissolve many chemicals and polymers.

When Professor Thomas Andrews reported the measurement of the critical properties of carbon dioxide as part of his 1876 Bakerian Lecture "On the Gaseous State of Matter", he probably could not have envisaged that this important industrial gas would also become very popular in supercritical fluid technology. In fact carbon dioxide's popularity stems from the fact that it is nontoxic and nonflammable, it has a near ambient critical temperature of 31.1 °C, and that it is the second least expensive solvent after water. The most widespread use of supercritical carbon dioxide has been in Supercritical Fluid Extraction processes for the food and pharmaceutical industries with several large extraction units in operation in the United States and in Europe for decaffeinating coffee and tea and extracting flavors and essential oils from hops, spices, and herbs. Other applications have been reported in recrystallization of pharmaceuticals, purification of surfactants, cleaning and degreasing of products in the fabrication of printed circuit boards, and as a substitute for organic diluents in spray painting and coating processes.

The potential of supercritical carbon dioxide in polymer processes has been recently a focus of research and development both in academia and in industry. The main driver behind this effort is the chemical industry's pursuit of sustainable growth strategies, which aim to reduce the environmental footprint of existing or new polymer processes. The objective of the research and development effort has been to demonstrate whether carbon dioxide can be applied as an environmentally friendly substitute for many halogenated and other organic solvents used in polymer processes thereby reducing atmospheric pollution and eliminating solvent residues in products. Supercritical carbon dioxide could be most advantageously applied in developing improved polymer processes and

Supercritical Carbon Dioxide: in Polymer Reaction Engineering
Edited by Maartje F. Kemmere and Thierry Meyer
Copyright © 2005 WILEY-VCH Verlag GmbH & Co. KGaA, Weinheim
ISBN: 3-527-31092-4

products when environmental compliance pressures would require a process change, when regulatory requirements could require changes in product purity, and when improved products in terms of performance can result from substituting the traditional solvent with carbon dioxide.

This book edited by Professors M. Kemmere and Th. Meyer provides both academic researchers and industrial practitioners a thorough overview of the state of the art of the application of supercritical carbon dioxide in polymer processes by carefully balancing the exposition of recent research results and emerging commercial applications with the discussion of the special challenges and needs of this exciting new technology. Written mainly by prominent American and European academic researchers in the field, the book is comprised of three parts, which focus on the fundamentals aspects of this technology (thermodynamics, transport phenomena, and polymerization kinetics), and its application in polymerization reactions (including dispersion and emulsion systems as well as fluoropolymers synthesis) and polymer processing operations (including extrusion and reduction of residual monomer).

We hope that the publication of this book, which will surely become a standard reference in the field, will spur the interest in further exploring the potential of supercritical carbon dioxide applications in polymer technology both in terms of fundamental understanding of the relevant physico-chemical phenomena and in advancing the state of the design and commercialization of environmentally friendly polymer processes producing products with unique performance characteristics.

June, 2005

Harold L. Snyder
Technology Director
DuPont Fluoroproducts

John P. Congalidis
Senior Research Planning Associate
DuPont Central Research and Development

John R. Richards
Senior Research Associate
DuPont Engineering Research and Technology

E. I. du Pont de Nemours and Company
Wilmington, Delaware 19880, USA

Preface

The idea of producing a book on the application of supercritical carbon dioxide in polymer processes was born on a fine November evening in Barcelona during the meeting of the European Working Party on Polymer Reaction Engineering in 2002. As the idea still seemed reasonable the next morning, we decided to put words into action, and two years later the book was complete. From the outset, we were determined to give the manuscript a chemical engineering focus because of the increasing number of supercritical polymer processes on the verge of industrial application.

Our aim has been to present a state-of-the-art overview of polymer processes in high-pressure carbon dioxide using a multidisciplinary and synergetic approach that starts from fundamentals, goes through polymerization processes, and ends with post-processing. The contributors to this book are internationally recognized experts from different fields of CO_2-based polymer processes from Europe and the United States. We would like to express our gratitude to all the authors for the high quality of every contribution, and we are convinced that this compilation will become a reference book in the field.

Editing a book has resulted in strong links between Eindhoven and Lausanne, enabling us to adopt the good habits of both countries. In particular, the happy evenings spent with Francine, Jos, Morgane, and Quentin were a real pleasure, not only due to the presence of "tarte à la crème", "stroopwafels", "crème brulée", and too many chocolates, but also by the sealing of a strong friendship. This home support and understanding, also when we were traveling, certainly facilitated the editing process by introducing fun and fresh air into a hard job.

Furthermore, many thanks are due to our collaborators in the Process Development Group in Eindhoven and the Polymer Reaction Engineering Group in Lausanne for their creativeness and enthusiasm in the field of polymer science in supercritical carbon dioxide. Finally, we would like to thank Karin Sora and her team from Wiley-VCH for their great help in producing this book.

Eindhoven and Lausanne, July 2005

Maartje F. Kemmere
Thierry Meyer

Supercritical Carbon Dioxide: in Polymer Reaction Engineering
Edited by Maartje F. Kemmere and Thierry Meyer
Copyright © 2005 WILEY-VCH Verlag GmbH & Co. KGaA, Weinheim
ISBN: 3-527-31092-4

Contents

Foreword *V*

Preface *VII*

List of Contributors *XVII*

1 **Supercritical Carbon Dioxide for Sustainable Polymer Processes** *1*
Maartje Kemmere

1.1 Introduction *1*
1.2 Strategic Organic Solvent Replacement *3*
1.3 Physical and Chemical Properties of Supercritical CO_2 *5*
1.4 Interactions of Carbon Dioxide with Polymers and Monomers *8*
1.5 Concluding Remarks and Outlook *11*
Notation *12*
References *13*

2 **Phase Behavior of Polymer Systems in High-Pressure Carbon Dioxide** *15*
Gabriele Sadowski

2.1 Introduction *15*
2.2 General Phase Behavior in Polymer/Solvent Systems *15*
2.3 Polymer Solubility in CO_2 *19*
2.4 Thermodynamic Modeling *27*
2.5 Conclusions *32*
Notation *33*
References *34*

3   **Transport Properties of Supercritical Carbon Dioxide**   *37*
            Frederic Lavanchy, Eric Fourcade, Evert de Koeijer, Johan Wijers,
            Thierry Meyer, and Jos Keurentjes

3.1   Introduction   *37*
3.2   Hydrodynamics and Mixing   *39*
3.2.1   Laser-Doppler Velocimetry and Computational Fluid Dynamics   *39*
3.2.2   Flow Characteristics   *41*
3.3   Heat Transfer   *44*
3.3.1   Specific for Near-Critical Fluids: the Piston Effect   *45*
3.3.2   Reaction Calorimetry   *46*
3.3.3   Heat Transfer in Stirred Vessel with SCFs   *48*
3.4   Conclusions   *53*
            Notation   *53*
            References   *54*

4   **Kinetics of Free-Radical Polymerization in Homogeneous Phase
            of Supercritical Carbon Dioxide**   *55*
            Sabine Beuermann and Michael Buback

4.1   Introduction   *55*
4.2   Experimental   *57*
4.3   Initiation   *57*
4.4   Propagation   *62*
4.4.1   Propagation Rate Coefficients   *62*
4.4.2   Reactivity Ratios   *67*
4.5   Termination   *69*
4.6   Chain Transfer   *73*
4.7   Conclusions   *75*
            Notation   *76*
            References   *79*

5   **Monitoring Reactions in Supercritical Media**   *81*
            Thierry Meyer, Sophie Fortini, and Charalampos Mantelis

5.1   Introduction   *81*
5.2   On-line Analytical Methods Used in SCF   *82*
5.2.1   Spectroscopic Methods   *82*
5.2.1.1   FTIR   *82*
5.2.1.2   Raman Spectroscopy   *84*
5.2.1.3   UV/Vis   *85*
5.2.1.4   NMR   *87*
5.2.2   Reflectometry   *89*
5.2.3   Acoustic Methods   *89*
5.3   Calorimetric Methods   *90*
5.3.1   Power Compensation Calorimetry   *90*

5.3.2   Heat Flow Calorimetry   *91*
5.3.2.1   Heat Balance Equations   *92*
5.3.2.2   Determination of Physico-Chemical Parameters   *95*
5.3.2.3   Calorimeter Validation by Heat Generation Simulation   *96*
5.4   MMA Polymerization as an Example   *97*
5.4.1   Calorimetric Results   *97*
5.4.2   The Coupling of Calorimetry and On-Line Analysis   *100*
5.5   Conclusions   *101*
           Notation   *102*
           References   103

6   **Heterogeneous Polymerization in Supercritical Carbon Dioxide**   *105*
           Philipp A. Mueller, Barbara Bonavoglia, Giuseppe Storti,
           and Massimo Morbidelli

6.1   Introduction   *105*
6.2   Literature Review   *106*
6.3   Modeling of the Process   *108*
6.4   Case Study I: MMA Dispersion Polymerization   *115*
6.5   Case Study II: VDF Precipitation Polymerization   *124*
6.6   Concluding Remarks and Outlook   *132*
           Notation   *133*
           References   136

7   **Inverse Emulsion Polymerization in Carbon Dioxide**   *139*
           Eric J. Beckman

7.1   Introduction   *139*
7.2   Inverse Emulsion Polymerization in CO_2: Design Constraints   *141*
7.3   Surfactant Design for Inverse Emulsion Polymerization   *142*
7.3.1   Designing CO_2-philic Compounds: What Can We Learn
           from Fluoropolymer Behavior?   *143*
7.3.2   Non-Fluorous CO_2-Philes: the Role of Oxygen   *144*
7.4   Inverse Emulsion Polymerization in CO_2: Results   *148*
7.5   Future Challenges   *154*
           References   154

8   **Catalytic Polymerization of Olefins in Supercritical Carbon
           Dioxide**   *157*
           Maartje Kemmere, Tjerk J. de Vries, and Jos Keurentjes

8.1   Introduction   *157*
8.2   Phase Behavior of Polyolefin-Monomer-CO_2 Systems   *158*
8.2.1   Cloud-Point Measurements on the PEP-Ethylene-CO_2 System   *158*
8.2.2   SAFT Modeling of the PEP-Ethylene-CO_2 System   *161*
8.3   Catalyst System   *162*
8.3.1   Solubility of the Brookhart Catalyst in scCO_2 *163*

8.3.2   Copolymerization of Ethylene and Norbornene Using a Neutral
          Pd-Catalyst   165
8.3.3   Ring-Opening Metathesis Polymerization of Norbornene
          Using an MTO Catalyst   166
8.4   Polymerization of Olefins in Supercritical CO_2 Using Brookhart
          Catalyst   168
8.4.1   Catalytic Polymerization of 1-Hexene in Supercritical CO_2 168
8.4.2   Catalytic Polymerization of Ethylene in Supercritical CO_2 170
8.4.2.1   Experimental Procedure for Polymerization Experiments   170
8.4.2.2   Determination of Reaction Rate   171
8.4.2.3   Results of the Ethylene Polymerizations   173
8.4.2.4   Monitoring Reaction Rate Using SAFT-LKP and SAFT-PR   175
8.4.2.5   Topology of Synthesized Polyethylenes   177
8.4.3   Copolymerization of Ethylene and Methyl Acrylate
          in Supercritical CO_2 180
8.5   Concluding Remarks and Outlook   183
          Notation   185
          References   186

9   **Production of Fluoropolymers in Supercritical Carbon Dioxide**   189
          Colin D. Wood, Jason C. Yarbrough, George Roberts,
          and Joseph M. DeSimone

9.1   Introduction   189
9.2   Fluoroolefin Polymerization in CO_2 189
9.2.1   Overview   189
9.2.2   TFE-based Materials   192
9.2.3   Ionomer Resins and Nafion®   195
9.2.4   VF2-based Materials   195
9.2.5   VF2 and TFE Telomerization   196
9.3   Fluoroalkyl Acrylate Polymerizations in CO_2 197
9.4   Amphiphilic Poly(alkylacrylates)   199
9.5   Photooxidation of Fluoroolefins in Liquid CO_2 200
9.6 CO_2/Aqueous Hybrid Systems   202
9.7   Conclusions   202
          References   203

10   **Polymer Processing with Supercritical Fluids**   205
          Oliver S. Fleming and Sergei G. Kazarian

10.1   Introduction   205
10.2   Phase Behavior of CO_2/Polymer Systems and the Effect of CO_2
          on Polymers   206
10.2.1   Solubility of CO_2 in Polymers   206
10.2.2 CO_2-Induced Plasticization of Polymers   207
10.2.3 CO_2-Induced Crystallization of Polymers   208

10.2.4   Interfacial Tension in CO_2/Polymer Systems   *211*
10.2.5   Diffusion of CO_2 in Polymers and Solutes in Polymers
          Subjected to CO_2 *213*
10.2.6   Foaming   *215*
10.3   Rheology of Polymers Under High-Pressure CO_2 *218*
10.3.1   Methods for the Measurements of Polymer Viscosity
          Under High-Pressure CO_2 *218*
10.3.2   Viscosity of Polymer Melts Subjected to CO_2 *219*
10.3.3   Implications for Processing: Extrusion   *220*
10.4   Polymer Blends and CO_2 *222*
10.4.1 CO_2-Assisted Blending of Polymers   *222*
10.4.2 CO_2-Induced Phase Separation in Polymer Blends   *224*
10.4.3   Imaging of Polymeric Materials Subjected to High-Pressure CO_2 *226*
10.5   Supercritical Impregnation of Polymeric Materials   *228*
10.5.1   Dyeing of Polymeric Materials   *229*
10.5.2   Preparation of Materials for Optical Application   *230*
10.5.3   Preparation of Biomaterials and Pharmaceutical Formulations   *230*
10.6   Conclusions and Outlook   *232*
          Notation   *233*
          References   *234*

11   Synthesis of Advanced Materials Using Supercritical Fluids   *239*
          Andrew I. Cooper

11.1   Introduction   *239*
11.2   Polymer Synthesis   *239*
11.2.1   Reaction Pressure   *240*
11.2.2   Inexpensive Surfactants   *240*
11.3   Porous Materials   *243*
11.3.1   Porous Materials by SCF Processing   *243*
11.3.2   Porous Materials by Chemical Synthesis   *245*
11.4   Nanoscale Materials and Nanocomposites   *247*
11.4.1   Conformal Metal Films   *247*
11.4.2   Synthesis of Nanoparticles   *247*
11.4.3   Synthesis of Nanowires   *248*
11.5   Lithography and Microelectronics   *249*
11.5.1   Spin Coating and Resist Deposition   *249*
11.5.2   Lithographic Development and Photoresist Drying   *250*
11.5.3   Etching Using SCF Solvents   *250*
11.5.4   "Dry" Chemical Mechanical Planarization   *251*
11.6   Conclusion and Future Outlook   *251*
          References   *253*

12   **Polymer Extrusion with Supercritical Carbon Dioxide**   *255*
           Leon P. B. M. Janssen and Sameer P. Nalawade

12.1   Introduction   *255*
12.2   Practical Background on Extrusion   *256*
12.3   Supercritical CO_2-Assisted Extrusion   *257*
12.4   Mixing and Homogenization   *260*
12.4.1   Dissolution of Gas into Polymer Melt   *260*
12.4.2   Diffusion into the Polymer Melt   *261*
12.5   Applications   *262*
12.5.1   Polymer Blending   *262*
12.5.2   Microcellular Foaming   *265*
12.5.3   Particle Production   *268*
12.5.4   Reactive Extrusion   *269*
12.6   Concluding Remarks   *270*
           Notation   *271*
           References   *271*

13   **Chemical Modification of Polymers in Supercritical Carbon
           Dioxide**   *273*
           Jesse M. de Gooijer and Cor E. Koning

13.1   Introduction   *273*
13.2   Brief Review of the State of the Art   *275*
13.3   End-group Modification of Polyamide 6 in Supercritical CO_2 *277*
13.3.1   Background   *277*
13.3.1.1   Sorption and Diffusion   *278*
13.3.2   Amine End-Group Modification with Succinic Anhydride   *279*
13.3.2.1   Sorption Measurements   *281*
13.3.2.2   Melt Stability of Modified and Unmodified PA-6   *285*
13.3.2.3   Conclusions   *286*
13.3.3   Amine and Carboxylic Acid End-Group Modification
           with 1,2-Epoxybutane   *286*
13.3.3.1   Melt Stability of Modified and Unmodified PA-6   *289*
13.3.3.2   Conclusions   *289*
13.3.4   Amine End-group Modification with Diketene and Diketene
           Acetone Adduct   *289*
13.3.4.1   Modification of PA-6 Granules with Diketene and Diketene
           Acetone Adduct in Supercritical and Subcritical CO_2 *290*
13.3.4.2   Molecular Characterization   *291*
13.3.4.3   Conclusions   *292*
13.3.5   General Conclusions on Polyamide Modification   *292*
13.4   Carboxylic Acid End-group Modification of Poly(Butylene
           Terephthalate) with 1,2-Epoxybutane in Supercritical CO_2 *292*
13.4.1   Background   *292*
13.4.2   Chemical Modification of PBT with 1,2-Epoxybutane   *293*

13.4.2.1   Influence of Acid End-Group Concentration on Hydrolytic
           Stability   296
13.4.2.2   Determination of Molecular Weights   296
13.4.3   General Conclusions Concerning PBT Modification   297
13.5   Concluding Remarks and Outlook   297
           Notation   298
           References   299

14   **Reduction of Residual Monomer in Latex Products**
           Using High-Pressure Carbon Dioxide   303
           Maartje F. Kemmere, Marcus van Schilt, Marc Jacobs, and Jos Keurentjes

14.1   Introduction   303
14.2   Overview of Techniques for Reduction of Residual Monomer   304
14.2.1   Conversion of Residual Monomer   305
14.2.2   Removal of Residual Monomer   305
14.2.3   Alternative Technology: High-Pressure Carbon Dioxide   307
14.3   Enhanced Polymerization in High-Pressure Carbon Dioxide   307
14.3.1   Procedure for Pulsed Electron Beam Experiments   308
14.3.2   Results and Discussion   308
14.4   Extraction Capacity of Carbon Dioxide   310
14.4.1   Modeling Phase Behavior with the Peng-Robinson Equation
           of State   311
14.4.2   Procedure for Measuring Monomer Partition Coefficients   313
14.4.3   Validation of the Experimental Determination of Partition
           Coefficients   315
14.4.4   Measured Partition Coefficients of MMA over Water and CO_2 316
14.4.5   Prediction of Partition Coefficients of MMA over Water
           and CO_2 319
14.4.5.1   Modeling the Two-Component Systems CO_2-H_2O, MMA-CO_2
           and MMA-H_2O 320
14.4.5.2   Modeling the Three-Component System CO_2-H_2O-MMA   320
14.5   Process Design for the Removal of MMA from a PMMA Latex
           Using CO_2 323
14.5.1   Extraction Model   323
14.5.1.1   Diffusion and Mass Transfer Coefficients   324
14.5.1.2   Partition Coefficients   326
14.5.1.3   Interfacial Surface Areas   326
14.5.2   Process Flow Diagram, Equipment Selection, and Equipment
           Sizing   326
14.5.3   Economic Evaluation   328
14.6   Conclusion and Future Outlook   330
           Notation   330
           References   332

           Subject Index   335

List of Contributors

Prof. Eric J. Beckman
Chemical Engineering Department
University of Pittsburgh
Benedum Hall 1249
Pittsburgh, PA 15261
USA

Dr. Sabine Beuermann
Institute of Physical Chemistry
Georg-August University Göttingen
Tammannstrasse 6
37077 Göttingen
Germany

Barbara Bonavoglia
Institute for Chemistry
and Bioengineering
Group Morbidelli
Swiss Federal Institute of Technology
Zurich, ETHZ
ETH Hoenggerberg/HCI F125
8093 Zurich
Switzerland

Prof. Michael Buback
Institute of Physical Chemistry
Georg-August University Göttingen
Tammannstraße 6
37077 Göttingen
Germany

Prof. Andrew I. Cooper
Department of Chemistry
University of Liverpool
Liverpool
Merseyside, L69 3BX
UK

Dr. Jesse M. de Gooijer
Laboratory of Polymer Chemistry
Eindhoven University of Technology
PO Box 513
5600 MB Eindhoven
The Netherlands

Evert de Koeijer
Eindhoven University of Technology
Process Development Group
PO Box 513
5600 MB Eindhoven
The Netherlands

Prof. Joseph DeSimone
Department of Chemical Engineering
North Carolina State University
Raleigh, NC 27695
USA

Dr. Tjerk J. de Vries
Process Development Group
Eindhoven University of Technology
PO Box 513
5600 MB Eindhoven
The Netherlands

Supercritical Carbon Dioxide: in Polymer Reaction Engineering
Edited by Maartje F. Kemmere and Thierry Meyer
Copyright © 2005 WILEY-VCH Verlag GmbH & Co. KGaA, Weinheim
ISBN: 3-527-31092-4

Oliver S. Fleming
Department of Chemical Engineering
South Kensington Campus
Imperial College London
London, SW7 2AZ
UK

Sophie Fortini
Swiss Federal Institute of Technology
Institute of Chemical Sciences
& Engineering
EPFL, ISIC-GPM
Station 6
1015 Lausanne
Switzerland

Dr. Eric Fourcade
Eindhoven University of Technology
Process Development Group
PO Box 513
5600 MB Eindhoven
The Netherlands

Dr. Marc A. Jacobs
Process Development Group
Eindhoven University of Technology
PO Box 513
5600 MB Eindhoven
The Netherlands

Prof. Leon P. B. M. Janssen
Process Development Group
Department of Chemical Engineering
University of Groningen
Nijenborgh 4
9747 AG Groningen
The Netherlands

Dr. Sergei G. Kazarian
Department of Chemical Engineering
South Kensington Campus
Imperial College London
London, SW7 2AZ
UK

Dr. Maartje F. Kemmere
Process Development Group
Eindhoven University of Technology
PO Box 513
5600 MB Eindhoven
The Netherlands

Prof. Jos T. F. Keurentjes
Process Development Group
Eindhoven University of Technology
PO Box 513
5600 MB Eindhoven
The Netherlands

Prof. Cor E. Koning
Laboratory of Polymer Chemistry
Eindhoven University of Technology
PO Box 513
5600 MB Eindhoven
The Netherlands
and
Department of Physical and Colloidal
Chemistry
Free University of Brussels
1050 Brussels
Belgium

Dr. Frederic Lavanchy
Institute of Chemical Sciences and
Engineering
Swiss Federal Institute of Technology
EPFL, ISIC-GPM
Station 6
1015 Lausanne
Switzerland

Charalampos Mantelis
Swiss Federal Institute of Technology
Institute of Chemical Sciences
& Engineering
EPFL, ISIC-GPM
Station 6
1015 Lausanne
Switzerland

MER Dr. Thierry Meyer
Swiss Federal Institute of Technology
Institute of Chemical Sciences
& Engineering
EPFL, ISIC-GPM
Station 6
1015 Lausanne
Switzerland

Prof. Massimo Morbidelli
Institute for Chemistry
and Bioengineering
Group Morbidelli
Swiss Federal Institute of Technology
Zurich, ETHZ
ETH Hoenggerberg/HCI F125
8093 Zurich
Switzerland

Philipp A. Mueller
Institute for Chemistry
and Bioengineering
Group Morbidelli
Swiss Federal Institute of Technology
Zurich, ETHZ
ETH Hoenggerberg/HCI F125
8093 Zurich
Switzerland

Sameer P. Nalawade
Process Development Group
Department of Chemical Engineering
University of Groningen
Nijenborgh 4
9747 AG Groningen
The Netherlands

Prof. George Roberts
Department of Chemical Engineering
North Carolina State University
Raleigh, NC 27695
USA

Prof. Gabriele Sadowski
Department of Biochemical
and Chemical Engineering
Chair for Thermodynamics
University of Dortmund
Emil-Figge-Strasse 70
44227 Dortmund
Germany

Prof. Giuseppe Storti
Institute for Chemistry
and Bioengineering
Group Morbidelli
Swiss Federal Institute of Technology
Zurich, ETHZ
ETH Hoenggerberg/HCI F125
8093 Zurich
Switzerland

Marcus A. van Schilt
Process Development Group
Eindhoven University of Technology
PO Box 513
5600 MB Eindhoven
The Netherlands

Johan Wijers
Process Development Group
Eindhoven University of Technology
PO Box 513
5600 MB Eindhoven
The Netherlands

Colin Wood
Department of Chemistry
University of North Carolina
at Chapel Hill
Chapel Hill, NC 27599
USA

Dr. Jason C. Yarbrough
Department of Chemistry
University of North Carolina
at Chapel Hill
Chapel Hill, NC 27599
USA

1
Supercritical Carbon Dioxide for Sustainable Polymer Processes *

Maartje Kemmere

1.1
Introduction

Environmental and human safety concerns have become determining factors in chemical engineering and process development. Currently, there is a strong emphasis on the development of more sustainable processes, particularly in the polymer industry. Many conventional production routes involve an excessive use of organic solvents, either as a reaction medium in the polymerization step or as a processing medium for shaping, extraction, impregnation, or viscosity reduction. In each of these steps, most of the effort of the process is put into the solvent recovery, as schematically indicated in Fig. 1.1 for the polymerization step.

Illustrative examples include the production of butadiene rubber, with a product/solvent ratio of 1:6 [1], and the production of elastomers such as EPDM (ethylene-propylene-diene copolymer) in an excess of hexane [2]. Annually, these types of processes add substantially to the total emissions of volatile organic

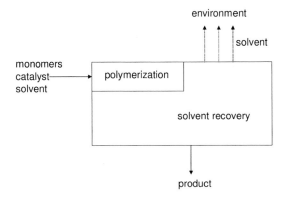

Fig. 1.1 Visualization of the relative effort required for polymerization and solvent recovery in conventional catalytic polymerization processes based on organic solvents.

---

* The symbols used in this chapter are listed at the end of the text, under "Notation".

Supercritical Carbon Dioxide: in Polymer Reaction Engineering
Edited by Maartje F. Kemmere and Thierry Meyer
Copyright © 2005 WILEY-VCH Verlag GmbH & Co. KGaA, Weinheim
ISBN: 3-527-31092-4

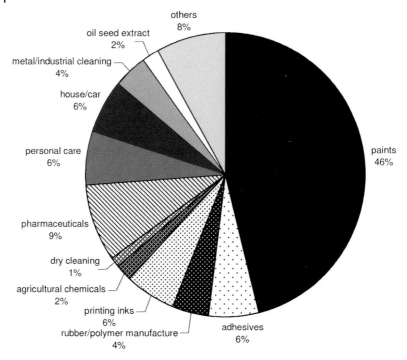

Fig. 1.2 Annual European solvent sales: 5 million tonnes, for which rubber and polymer manufacture accounts for 56% [4].

(VOCs). Approx. 20 million tonnes of VOCs are emitted into the atmosphere each year as a result of industrial activities [3]. According to Fig. 1.2, the annual European solvent sales to the rubber and polymer manufacturing industries, including polymer industries such as paints and adhesives, amount to 2.8 million tonnes.

Based on these facts, it is highly desirable from an environmental, safety, and economical point of view to develop alternative routes to reducing the use of organic solvents in polymer processes. Two obvious solutions to the organic solvents problem are the development of solvent-free processes and the replacement of solvents by environmentally benign products. Solvent-free polymerizations generally suffer from processing difficulties as a result of increased viscosities and mass transfer limitations, for instance in melt phase polymerization [5]. Solvent replacement, on the other hand, although it prevents the loss of dangerous organic solvents, still necessitates an energy-intensive solvent removal step. Using a "volatile" solvent makes the solvent removal step relatively easy. An intermediate solution is using one of the reactants in excess, as a result of which it partly acts as a solvent or plasticizer. In this case the excess of reactant still needs to be removed. Again, this becomes easier when the reactant involved is more volatile or, even better, gaseous.

Currently, the possibilities of green alternatives to replace organic solvents are being explored for a wide variety of chemical processes.

1.2
Strategic Organic Solvent Replacement

Solvents that have interesting potential as environmentally benign alternatives to organic solvents include water, ionic liquids, fluorous phases, and supercritical or dense phase fluids [5, 6]. Obviously, each of these approaches exhibits specific advantages and potential drawbacks. Ionic liquids (room-temperature molten organic salts), for example, have a vapor pressure that is negligible. Because they are non-volatile, commercial application would significantly reduce the VOC emission. In general, ionic liquids can be used in existing equipment at reasonable capital cost [7]. Nevertheless, the cost of a room-temperature molten salt is substantial. In addition, the separation of ionic liquids from a process stream is another important point of concern.

With respect to dense phase fluids, supercritical water has been shown to be a very effective reaction medium for oxidation reactions [8, 9]. Despite extensive research efforts, however, corrosion and investment costs form major challenges in these processes because of the rather extreme operation conditions required (above 647 K and 22.1 MPa) [10]. Still, several oxidation processes for waste water treatment in chemical industries are based on supercritical water technology (see, e.g., [11]).

In Table 1.1, the critical properties of some compounds which are commonly used as supercritical fluids are shown. Of these, carbon dioxide and water are the most frequently used in a wide range of applications. The production of polyethylene in supercritical propane is described in a loop reactor [13]. Supercritical ethylene and propylene are also applied, where they usually act both as a solvent and as the reacting monomer. In the field of polymer processing, the Dow Chemical Company has developed a process in which carbon dioxide is used to replace chlorofluorocarbon as the blowing agent in the manufacture of polystyrene foam sheet [14, 15].

Table 1.1 Critical conditions of several substances [12].

Solvent	T_c (K)	P_c (MPa)	Solvent	T_c (K)	P_c (MPa)
Acetone	508.1	4.70	Hexafluoroethane	293.0	3.06
Ammonia	405.6	11.3	Methane	190.4	4.60
Carbon dioxide	304.1	7.38	Methanol	512.6	8.09
Cyclohexane	553.5	4.07	n-hexane	507.5	3.01
Diethyl ether	466.7	3.64	Propane	369.8	4.25
Difluoromethane	351.6	5.83	Propylene	364.9	4.60
Difluoroethane	386.7	4.50	Sulfur hexafluoride	318.7	3.76
Dimethyl ether	400.0	5.24	Tetrafluoromethane	227.6	3.74
Ethane	305.3	4.87	Toluene	591.8	41.1
Ethylene	282.4	5.04	Trifluoromethane	299.3	4.86
Ethyne	308.3	6.14	Water	647.3	22.1

The interest in CO_2-based processes has strongly increased over the past decades. Fig. 1.3 shows the number of papers and patents that have been published over the years concerning polymerizations in supercritical carbon dioxide ($scCO_2$). In the last ten years, a substantial rise in publications can be observed, which illustrates the increasing interest in $scCO_2$ technology for polymer processes.

Carbon dioxide is considered to be an interesting alternative to most traditional solvents [17, 18] because of its practical physical and chemical properties: it is a solvent for monomers and a non-solvent for polymers, which allows for easy separation. To a somewhat lesser extent, it can also be a sustainable source of carbon [19]. The use of CO_2 as a reactant is considered to contribute to the solution of the depletion of fossil fuels and the sequestration of the greenhouse gas CO_2. One example in this area is the copolymerization of carbon dioxide with oxiranes to aliphatic polycarbonates [19–22].

Since *sustainability* is expected to become the common denominator of all polymer processes [23], it is important to consider this topic in relation to supercritical fluids, and $scCO_2$ in particular. To develop sustainable processes, process intensification is essential. The following requirements have been defined to be important for process intensification [24–26]:

- to match heat and mass transfer rates with the reaction rate,
- to enhance selectivity and specificity of reactions,
- to have no net consumption of auxiliary fluids,
- to achieve a high conversion of raw material,
- to improve product quality.

The present status of the sustainability of chemical processes in general has recently been reviewed [27]. Although there have been remarkable gains in energy effectiveness for the chemical industry both in Europe and the USA, it is a necessity to introduce sustainable development priorities in chemical engineering

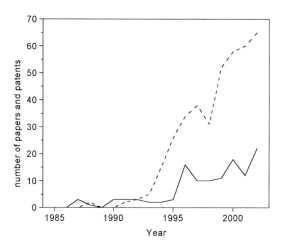

Fig. 1.3 Number of publications concerning polymerization in $scCO_2$; papers (dashed line), patents (solid line) [16].

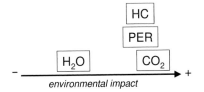

Fig. 1.4 Relative environmental impact of four dry cleaning technologies on a system level [35].

education in order to cope with future challenges. Moreover, new methodologies and design tools are being developed to implement the theme of sustainability in the conceptual process design of chemical process innovation, as illustrated in Fig. 1.4 [28].

Closely related to sustainability is the term *green chemistry*, which is defined as the utilization of a set of principles that reduces or eliminates the use or generation of hazardous substances in the design, manufacture, and applications of chemical products [6, 29, 30]. Life-cycle assessment (LCA) has been shown to be a useful tool to identify the more sustainable products and processes [31–33], including an environmental assessment of organic solvents as reported by Hellweg et al. [34]. The LCA-comparison of four dry cleaning technologies, i.e. based on perchloroethylene (PER), hydrocarbon (HC), wet-cleaning (H$_2$O), and liquid CO$_2$ [35], including a wide range of scientifically-based and known environmental impacts, forms an interesting case study. Based on the tendencies in the results, the wet-cleaning process does not look favorable as compared to the other three technologies (see Fig. 1.4). Various LCA studies emphasize that each specific process has to be considered individually, including analysis on energy consumption, emissions, material consumption, risk potential, and toxicity potential [33]. It is impossible to discuss in general whether polymer processes based on supercritical CO$_2$ can be sustainable or not.

Nevertheless, it is evident that the chemical process industry has to comply with regulatory issues and more stringent quality demands, which necessitates focusing on green chemistry and green engineering. Therefore, there is an increasing demand for innovative products and processes. In the past, polymer reaction engineering (PRE) was strongly based on engineering sciences. Currently, the focus is changing toward an integrated, multidisciplinary approach that is strongly driven by sustainability [36]. In the near future, a changeover will occur from technology-based PRE toward product-inspired PRE, for which it is expected that supercritical technology will play an important role [37].

1.3
Physical and Chemical Properties of Supercritical CO$_2$

In 1822, Baron Cagniard de la Tour discovered the critical point of a substance in his famous cannon barrel experiments [38]. Listening to discontinuities in the sound of a rolling flint ball in a sealed cannon, he observed the critical temperature. Above this temperature, the distinction between the liquid phase and the gas phase disappears, resulting in a single supercritical fluid phase behavior. In 1875, Andrews discovered the critical conditions of CO_2 [39]. The reported values were a critical temperature of 304.05 K and a critical pressure of 7.40 MPa, which are in close agreement with today's accepted values of 304.1 K and 7.38 MPa. In the early days, supercritical fluids were mainly used in extraction and chromatography applications. A well-known example of supercritical fluid extraction is caffeine extraction from tea and coffee [40]. Supercritical chromatography was frequently used to separate polar compounds [41, 42]. Nowadays, an increasing interest is being shown in supercritical fluid applications for reaction, catalysis, polymerization, polymer processing, and polymer modification [43]. More detailed historical overviews are given by Jessop and Leitner [12] and by McHugh and Krukonis [40].

A supercritical fluid is defined as a substance for which the temperature and pressure are above their critical values and which has a density close to or higher than its critical density [44–46]. Above the critical temperature, the vapor-liquid coexistence line no longer exists. Therefore, supercritical fluids can be regarded as "hybrid solvents" because the properties can be tuned from liquid-like to gas-like without crossing a phase boundary by simply changing the pressure or the temperature. Although this definition gives the boundary values of the supercritical state, it does not describe all the physical or thermodynamic properties. Baldyga [47] explains the supercritical state differently by stating that on a characteristic microscale of approximately 10–100 Å, statistical clusters of augmented density define the supercritical state, with a structure resembling that of liquids, surrounded by less dense and more chaotic regions of compressed gas. The number and dimensions of these clusters vary significantly with pressure and temperature, resulting in high compressibility near the critical point.

To illustrate the "hybrid" properties of supercritical fluids, Table 1.2 gives some characteristic values for density, viscosity, and diffusivity. The unique properties of supercritical fluids as compared to liquids and gases provide opportunities for a variety of industrial processes.

In Fig. 1.5, two projections of the phase behavior of carbon dioxide are shown. In the pressure-temperature phase diagram (Fig. 1.5 a), the boiling line is observed, which separates the vapor and liquid regions and ends in the critical point. At the critical point, the densities of the equilibrium liquid phase and the saturat-

Table 1.2 Comparison of typical values of physical properties of gases, supercritical fluids and liquids [48], where ρ, η and $\mathcal{D}$ stand for density, viscosity and diffusivity, respectively.

Properties	Gas	Supercritical fluid	Liquid
ρ (kg m^{-3})	1	100–800	1000
η (Pa s)	0.001	0.005–0.01	0.05–0.1
$\mathcal{D}$ (m^2 s^{-1})	$1 \cdot 10^{-5}$	$1 \cdot 10^{-7}$	$1 \cdot 10^{-9}$

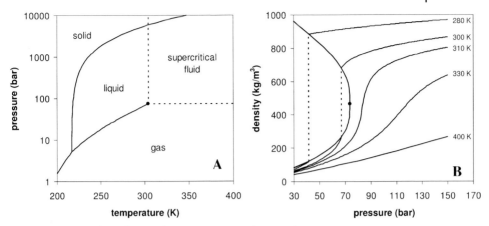

Fig. 1.5 Schematic phase diagram for a pure CO$_2$. (a) The critical point at the critical temperature, T_{cr} and the critical pressure, P_{cr}, marks the end of the vapor-liquid equilibrium line and the beginning of the supercritical fluid region. (b) Density of CO$_2$ as a function of pressure at different temperatures (solid lines) and at the vapor-liquid equilibrium line (dashed line) [44, 45].

ed vapor phases become equal, resulting in the formation of a single supercritical phase. This can be observed in the density-pressure phase diagram (Fig. 1.5 b). The transition from the supercritical state to liquid CO$_2$ is illustrated in Fig. 1.6.

In general, supercritical carbon dioxide can be regarded as a viable alternative solvent for polymer processes. Besides the obviously environmental benefits, supercritical carbon dioxide has also desirable physical and chemical properties

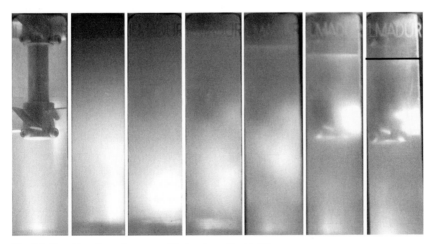

Fig. 1.6 Transition from the supercritical state to liquid CO$_2$.
The line indicates the liquid-vapor interface.

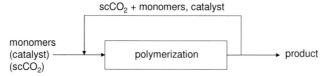

Fig. 1.7 Schematic view of a catalytic polymerization based on CO_2 technology, in which the catalyst and monomers can be recycled in a closed-loop process.

from a process point of view. These include its relatively chemical inertness, readily accessible critical point, excellent wetting characteristics, low viscosity, and highly tunable solvent behavior, facilitating easy separation. The use of such a "volatile" solvent makes the solvent removal step relatively easy. In principle, this allows for a closed-loop polymer process, in which the components like catalyst and monomers can be recycled. Fig. 1.7 schematically illustrates the efficiency of a CO_2-based polymerization as compared to a conventional process shown in Fig. 1.1.

Moreover, supercritical carbon dioxide is a non-toxic and non-flammable solvent with a low viscosity and high diffusion rate and no surface tension. A drawback of CO_2, however, is that only volatile or relatively non-polar compounds are soluble, as CO_2 is non-polar and has low polarizability and a low dielectric constant, as discussed in Section 1.4.

1.4
Interactions of Carbon Dioxide with Polymers and Monomers

For application of supercritical CO_2 as a medium in polymer processes, it is important to consider its interactions with polymers and monomers. In general, the thermodynamic properties of pure substances and mixtures of molecules are determined by intermolecular forces acting between the molecules or polymer segments. By examining these potentials between molecules in a mixture, insight into the solution behavior of the mixture can be obtained. The most commonly occurring interactions are dispersion, dipole-dipole, dipole-quadrupole, and quadrupole-quadrupole (Fig. 1.8).

For small molecules, the contribution of each interaction to the intermolecular potential energy $\Gamma_{ij}(r,T)$ is given by the polarizability a, the dipole moment μ, the quadruple moment Q, and in some cases specific interactions such as complex formation or hydrogen bonding [49]. The interactions work over different distances, with the longest range for dispersion and dipole interactions. Note that the dispersion interaction depends on the polarizability only and not on the temperature. Consequently, an increased polarizability of the supercritical solvent is expected to decrease the pressures needed to dissolve a nonpolar solute or polymer. Furthermore, at elevated temperatures, the configurational

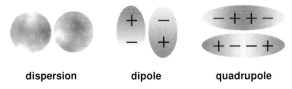

dispersion   **dipole**   **quadrupole**

Fig. 1.8 Charge distributions for various molecular interactions.

alignment of directional interactions as dipoles or quadrupoles is disrupted by the thermal energy, leading to a nonpolar behavior. Hence, it may be possible to dissolve a nonpolar solute or a polymer in a polar supercritical fluid. However, to obtain sufficient density for dissolving the solutes at these elevated temperatures, substantially higher pressures need to be applied. Additionally, specific interactions such as complex formation and hydrogen bonding can increase the solvent strength of the supercritical fluid. These interactions are also highly temperature sensitive.

The solvent strength of carbon dioxide for solutes is dominated by low polarizability and a strong quadrupole moment (Table 1.3). Consequently, carbon dioxide is difficult to compare to conventional solvents because of this ambivalent character. With its low polarizability and nonpolarity, carbon dioxide is similar to perfluoromethane, perfluoroethane, and methane.

In general, carbon dioxide is a reasonable solvent for small molecules, both polar and nonpolar. With the exception of water, for many compounds, including most common monomers, complete miscibility can be obtained at elevated pressures. However, the critical point of the mixture, i.e. the lowest pressure at a given temperature where CO_2 is still completely miscible, rises sharply with increasing molecule size. Consequently, most larger components and polymers exhibit very limited solubility in carbon dioxide. Polymers that do exhibit high solubility in carbon dioxide are typically characterized by a flexible backbone and high free volume (hence a low glass transition temperature T_g), weak interactions between the polymer segments, and a weakly basic interaction site such

Table 1.3 Physical properties of various solvents [48–50], where a is the polarizability, μ is the dipole moment and Q is the quadrupole moment.

Solvent	$a \cdot 10^{25}$ (cm^3)	μ (D)	$Q \times 10^{26}$ (erg$^{1/2}$cm$^{5/2}$)
Methane	26	0.0	
Ethane	45.0	0.0	−0.7
Ethyne	33.3	0.0	+3.0
Hexafluoroethane	47.6	0.0	−0.7
Carbon dioxide	27.6	0.0	−4.3
n-hexane	118.3	0.0	
Methanol	32.3	1.7	
Acetone	63.3	2.9	

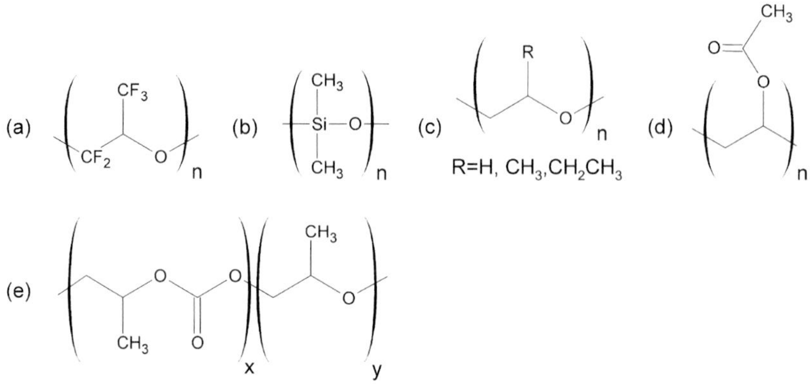

Fig. 1.9 Polymeric structures soluble in scCO$_2$. (a) Perfluoropoly(propylene oxide), (b) polydimethylsiloxane, (c) poly(ethylene, propylene and butylene oxide), (d) polyvinylacetate, (e) poly(ether carbonate).

as a carbonyl group [51–54]. Carbon dioxide-soluble polymers incorporating these characteristics include, e.g., polyalkene oxides, perfluorinated polypropylene oxide, polymethyl acrylate, polyvinyl acetate, polyalkyl siloxanes, and polyether carbonate (Fig. 1.9).

Although the solubility of polymers in CO$_2$ is typically very low, the solubility of carbon dioxide in many polymers is substantial. The sorption of carbon dioxide by the polymers and the resulting swelling of the polymer influence the mechanical and physical properties of the polymer. The most important effect is plasticization, i.e. the reduction of the T_g of glassy polymers. The plasticization effect, characterized by increased segmental and chain mobility as well as an increase in interchain distance, is largely determined by polymer–solvent interactions and solvent size [55]. The molecular weight of the polymer is of little influence on the swelling once the entanglement molecular weight has been exceeded.

The interaction of CO$_2$ and polymers can be divided into three application areas: processing of swollen or dissolved polymers and applications where carbon dioxide does not interact with the polymer. An extensive review on polymer processing using supercritical fluids has been written by Kazarian [55], including possible applications based on the specific interaction of CO$_2$ and the polymer system involved.

Obviously, the sorption and swelling of polymers by CO$_2$ are crucial effects in designing polymer processes based on high-pressure technology, because important properties such as diffusivity, viscosity, glass transition, melting point, compressibility, and expansion will change. The plasticization effect of CO$_2$ facilitates mass transfer properties of solutes into and out of the polymer phase, which leads to many applications: increased monomer diffusion for polymer synthesis, enhanced diffusion of small components in polymers for impregnation and extraction purposes, polymer fractionation, and polymer extrusion.

Another important requirement for the development of new polymer processes based on $scCO_2$ is knowledge about the phase behavior of the mixture involved, which enables the process variables to be tuned properly to achieve maximum process efficiency. Determining parameters in the phase behavior of a system are the solvent quality, the molecular weight, chain branching, and chemical architecture of the polymer, as well as the effect of endgroups and the addition of a cosolvent or an antisolvent. An overview of the available literature on the phase behavior of polymers in supercritical fluids has been published by Kirby and McHugh [50]. In addition, the possibilities of carbon dioxide as a medium for polymerization reactions and polymer processing have been reviewed [56–60].

1.5
Concluding Remarks and Outlook

A steady stream of emerging technologies has brought carbon dioxide all the way from a potential alternative solvent in the early 1970s to its use in industry [61]. The most promising applications of supercritical fluids are those in which their unusual properties can be exploited for manufacturing products with characteristics and specifications that are difficult to obtain by other processes.

Although there have been many interesting developments over the past twenty years, technical issues sometimes seem to hinder the progress of certain new processes toward commercialization [37, 62]. Applying carbon dioxide as a clean solvent in polymer processes is not the simplest route, because it involves, amongst others complications, high-pressure equipment, complex phase behavior, new measurement techniques, and the development of novel process concepts rather than extending conventional technologies. The development trajectory (see Fig. 1.10) from the concept idea via the laboratory bench and pilot scale to industrial implementation is often long. Currently, there exists a lack of facilities between laboratory scale research (5–500 mL) and the industrial scale application, mainly caused by the absence of pilot scale facilities. To break down the boundaries between the academic approach and industrial practice, close collaboration between industrial R & D, research institutes, and universities is essential to reduce costs, to exploit existing know-how and experimental facilities, and to reduce the development time. Bearing in mind the economics of an emerging technology as compared to long existing processes, it is a challenge to implement new process concepts at reasonable costs. For these reasons, the number of large-scale industrial polymer processes based on supercritical fluids will be limited in the short term. However, stimulation from government and research consortia should contribute substantially to the progress of development.

Several process design calculations [64, 65] have shown that polymer processes based on $scCO_2$ technology can be economically feasible, depending on the value of the product and the process conditions. Moreover, further developments will reduce costs of supercritical application substantially. It is expected that the major application of supercritical carbon dioxide will first be in the food

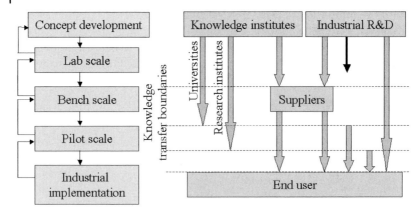

Fig. 1.10 Development trajectory of an emerging technology [63].

and pharmaceuticals industry because of additional marketing advantages, such as the GRAS (generally regarded as safe) status. However, the fact that DuPont is commercializing the production of fluoropolymers in $scCO_2$ [66] illustrates the application possibility of supercritical fluid technology in polymer processes also. In addition, the long-existing ldPE tubular process (ca. 250 MPa, 600 K) proves that a high-pressure polymerization process performed on a large scale can survive in a highly competitive field.

Nevertheless, the progress made in research today will enable the development of sustainable industrial polymer processes for the future. For this reason, the various subjects in this book have been addressed from an engineering point of view. The book is divided into three parts: an overview of polymer fundamentals, polymerization reactions, and polymer processing in supercritical carbon dioxide. It covers topics in a multidisciplinary approach starting in Part I with thermodynamics (Chapter 2), mass and heat transfer (Chapter 3), polymerization kinetics (Chapter 4), and monitoring (Chapter 5). In Part II, different types of polymerization processes (Chapters 6 to 9) will be discussed, and Part III describes the possibilities for polymer post-processing (Chapters 10 and 11), including reactive extrusion (Chapter 12), end group modification (Chapter 13), and residual monomer removal (Chapter 14).

Notation

P_c	critical pressure	[MPa]
Q	quadrupole moment	[erg$^{1/2}$ cm$^{5/2}$]
T_c	critical temperature	[K]
T_g	glass transition temperature	[K]
α	polarizability	[cm^3]
Γ_{ij}	potential energy	[J]

μ	dipole moment	[D]
ρ	density	[kg m^{-3}]
$\mathcal{D}$	diffusivity	[m^2 s^{-1}]
η	viscosity	[Pa s]
σ	surface tension	[N m^{-1}]

References

1 International Institute of Synthetic Rubber Producers, Inc., Internet publication, http://www.iisrp.com/WebPolymers/01FinalPolybutadieneVer2.pdf.

2 Condensed Course on Polymer Reaction Engineering, 2002, September 24–26, Eindhoven University of Technology, The Netherlands.

3 D. T. Allen, D. R. Shonnard, *Green Engineering, Environmentally conscious design of chemical processes*, Prentice Hall PTR, Upper Saddle River, 2002.

4 European Solvents Industry Group, http://www.esig.info/solvents_usesbenefits_details.php.

5 J. M. DeSimone, *Science*, 2002, *297*, 799.

6 P. T. Anastas, *Green chemistry as applied to solvents*, in *Clean solvents alternative media for chemical reactions and processing*, eds.: M. A. Abraham, L. Moens, ACS Symposium Series 819, Washington, 2002.

7 J. F. Brennecke, E. J. Maginn, *AIChE J.*, 2001, *47*(11), 2384.

8 M. Modell, US Patent 4338199, 1982.

9 T. B. Thomason, M. Modell, *Hazard Waste*, 1984, *1*(4), 453.

10 T. M. Hayward, I. M. Svishchev, R. C. Makhija, *J. Supercritical Fluids*, 2003, *27*, 275.

11 J. W. Griffith, D. H. Raymond, *Waste Management*, 2002, *22*, 453.

12 W. Leitner, P. G. Jessop, *Chemical synthesis using supercritical fluids*, Wiley-VCH, Weinheim, 1999.

13 A. Ahvenainen, K. Sarantila, H. Andtsjo, *WO Patent 92/12181*, 1992.

14 G. C. Welsh, US Patent 5266605, 1993.

15 G. C. Welsh, US Patent 5250577, 1993.

16 SciFinder Scolar 2002, http://www.cas.org/SCIFINDER/SCHOLAR/index.html.

17 E. J. Beckman, *J. Supercritical Fluids*, 2004, *28*, 121.

18 K. P. Johnston, R. M. Lemert, *Supercritical fluid technology: theory and applications*, in: *Encyclopedia of Chemical Processes and Design*, ed.: J. J. McKetta, Marcel Dekker, New York, 1996.

19 D. J. Darensbourg, M. W. Holtecamp, *Coordination Chemistry Reviews*, 1996, *153*, 155.

20 S. Inoue, H. Koinuma, T. Tsuruta, *J. Polym. Sci. B: Polymer Lett.*, 1969, *7*, 287.

21 H. Sugimoto, S. Inoue, *J. Polym. Sci. A*, 2004, *42*, 5561.

22 W. J. van Meerendonk, R. Duchateau, C. E. Koning, G. M. Gruter, *Macromolecular Rapid Commun.*, 2004, *25*, 382.

23 Th. Meyer, J. T. F. Keurentjes, *Polymer reaction engineering, an integrated approach*, in *Handbook for polymer reaction engineering*, eds.: Th. Meyer, J. T. F. Keurentjes, Wiley-VCH, Weinheim, 2005.

24 C. Ramshaw, *Chem. Eng.*, 1985, *415*, 30.

25 A. I. Stankiewicz, J. A. Moulijn, *Chem. Eng. Prog.*, 2000, *96*, 22.

26 R. Jachuck, *Process intensification for green chemistry*, in *Handbook of Green Chemistry & Technology*, eds.: J. Clark, D. Macquarrie, Blackwell Publishing, Oxford, 2002.

27 R. J. Batterham, *Chem. Eng. Sci.*, 2003, *58*, 2167.

28 G. Korevaar, *Sustainable chemical processes and products, new design methodology and design tools*, Ebrun Academic Publishers, Delft, 2004.

29 P. T. Anastas, J. C. Warner, *Green Chemistry, Theory and Practice*, Oxford University Press, Oxford, 2000.

30 M. Lancaster, *Principles of sustainable and green chemistry*, in *Handbook of Green Chemistry & Technology*, eds.: J. Clark, D. Macquarrie, Blackwell Publishing, Oxford, 2002.

31 M. Stewart, O. Jolliet, *Int. J. LCA*, **2004**, *9*, 153.

32 D. W. Pennington, J. Potting, G. Finnveden, E. Lindeijer, O. Jolliet, T. Rydberg, G. Rebitzer, *Environ. Int.*, **2004**, *30*, 721.

33 P. Saling, A. Kicherer, B. Dittrich-Krämer, R. Wittlinger, W. Zombik, I. Schmidt, W. Schrott, S. Schmidt, *Int. J. LCA*, **2002**, *4*, 203.

34 S. Hellweg, U. Fischer, M. Scheringer, K. Hungerbühler, *Green Chem.*, **2004**, *6*, 418.

35 P. H. Flückiger, *The use of life-cycle assessment and product risk assessment within application development of chemicals, a case study of perchloroethylene use in dry cleaning*, PhD thesis, Swiss Federal Institute of Technology Zürich, **1999**.

36 Th. Meyer, J. T. F. Keurentjes, *Trans. IChemE A*, **2004**, *82(A12)*, 1.

37 E. J. Beckman, *Ind. Eng. Chem. Res.*, **2003**, *42*, 1598.

38 C. Cagniard de la Tour, *Ann. Chim. Phys.*, **1822**, *22*, 127.

39 T. Andrews, *Proc. R. Soc.*, **1875**, *24*, 455.

40 M. A. McHugh, V. J. Krukonis, *Supercritical fluid extractions: principles and practice*, 2nd edn., Butterworth-Heinemann, Stoneham, MA, **1994**.

41 T. A. Berger, *J. Chromatogr. A*, **1997**, *785*, 3.

42 G. O. Cantrell, J. A. Blackwell, *J. Chromatogr. A*, **1997**, *782*, 237.

43 C. A. Eckert, B. L. Knutson, P. G. Debenedetti, *Nature*, **1996**, *383*, 313.

44 S. Angus, B. Armstrong, K. M. de Reuck, *International Thermodynamic Tables of the Fluid State. Carbon Dioxide*, Pergamon Press, Oxford, **1976**.

45 R. Span, W. Wagner, *J. Phys. Chem. Ref. Data*, **1996**, *25*, 1509.

46 J. A. Darr, M. Poliakoff, *Chem. Rev.*, **1999**, *99*, 495.

47 J. Baldyga, M. Henczka, B. Y. Shekunov, *Fluid dynamics, mass transfer and particle formation in supercritical fluids*, in Supercritical fluid technology for drug product development, eds.: P. York, U. B. Kompella, B. Shekunov, Marcel Dekker, New York, **2004**.

48 R. C. Reid, J. M. Prausnitz, B. E. Poling, *The Properties of Gases and Liquids*, 4th edn., McGraw-Hill, New York, **1987**.

49 J. M. Prausnitz, R. N. Lichtenthaler, E. G. de Azevedo, *Molecular Thermodynamics of Fluid Phase Equilibria*, 2nd edn., Prentice-Hall, Englewood Cliffs, **1986**.

50 C. F. Kirby, M. A. McHugh, *Chem. Rev.*, **1999**, *99*, 565.

51 F. Rindfleisch, T. P. DiNoia, M. A. McHugh, *J. Phys. Chem.*, **1996**, *100*, 15581.

52 T. Sarbu, T. Styranec, E. J. Beckman, *Ind. Eng. Chem. Res.*, **2000**, *39*, 1678.

53 T. Sarbu, T. Styranec, E. J. Beckman, *Nature*, **2000**, *405*, 165.

54 S. G. Kazarian, M. F. Vincent, F. V. Bright, C. L. Liotta, C. A. Eckert, *J. Am. Chem. Soc.*, **1996**, *118*, 1729.

55 S. G. Kazarian, *Polym. Sci., Ser. C*, **2000**, *42(1)*, 78.

56 K. M. Scholsky, *J. Supercrit. Fluids*, **1993**, *6*, 103.

57 K. A. Shaffer, J. M. DeSimone, *Trends Polym. Sci.*, **1995**, *3(5)*, 146.

58 J. L. Kendall, D. A. Canelas, J. L. Young, J. M. DeSimone, *Chem. Rev.*, **1999**, *99*, 543.

59 A. I. Cooper, *J. Mater. Chem.*, **2000**, *10*, 207.

60 M. F. Kemmere, *Recent developments in polymer processes*, in Handbook for polymer reaction engineering, eds.: Th. Meyer, J. T. F. Keurentjes, Wiley-VCH, Weinheim, **2005**.

61 D. Adam, *Nature*, **2000**, *407*, 938.

62 L. Chordia, R. Robey, *Proceedings of the 5th International Symposium on Supercritical Fluids*, Atlanta, **2000**.

63 J. M. K. Timmer, *Development of large scale applications (Workshop): Application of high-pressure carbon dioxide in polymer processes*, Eindhoven University of Technology, **2003**.

64 G. Brunner, *Nato Science Ser. E, Appl. Sci.*, **2000**, *366*, 517.

65 M. F. Kemmere, M. H. W. Cleven, M. A. van Schilt, J. T. F. Keurentjes, *Chem. Eng. Sci.*, **2002**, *57*, 3929.

66 G. Parkinson, *Chem. Eng.*, **1999**, *106*, May, 17.

2
Phase Behavior of Polymer Systems in High-Pressure Carbon Dioxide *

Gabriele Sadowski

2.1
Introduction

The investigation of phase behavior in polymer/supercritical fluid systems started with the development of the high-pressure polyethylene process, where LDPE has to be dissolved in supercritical ethylene. However, only in the last few decades has the phase behavior of polymers in other supercritical solvents, in particular carbon dioxide, attracted increasing research interest. The major reason is that most polymers are soluble in supercritical gases to only a very limited extent unless tremendous pressures are applied. Thus, even today the vision of using the unique solvent properties of supercritical fluids at moderate conditions in polymer processing still remains a challenge for polymer chemists and engineers.

Meanwhile, thermodynamics can provide a powerful tool for understanding the underlying phenomena and can thus help to develop a firm basis for the successful purification and application of supercritical solvents as polymer reaction media, as well as for the modification of the mechanical properties and morphology of polymers.

2.2
General Phase Behavior in Polymer/Solvent Systems

Polymers very often show only limited solubility in liquid or supercritical solvents. Moreover, solubility is not only a function of temperature, pressure, and concentration. For polymer systems, it also depends on the molecular weight and the molecular-weight distribution of the polymer. In the case of copolymers, it is moreover a function of the comonomer composition in the backbone.

---

* The symbols used in this chapter are listed at the end of the text, under "Notation".

Supercritical Carbon Dioxide: in Polymer Reaction Engineering
Edited by Maartje F. Kemmere and Thierry Meyer
Copyright © 2005 WILEY-VCH Verlag GmbH & Co. KGaA, Weinheim
ISBN: 3-527-31092-4

Fig. 2.1 a and b shows the phase behavior of a polymer/solvent system. At low temperatures, this typically demixes into two liquid phases (LL): one very dilute solvent-rich phase and the other a more concentrated polymer-rich phase.

In this region, increasing temperature leads to improved miscibility. Above the critical temperature (*Upper Critical Solution Temperature*; UCST) the system is at first completely miscible and forms a homogeneous liquid solution (L). However, polymer/solvent systems typically show a second region of liquid-liquid demixing at high temperatures. The reason is the so-called free-volume effect: at high temperatures, especially when approaching the critical temperature of the solvent, large differences in the thermal expansion of the polymer and the solvent are observed. Therefore, the density (reverse of "free volume") of the solvent decreases much more than that of the polymer. This leads to a separation of polymer and solvent molecules from each other and thereby reduces the solvent power. This effect becomes even more pronounced for increasing temperatures and thus leads to a demixing region that shows a *Lower Critical Solution Temperature* (LCST). Although some polymer/solvent mixtures do not show UCST behavior, the LCST demixing does typically occur in polymer/solvent systems because of the large differences in the thermal expansion coefficients.

Moreover, polymer solubility is strongly affected by the polymer molecular weight (Fig. 2.1 a). It is well known that, irrespective of the chemical structure, polymers of high molecular weight show much lower solubilities than those of lower molecular weight or oligomers. However, this effect decreases with increasing molecular weight and tends to vanish for polymers of molecular weight higher than about 100 kg/mol.

For polydisperse polymers, the solubility of a polymer is not only a function of the average molecular weight but also of the polydispersity. A polymer having a very broad molecular-weight distribution behaves qualitatively like a mixture of short and long polymer molecules. Whereas the longer molecules dissolve only very little, the shorter ones can act as co-solvents and thus enhance the solubility of the longer ones. Moreover, the phase equilibrium curves are no longer binodals (as in Fig. 2.1) but split into a cloud point curve, a shadow curve, and an infinite number of coexistence curves (for further details see, e.g., [1, 2]). Therefore, for polydisperse polymers (e.g., $M_w/M_n > 5$) the molecular-weight distribution of the polymer has also to be explicitly considered in the modeling (see, e.g., [3, 4]).

The influence of pressure on demixing is illustrated in Fig. 2.1 b. As expected for incompressible liquids, the UCST demixing is only very slightly influenced by pressure. However, the LCST demixing shows a much more pronounced pressure dependence. Here, the system pressure has a direct impact on the free-volume difference of solvent and polymer, which causes the demixing behavior in this region. Thus, in most cases, the polymer solubility can be improved by increasing the pressure in the system. This applies naturally in particular to systems with high differences in free volume, i.e. to mixtures where the system temperature is close to or even above the critical temperature of the solvent.

At that point it becomes obvious that from thermodynamic point of view there is no qualitative difference between the so-called "normal" (liquid) solvents

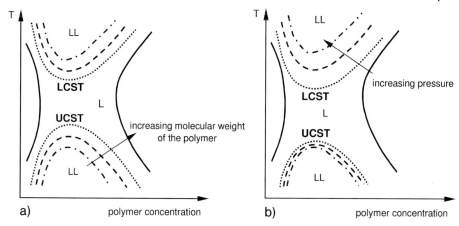

Fig. 2.1 Phase behavior of polymer-solvent systems as function of temperature and concentration: (a) influence of polymer molecular weight, (b) influence of pressure.

on the one hand and supercritical solvents, like carbon dioxide, on the other. In both cases, the solvent power is determined by the chemical nature and structure (implying enthalpic and entropic contributions) and by density (free-volume contribution).

The similarity of the phase behavior in liquids and supercritical solvents also becomes very evident from the p,T projection, which is often used for polymer/solvent systems.

Fig. 2.2 shows a typical p,T projection of a polymer/solvent system, where the solid lines for a given polymer concentration denote the transition from a homogeneous solution (L) to a demixed system (LL) and to a vapor-liquid system (VL), respectively.

The UCST branch depends only slightly on pressure and has a (mostly) negative slope. The LCST branch, which is much more pressure dependent, passes through a maximum and finally disembogues at the hypothetical critical point of the polymer. With increasing differences in chemical nature and size of polymer and solvent, the homogeneous region L becomes smaller and is shifted to higher pressures. Finally, UCST and LCST curves merge to give the so-called U-LCST behavior, which is typical for polymer/supercritical solvent mixtures. The experimentally accessible range of such a phase diagram depends on the particular temperature and pressure conditions for the system of interest. Typical windows for liquid systems as well as for supercritical solvents are marked in Fig. 2.2.

Fig. 2.3 gives examples of p,T projections for the systems polyethylene/ethylene (Fig. 2.3a) and poly(butyl methacrylate)/carbon dioxide (Fig. 2.3b).

In both cases, the cloud point curves were measured for different molecular weights of the polymers. In analogy to liquid solvents (Fig. 2.1a), shorter polymer molecules have better solubility in supercritical gases, such as ethylene and carbon dioxide, than larger ones, and thus dissolve at lower pressures.

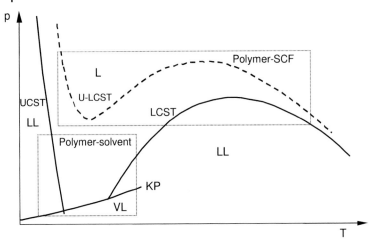

Fig. 2.2 Phase behavior of polymer-solvent systems as function of temperature and pressure. The lines indicate the two-phase boundaries at constant polymer concentration. Solid line is for normal solvents, dashed line indicates the behavior in supercritical fluids (SCF).

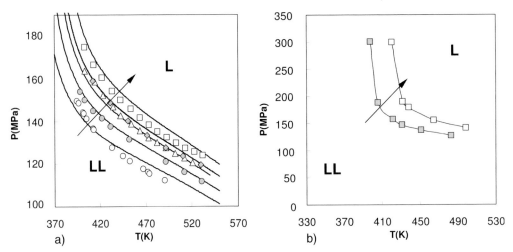

Fig. 2.3 Impact of the molecular weight on polymer solubility. Arrows indicate increasing molecular weight of the polymer:
(a) Polyethylene in supercritical ethylene. Symbols are experimental cloud point data: open squares 129 kg/mol, filled diamonds 58.3 kg/mol, open triangles 45.3 kg/mol, filled circles 30 kg/mol, open circles 19.3 kg/mol. Lines are predicted using the PC-SAFT model [5]. (b) Poly(butyl methacrylate) in supercritical carbon dioxide. Symbols are experimental cloud point data: open squares 320 kg/mol, filled squares 100 kg/mol [1].

2.3
Polymer Solubility in CO_2

As can be seen from Fig. 2.3, very high pressures are often needed to dissolve polymers in supercritical CO_2. This can partly be understood from the tremendous free-volume differences of polymer and CO_2 at low pressures and high temperatures, or, in other words, at low densities of CO_2. At high pressures, the CO_2 density is increased considerably, leading to an increase in solvent power. Secondly, as mentioned above, the mutual solubility is a question of intermolecular interactions, here in particular of the polymer and the CO_2. CO_2 does have a remarkable quadrupole moment, which substantially determines its solvent properties. Therefore, it can favorably interact with polar molecules but is, on the other hand, only a weak solvent for nonpolar polymers. Thus, CO_2 does not dissolve polyolefins particularly well unless their molecular weight is extremely low [6–8].

Much research has been done to determine how the solubility of polymers in CO_2 can be improved. One obvious way is to increase the polarity of the polymer (see, e.g., [6, 9–17]).

Rindfleisch et al. [6] determined the solubility of different poly(acrylates) in CO_2 (Fig. 2.4). With decreasing length of monomer units, from octadecyl acrylate to ethyl acrylate, their polarity increases. The dipole-quadrupole interactions between these groups and CO_2 promote the mutual solubility, which leads to a

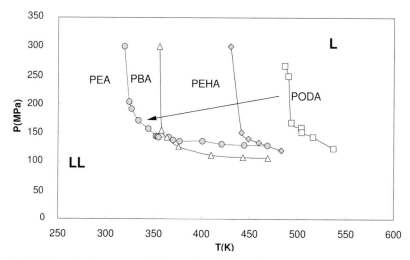

Fig. 2.4 Impact of monomer polarity on the solubility of various poly(acrylates). The arrow indicates increasing polarity of the acrylate group. Filled circles: poly(ethyl acrylate)(PEA), open triangles: poly(butyl acrylate)(PBA), filled diamonds: poly(ethyl hexyl acrylate) (PEHA), open squares: poly(octadecyl acrylate)(PODA). Experimental data from [6].

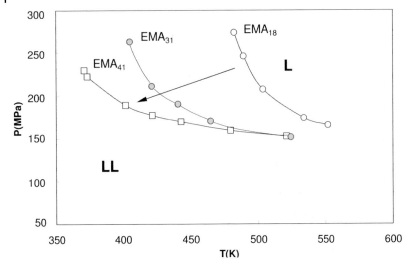

Fig. 2.5 Impact of the copolymer composition on the solubility
of ethylene/methylacrylate-copolymers (EMA) in supercritical carbon
dioxide. Subscripts indicate the amount of methylacrylate monomers
in the copolymer in mol%. Experimental data from [6].

growth of the homogeneous region by shifting the cloud point curves to lower
temperatures. However, in all cases, extremely high pressures are needed to dissolve the polymers.

Another possibility to increase the polarity of a polymer is the incorporation
of polar units into the polymer backbone via the synthesis of copolymers.
Fig. 2.5 shows the CO_2 solubility of poly(ethylene-co-methyl acrylate)s with varying amounts of the methyl acrylate monomers in the copolymer molecules. As
the methyl acrylate content increases, the favorable dipole-quadrupole interactions between the methyl acrylate units and the CO_2 lead to enhanced solubility
and shift the cloud point curves to lower temperatures and pressures.

However, the influence of polar comonomer units on polymer solubility is in
general neither linear nor necessarily monotonic. Fig. 2.6a shows the ethylene
solubility of poly(ethylene-co-methyl acrylate) copolymers for different amounts
of the methyl acrylate monomer in the copolymer from 0 mol% (corresponds to
LDPE) to 44 mol%. For small amounts of the methyl acrylate monomer, favorable interactions of the methyl acrylate units of the copolymer with the quadrupole moment of the ethylene enhance the solubility of the copolymer. Here, the
copolymers first show a decreasing cloud point pressure. However, upon further
increase of the methyl acrylate contents (above 13 mol%), the importance of the
polar intermolecular interactions between the different methyl acrylate units of
the copolymer molecules becomes dominant, leading to decreasing solubility.
However, for the similar system poly(ethylene-co-propyl acrylate), very different
behavior is observed. Here, the solubility of the copolymer increases with in-

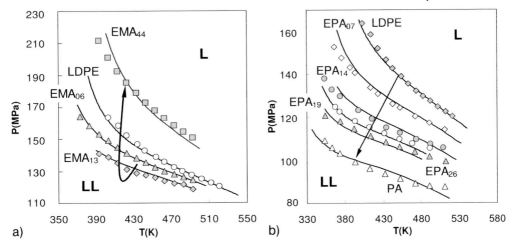

Fig. 2.6 Impact of the copolymer composition on the solubility in supercritical ethylene. Arrows indicate an increasing amount of the acrylate comonomer(s) in the polymer backbone. Symbols are experimental data. Lines are modeling results with the PC-SAFT model [18]. (a) Ethylene/methylacrylate-copolymers (EMA). Subscripts indicate the amount of methylacrylate monomers in the copolymer in mol% (LDPE = EMA$_{00}$). (b) Ethylene/propylacrylate-copolymers (EPA) (LDPE = EPA$_{00}$, PA = EPA$_{100}$).

creasing amounts of the acrylate monomers in the backbone over the whole range of copolymer compositions (Fig. 2.6 b). However, as observed in general, in this case also the copolymer solubility is a strongly non-linear function of the comonomer contents.

Finally, co-solvents are often used to enhance the polymer solubility in CO$_2$ (see numerous examples in Table 2.1). These are usually organic liquids that are completely soluble in CO$_2$ at moderate pressures and are also good solvents for the polymers considered. Adding a co-solvent can considerably decrease the pressures that are needed to dissolve a polymer in CO$_2$. However, it also means that this component has to be removed (laboriously) later in the process. Therefore, systems which, for different reasons, already contain volatile substances other than CO$_2$, e.g., as reactants or as comonomers, are of particular interest (see, e.g., [19–22]).

Conversely, because of its weak solvent properties, CO$_2$ can be considered as an antisolvent for polymer/solvent separations and polymer precipitations (see, e.g., [23–26]). Adding CO$_2$ to an initially homogeneous polymer solution usually reduces the overall solvent power and therewith causes a demixing into two liquid phases. Fig. 2.7 illustrates this phase behavior for the solubility of polypropylene in pentane. With increasing amounts of CO$_2$ in the system, the LCST demixing of the polypropylene/pentane solution is shifted to lower temperatures. At the same time, higher pressures are needed to keep the CO$_2$ dissolved in the liquid, and thus the cloud point pressures rise with increasing amounts of CO$_2$.

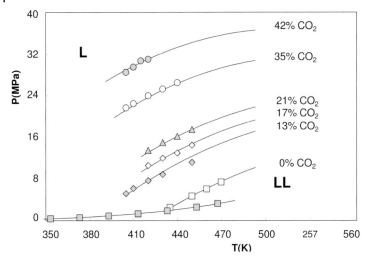

Fig. 2.7 The influence of added carbon dioxide on the solubility of polypropylene in pentane. Numbers indicate the wt% of carbon dioxide in the ternary system. Symbols are experimental data from [24]. Lines are modeling results using the PC-SAFT model [27].

Fig. 2.8 summarizes the impact of various polymer properties on the solubility in CO_2 and in supercritical solvents in general. Whereas an increase in molecular weight always causes a decrease in solubility and thereby leads to a shrinking of the homogeneous region (L), increasing branching and polydispersity of the polymer have a converse effect. Increasing polarity of the polymer mostly leads to improved solubility in CO_2 and ethylene. However, as indicated in Fig. 2.6, depending on the particular system, it might also have the opposite effect. Finally, for a given polymer/solvent system, the solubility can be significantly improved by adding a (liquid) cosolvent.

Much work has been done – especially during the last decade – on the measurement of polymer solubilities in CO_2, with the particular aim of influencing

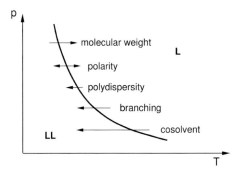

Fig. 2.8 Qualitative impact of various system properties on the (co)polymer solubility in carbon dioxide. The arrows indicate an increase of the corresponding property. The length of an arrow corresponds qualitatively with the strength of the property influence.

Table 2.1 Selected experimental data for co(polymer) solubilities in carbon dioxide (+ co-solvents).

Polymer	M (kg/mol)	Pressure range (MPa)	Temperature range (K)	Co-solvent	Reference
Poly(olefine)s					
Poly(ethylene)	420	20–80	380–480	Butane	[28]
Poly(ethylene)	125	0–23	383–503	*n*-Heptane	[26]
Poly(propylene)	50	5–32	400–450	Pentane	[24]
Poly(ethylene-co-propylene)	120	172–370	313–343	Ethylene	[29]
Poly(ethylene-co-propylene)	145	8–12	388–403	Hydrocarbon mixture	[30]
Poly(isobutylene)	0.2, 1	0–200	323–573		[7]
Poly(isobutylene)	1	0–200	323–573		[8]
Poly(isobutylene)	1000	0–25	323–493	*n*-Heptane	[26]
Poly(butadiene)	420	0–20	293–353	Tetrahydro-furane	[31]
Poly(butadiene)	5	0–20	293–353	Toluene	[31]
Poly(butadiene)	420	0–20	353–473	Cyclohexane	[26]
		0–24	323–473	Toluene	
Poly(styrene)					
Polystyrene	235	0–14	293–353	Toluene	[31]
Polystyrene	235	0–14	293–353	Tetrahydro-furane	[31]
Polystyrene	40–160	0–16	413–513	Cyclohexane	[23]
Poly([meth]acrylic)s					
Poly(methyl acrylate)	31	170–220	293–473		[6]
Poly(methyl acrylate)	1.4 –31	65–100	298–323		[51]
Poly(ethyl acrylate)	119	120–300	323–573		[6]
Poly(propyl acrylate)	140	120–150	373–353		[6]
Poly(butyl acrylate)	62	100–300	353–473		[6]
Poly(butyl acrylate)	62	50–120	300–573	Butyl acrylate	[19]
Poly(hexyl acrylate)	90	42–255	317–428	Hexyl acrylate	[22]
Poly(ethyl hexyl acrylate)	113	110–300	423–493		[6]
Poly(ethyl hexyl acrylate)	113	30–120	300–483	Ethyl hexyl acrylate	[19]
Poly(octadecyl acrylate)	23	100–260	483–533		[6]
Poly(octadecyl acrylate)	93	31–210	309–467	Octadecyl acrylate	[21]
Poly(methyl methacrylate)	120	0–20	293–373	Tetrahydro-furane	[31]
Poly(methyl methacrylate)	120	0–20	293–373	Toluene	[31]

Table 2.1 (continued)

Polymer	M (kg/mol)	Pressure range (MPa)	Temperature range (K)	Co-solvent	Reference
Poly(methyl methacrylate)	93	0–250	299–443	Methyl methacrylate	[20]
Poly(methyl methacrylate)	540	18	313	Acetone, ethanol, methylene chloride	[32]
Poly(ethyl methacrylate)	340	25–120	315–473	Ethyl methacrylate	[33]
Poly(butyl methacrylate)	320	6–202	313–485	Butyl methacrylate	[33]
Poly(hexyl methacrylate)	230	130–200	413–493		[15]
Poly(hexyl methacrylate)	400	50–220	334–473	Hexyl methacrylate	[22]
Poly(octyl methacrylate)	163	120–200	453–523		[15]
Poly(decyl methacrylate)	157	150–200	493–523		[15]
Poly(ethylene-co-methylacrylate)s	96–185	150–280	353–553		[6]
Poly(vinyl ester)s					
Poly(vinyl acetate)	125	50–100	303–423		[6]
Poly(vinyl acetate)	1–585	60–125	290–480		[51]
Poly(carbonate)s					
Poly(cyclohexene carbonate)	12–54	120–350	373–460	Cyclohexene oxide	[81]
Poly(ether)s					
Poly(propylene oxide)	2–3.5	90–135	323–343		[51]
Poly(ethylene glycol)	1–7.5	16	313	Ethanol	[34]
Poly(ethylene glycol)	7.5	16	313	Ethanol, toluene	[35]
Poly(ethylene glycol)	0.4	19–32	295		[36]
Poly(ethylene glycol)-diol	0.2–0.6	13–33	295		[36]
Poly(ethylene glycol)-mono-methylether	0.3–1	9–42	295		[36]
Poly(ethylene glycol)-di-methylether	∼ 0.6	13–26	295		[36]
Poly(ethylene glycol-co-propylene glycol)	1, 1.7	15–45	295		[36]
Poly(propylene glycol)	0.4	9–13	295		[36]

Table 2.1 (continued)

Polymer	M (kg/mol)	Pressure range (MPa)	Temperature range (K)	Co-solvent	Reference
Poly(propylene glycol)diol	0.4–2	6–25	295		[36]
Poly(propylene glycol)mono-methylether	1, 1.2	10–26	295		[36]
Poly(propylene glycol)mono-butyl ether	1	10–18	295–343		[36]
Poly(ethyl vinyl ether)	3.8	0–20	293–353	Toluene	[36]
Poly(ethyl vinyl ether)	1.5	28–36	295		[36]
Poly(ether-carbonate) copolymers	250 RU*	12–14	295		[13]
Poly(lactide)s					
Poly(lactide)	84–128	130–145	305–365		[37]
Poly(L-lactide)	2–100	0.5–76	318–373	Dichloro-methane	[38]
Poly(L-lactide)	2, 50, 100	3.6–71	303–373	Chlorodifluoro-methane	[39]
Poly(L-lactide)	2	3.6–71	305–393	Chlorodifluoro-methane	[40]
Poly(D,L-lactide)	30	2.5–72.5	303–373	Dimethyl ether	[41]
Poly(lactide-co-glycolide)	69–149	140–300	300–373		[37]
Poly(siloxane)s					
Poly(dimethyl-siloxane)	39–369	26–60	300–460		[52]
Poly(dimethyl-siloxane)	39, 94	28–52	323–424		[42]
Poly(dimethyl-siloxane)	2–486	15–75	298–373		[43]
Poly(dimethyl-siloxane-γ-propyl-acetate)	25 RU*	13.8–32.4	295		[10]
Poly(dimethylsiloxane) (hexyl functionalized)	25 RU*	13.8–32.4	295		[10]
Poly(dimethyl-siloxane)s (functionalized)	25 RU*	10–42	295		[17]
Poly(methylpropenoxy alkyl siloxane)	12	120–180	463–513		[15]
Poly(methylpropenoxy perfluoro alkyl siloxane)s	14.6, 17.7	10–40	298–383		[15]

Table 2.1 (continued)

Polymer	M (kg/mol)	Pressure range (MPa)	Temperature range (K)	Co-solvent	Reference
Fluoropolymers					
Teflon af	~ 400	50–100	323–453		[6]
Poly(vinyl fluoride)	125	0–220	398–523	Acetone, dimethyl ether, ethanol	[11]
Poly(vinylidene fluoride)	200	0–170	363–503	Acetone, dimethyl ether, ethanol	[11]
Poly(vinylidene fluoride-co-hexafluoro propylene)	85	50–230	0–503		[12]
Poly(vinylidene fluoride-co-hexafluoro propylene)	85	40–90	273–503		[44]
Poly(vinylidene fluoride-co-hexafluoro propylene)	85, 210	70–300	373–523		[6]
Poly(vinylidene fluoride-co-hexafluoro propylene)	210	100–300	458–518		[45]
Poly(tetrafluoroethy-lene-co-hexafluoro propylene)	190, 210	100–300	443–518		[46]
Fluorinated poly-(butadiene)	31–222	120–280	353–433		[14]
Fluorinated poly-(isoprene)	28–49	100–300	333–513		[14]
Poly(dihydroperfluoro octyl acrylate)	1200	14–27.5	303–353		[9]
Poly(dihydroperfluoro octyl acrylate)	1000	10–35	303–353		[47]
Poly(tetrahydro-perfluoro hexyl methacrylate)	200	30–60	313–403		[15]
Poly(tetrahydro-perfluoro octyl methacrylate)	292	30–50	203–403		[15]
Poly(tetrahydro perfluoro decyl acrylate)	–	5–27	283–507		[48]

Table 2.1 (continued)

Polymer	M (kg/mol)	Pressure range (MPa)	Temperature range (K)	Co-solvent	Reference
Poly(tetrahydro-perfluoro decyl methacrylate)	196	30–50	203–403		[15]
Poly(perfluoro-monoitaconates)	~ 150	14–54	293–352		[16]
Poly(perfluoro-diitaconates)	~ 150	8–40	293–424		[16]
Poly(perfluoro propylene oxide)	175 RU[a]	16–20	295		[13]
Hyperbranched polymers					
Hyperbranched poly(ester)	2, 5	1–17	332–370	Water, ethanol	[49]
Terpolymers					
Poly(styrene-co-methyl methyacry-late-co-glycidyl methacrylate)	4–5	4–80	310–393	Acetone	[50]

a) RU Repeat Units

and improving these solubilities. Table 2.1 is a summary of experimental studies to be found in the literature.

2.4
Thermodynamic Modeling

Thermodynamic modeling of the above-mentioned phase diagrams requires a
model that is able to account for the polymer chain-like structure, the polymer/
solvent interactions, and the influence of pressure on the phase behavior.
Whereas the first two issues can be at least qualitatively covered by using a lattice theory of the well-known Flory-Huggins type, such an approach is in general not able to describe the influence of pressure. Fulfillment of the third requirement requires a thermodynamic equation of state. Such a model naturally
accounts for density effects in a system.

There exist several approaches for the development of equations of state for
polymer systems. A possibility considered at an early stage was to extend the
Flory-Huggins theory by introducing holes into the lattice. Here, the number of
holes in the lattice is a measure of the system density. Equations of state based
on this idea are, for example, the Lattice-Fluid Theory (often called the Sanchez-

Lacombe model) [53] and the Mean-Field Lattice-Gas theory [54]. These two approaches were also successfully applied to polymer/carbon dioxide systems (see, e.g., [24, 28, 45, 52, 55, 56]). However, to achieve a quantitative description, a large set of parameters, most of them temperature dependent, has to be determined.

An alternative and very successful approach, which was pursued particularly over the last two decades, is the application of perturbation theories. The main assumption here is that the residual (difference from the ideal-gas state) part of the Helmholtz energy of a system A^{res} (and hence also the system pressure) can be written as sum of different terms. The main contribution is described by the Helmholtz energy of a chosen reference system A^{ref}. Contributions to the Helmholtz energy which are not covered by the reference system are considered as perturbations and are described by A^{pert}.

$$A^{res} = A - A^{id} = A^{ref} + A^{pert} \tag{1}$$

$$p = p^{ref} + p^{pert} \tag{2}$$

An appropriate reference system (at least for solvent molecules) is the hard-sphere (hs) system. Hard spheres are assumed to be spheres of a fixed diameter and not to have any attractive interactions. Such a reference system covers the repulsive interactions of the molecules, which are considered to mainly contribute to the thermodynamic properties. Moreover, for hard-sphere systems, analytical expressions for $A^{ref} = A^{hs}$ and $p^{ref} = p^{hs}$ are available (e.g., [57]).

Deviations of real molecules from the reference system may occur, e.g., due to attractive interactions (dispersion), non-spherical shape of the molecules (chain formation), and specific interactions (hydrogen bonding, dipole-dipole interactions). These contributions can be accounted for by using different perturbation terms. Depending on what kinds of perturbation are considered and which expressions are used for their description, different models based on perturbation theories have been developed in the literature.

The best-known model of this kind is the Statistical Associated Fluid Theory (SAFT) model [58–61]. Here, a non-spherical molecule (solvent or polymer) is assumed to be a chain of identical spherical segments. Starting from a reference system of m hard spheres (A^{hs}), this model considers three perturbation contributions, which are assumed to effect independently: attractive interactions of the (non-bonded) segments (A^{disp}), hard-sphere chain formation (A^{chain}), and association (A^{assoc}):

$$A^{res} = mA^{hs} + mA^{disp} + A^{chain} + A^{assoc} \tag{3}$$

The Carnahan-Starling formulation is used for A^{hs} and the segment-segment dispersion A^{disp} is described using a fourth-order perturbation term [62]; the contribution of chain formation as well as the association term is accounted for based on the work of Wertheim [63].

Subsequently, various perturbation theories were developed which are also based on Eq. (3) but differ in the use of specific expressions for the different types of perturbations. Examples are the Perturbed Hard-Sphere-Chain Theory (PHSC) [64], as well as the models proposed by Chang and Sandler [65], Gil-Villegas et al. [66], and Hino and Prausnitz [67].

Most of the perturbation theories require three pure-component parameters, which are physically meaningful: the number of segments (which is proportional to the molecular weight of a polymer), the size of the segments, and the energy related to the interaction of two segments. To describe a binary system, an additional binary parameter (k_{ij}) is used, which corrects for the deviations of solvent-polymer segment interactions from the geometric mean of those of the pure components. Applications of SAFT to various polymer/carbon dioxide systems can be found in the literature [7–9, 27, 29, 68, 69].

Although the above-mentioned perturbation theories account for the formation of chains in the repulsive contribution, the dispersion is still considered as resulting from the attraction of *unbonded* chain segments. This assumption is especially not justified in the case of polymer molecules where the segments do not interact independently but are influenced by the neighboring segments of the same molecule. Several attempts have been made to overcome this deficiency. Various models were suggested which use the square-well sphere (see, e.g., [66, 70, 71]) or the Lennard-Jones sphere (see, e.g., [71–74]) rather than the hard sphere as the reference to modify the chain contribution A^{chain}. However, the expressions finally obtained are lengthy, and thus these models were rarely used for engineering applications.

The recently proposed Perturbed-Chain SAFT (PC-SAFT) model [75, 76] adopts the opposite idea: here, a perturbation theory of second order is applied to the reference system of hard chains instead of hard spheres to develop a dispersion term A^{disp} for chain-like molecules:

$$A^{res} = mA^{hs} + A^{chain} + A^{disp}(m) + A^{assoc} \qquad (4)$$

This contribution now considers the attraction of chain molecules instead of that of unbonded segments and therefore becomes not only a function of reduced density η but also of chain length m:

$$\frac{A^{disp}(m)}{NkT} = -2\pi\rho \cdot m^2 \sigma^3 \left(\frac{\varepsilon}{kT}\right) I_1(m, \eta) - \pi\rho \, m \, kT \left(\frac{\partial\rho}{\partial p}\right)^{hc} m^2 \sigma^3 \left(\frac{\varepsilon}{kT}\right)^2 I_2(m, \eta)$$

$$(5)$$

where $I_1(m, \eta)$ and $I_2(m, \eta)$ are given by

$$I_1(\eta, m) = \sum_{i=0}^{6} a_i(m) \cdot \eta^i \quad \text{and} \quad I_2(\eta, m) = \sum_{i=0}^{6} b_i(m) \cdot \eta^i \qquad (6)$$

with

$$a_i(m) = a_{0i} + \frac{m-1}{m} a_{1i} + \frac{m-1}{m} \frac{m-2}{m} a_{2i} \; ;$$

$$b_i(m) = b_{0i} + \frac{m-1}{m} b_{1i} + \frac{m-1}{m} \frac{m-2}{m} b_{2i} \qquad (7)$$

The values of the a_{ki} and b_{ki} were determined from pure-component data of the n-alkane homologeous series and remain constant for all substances. Thus, the equation of state still requires three pure-component parameters just as much as the models mentioned earlier. However, from the physical point of view, the hard-chain system is a much better reference for a chain-like molecule than the hard-sphere system. Thus, the description of non-spherical molecules and in particular of polymer systems could be improved considerably.

For illustration, Fig. 2.9a and b give two examples which compare the results obtained by the original SAFT model with those obtained by PC-SAFT. As shown in Fig. 2.9a for the low-pressure vapor-liquid equilibrium in the polyethylene/toluene system, PC-SAFT gives superior results compared to the original SAFT and can predict the binary phase equilibrium without fitting binary parameters. It is even able to cover the correct molecular-weight dependence. A second example is illustrated in Fig. 2.9b, which shows the high-pressure carbon dioxide solubility in polyethylene. Whereas the original SAFT model is not

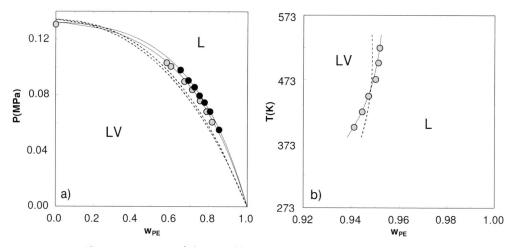

Fig. 2.9 Comparison of phase equilibrium calculations using SAFT (dashed lines) and PC-SAFT (solid lines) [76]. (a) Vapor-liquid phase equilibrium of polyethylene-toluene at $T=393$ K. Filled symbols are experimental data for polymer molecular weight of 6.2 kg/mol, open symbols are for 1.7 kg/mol. (b) Solubility of carbon dioxide in polyethylene ($M_n=87$ kg/mol) at $P=$ 9 MPa. Binary parameters were fitted for correlation: SAFT ($k_{ij}=0.242$) and PC-SAFT ($k_{ij}=0.181$).

able to describe the experimental data, the modeling with PC-SAFT even leads to quantitative results (one binary parameter was fitted in each case).

The PC-SAFT model was successfully applied to describe a whole variety of polymer solubilities in liquids as well as in supercritical solvents [18, 27, 76–78, 81]. The example in Fig. 2.10 illustrates the modeling results for the solubility of poly(methyl acrylate) and poly(methyl acetate) in supercritical carbon dioxide. Although the two components show great similarities from the chemical point of view, these polymers exhibit very different CO_2 solubility. Although the molecular weight of poly(methyl acrylate) (PMA) is much smaller than that of the considered poly(vinyl acetate) (PVA), much higher pressures are needed to dissolve the PMA. Moreover, the two systems show different slopes of the cloud point curves: whereas the solubility of PMA is improved at high temperatures, the PVA solubility decreases in the same temperature range. Using PC-SAFT, it was possible to model the very different phase behavior in the two systems without temperature-dependent parameters. This implies that the model could qualitatively predict the different temperature dependence of solubility just by using pure-component data. The binary parameters, which were fitted to each of the curves ($k_{\text{PMA–CO}_2}=0.052$ and $k_{\text{PVA–CO}_2}=0.04$), were only used to improve the quantitative description of the experimental solubilities.

In general, one of the most desirable capabilities of a thermodynamic model is to predict, rather than only to correlate, the impact of different substances and system properties on the solubility. Given a minimum of experimental data, the model should give an (at least qualitative) impression of how the phase behavior can be modified.

Thus, Fig. 2.3a gives an example of the ability of PC-SAFT to predict the molecular-weight dependence of polymer-solubility data. Although the binary parameter

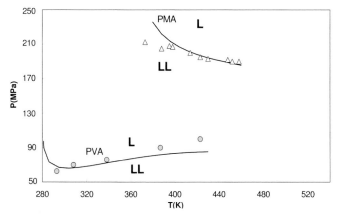

Fig. 2.10 Solubility of poly(vinyl acetate) (PVA; M_w=125 kg/mol) and poly(methyl acrylate) (PMA; M_w=31 kg/mol) in supercritical carbon dioxide. Symbols are experimental data (Rindfleisch et al. 1996). Lines are modeling results using the PC-SAFT model [79].

for the polymer/solvent system was fitted to data of one particular molecular weight only, the impact of molecular weight on polymer solubility could be predicted over a broad range of molecular weights, giving good agreement with experimental data.

A common approach to changing the solubility of a polymer is by modifying it by copolymerization. To get a first evaluation of promising comonomers and their impact on copolymer solubility, thermodynamic modeling can provide a powerful tool to predict the phase behavior of those systems. Figure 2.6a and b illustrate the modeling results for copolymer systems as obtained by the application of the PC-SAFT model. The lines in the two figures represent modeling results obtained for different copolymers that varied in comonomer composition. Using the pure-component information about the solvent, the corresponding homopolymers, and the homopolymer solubilities in the solvent, PC-SAFT is able to model the copolymer/solvent phase behavior over a wide range of comonomer compositions. The only parameter here that was fitted to the copolymer data is the binary parameter that describes the interactions between the unlike segments, which are not present in the homopolymer systems. Although this parameter, as well as all the other parameters, are just constants, the model could even predict that the solubility is a non-monotonous function of the acrylate content in the case of poly(ethylene-co-methyl acrylate) but a monotonous function in the case of poly(ethylene-co-propyl acrylate).

Finally, Fig. 2.7 shows, as an example, the modeling results obtained for the influence of the co-solvent/CO_2 ratio on the solubility of poly(propylene). Using parameters fitted only to pure-component and binary-system data, the phase behavior in the ternary system could be predicted, and were in good agreement with the experimental data.

2.5
Conclusions

Polymer supercritical-fluid systems show complex phase behavior (for a general overview see also [80]). Polymer solubility in these systems depends on (apart from temperature, pressure, and concentration) the chemical nature, molecular weight, and molecular-weight distribution of the polymer and on the comonomer composition in the case of copolymers.

The most interesting supercritical solvent is CO_2. However, in most cases the solubility of polymers in supercritical CO_2 is very limited, and great efforts have therefore been made to discover experimentally under which conditions the solubility of polymers could be increased.

Meanwhile, thermodynamic modeling has improved considerably, especially during the last decade. State-of-the art models were developed based on a molecular, physically meaningful approach. Based on a limited number of experiments, these models are not only able to describe but to a certain extent also to extrapolate or even predict the phase behavior in polymer systems. Thus, their application can considerably reduce the number of experiments needed and can support the understanding and development of supercritical-fluid applications in polymer processing.

Notation

Symbols

A	Helmholtz Energy
a_{ji}	model constants of the PC-SAFT equation of state
b_{ji}	model constants of the PC-SAFT equation of state
k	Boltzmann's constant
k_{ij}	binary interaction parameter
m	segment number
M_n	number average
M_w	weight average
N	total number of moles
P	pressure

Greek

ε	depth of the pair potential
η	reduced density
ρ	number density
σ	segment diameter

Superscripts

assoc	association
disp	dispersion
hs	hard sphere
id	ideal gas
pert	perturbation
ref	reference system
res	residual (deviation from ideal-gas state)

Abbreviations

L	Liquid
LDPE	Low Density Poly(ethylene)
LCST	Lower Critical Solution Temperature
LL	Liquid-Liquid System
PMA	poly(methyl acrylate)
SAFT	Statistical Associated Fluid Theory
UCST	Upper Critical Solution Temperature
VL	Vapor-Liquid System

References

1 R. Koningsveld, A. J. Staverman, *J. Polym. Sci. Part A-2* **1968**, *6*, 305.

2 R. Koningsveld, *Disc. Faraday Soc.* **1970**, *49*, 144.

3 T. Tork, G. Sadowski, W. Arlt, A. de Haan, G. Krooshof, *Fluid Phase Equilib.* **1999**, *163*, 79.

4 S. Behme, G. Sadowski, Y. Song, C.-C. Chen, *AIChE J.* **2003**, *49*, 258.

5 F. Becker, M. Buback, H. Latz, G. Sadowski, F. Tumakaka, Annual AIChE Meeting, Indianapolis, November **2002**.

6 F. Rindfleisch, T. P. DiNoia, M. A. McHugh, *J. Phys. Chem.* **1996**, *100*, 15581.

7 C. J. Gregg, F. P. Stein, M. Radosz, *Macromolecules* **1994**, *27*, 4972.

8 C. J. Gregg, F. P. Stein, M. Radosz, *Macromolecules* **1994**, *27*, 4981.

9 G. Luna-Barcenas, S. Mawson, S. Takishima, J. M. DeSimone, I. C. Sanchez, K. P. Johnston, *Fluid Phase Equilib.* **1998**, *146*, 325.

10 R. Fink, D. Hancu, R. Valentine, E. J. Beckman, *J. Phys. Chem. B* **1999**, *103*, 6441.

11 M. Lora, J. S. Lim, M. A. McHugh, *J. Phys. Chem. B* **1999**, *103*, 2818.

12 T. P. DiNoia, S. E. Conway, J. S. Lim, M. A. McHugh, *J. Polym. Sci.* **2000**, *38*, 2832.

13 T. Sarbu, T. Styranec, E. J. Beckman, *Nature* **2000**, *405*, 165.

14 M. A. McHugh, I.-H. Park, J. J. Reisinger, Y. Ren, T. P. Lodge, M. A. Hillmeyer, *Macromolecules* **2002**, *35*, 4653.

15 M. A. McHugh, A. Garach-Domech, I.-H. Park, D. Li, E. Barbu, P. Graham, J. Tsibouklis, *Macromolecules* **2002**, *35*, 6479.

16 D. Li, Z. Shen, M. A. McHugh, J. Tsibouklis, E. Barbu, *IEC Res.* **2003**, *42*, 6499.

17 S. Kilic, S. Michalik, Y. Wang, J. K. Johnson, R. M. Enick, E. J. Beckman, *IEC Res.* **2003**, *42*, 6415.

18 F. Becker, M. Buback, H. Latz, G. Sadowski, F. Tumakaka, *Fluid Phase Equilib.* **2004**, *215*, 263.

19 M. A. McHugh, F. Rindfleisch, P. T. Kuntz, C. Schmaltz, M. Buback, *Polymer* **1998**, *39*, 6049.

20 M. Lora, M. A. McHugh, *Fluid Phase Equilib.* **1999**, *157*, 285.

21 H.-S. Byun, T.-H. Choi, *J. Appl. Polym. Sci.* **2002**, *86*, 372.

22 H.-S. Byun, J.-G. Kim, J.-S. Yang, *IEC Res.* **2004**, *43*, 1543.

23 B. Bungert, G. Sadowski, W. Arlt, *Fluid Phase Equilib.* **1997**, *139*, 349.

24 T. M. Martin, A. A. Lateef, C. B. Roberts, *Fluid Phase Equilib.* **1999**, *154*, 241.

25 D. Li, Z. Liu, G. Yang, B. Han, H. Yan, *Polymer* **2000**, *41*, 5707.

26 S. N. Joung, J.-U. Park, S. Y. Kim, K.-P. Yoo, *J. Chem. Eng. Data* **2002**, *47*, 270.

27 J. Gross, G. Sadowski, Perturbed-Chain-SAFT: development of a new equation of state for simple, associating, multipolar and polymeric compounds. In: *Supercritical Fluids as Solvents and Reaction Media*, edited by G. Brunner, Elsevier, Amsterdam, **2004**.

28 Y. Xiong, E. Kiran, *J. Appl. Polym. Sci.* **1994**, *53*, 1179.

29 T. J. DeVries, P. J. A. Somers, T. W. deLoos, M. A. G. Vorstman, J. T. F. Keurentjes, *IEC Res.* **2000**, *39*, 4510.

30 M. A. McHugh, T. L. Guckes, *Macromolecules* **1985**, *18*, 674.

31 A. A. Kiamos, M. D. Donohue, *Macromolecules* **1994**, *27*, 357.

32 C. Domingo, A. Vega, M. A. Fonovich, C. Elvira, P. Subra, *J. Appl. Polym. Sci.* **2003**, *90*, 3652.

33 H.-S. Byun, M. A. McHugh, *IEC Res.* **2000**, *39*, 4658.

34 K. Mishima, T. Tokuyasu, K. Matsuyama, N. Komorita, T. Enjoji, M. Nagatani, *Fluid Phase Equilib.* **1998**, *144*, 299.

35 K. Mishima, K. Matsuyama, M. Nagatani, *Fluid Phase Equilib.* **1999**, *161*, 315.

36 C. Drohmann, E. J. Beckman, *J. Supercrit. Fluids* **2002**, *22*, 103.

37 S. E. Conway, H.-S. Byun, M. A. McHugh, J. D. Wang, F. S. Mandel, *J. Appl. Polm. Sci.* **2001**, *80*, 1155.

38 B. C. Lee, Y.-M. Kuk, *J. Chem. Eng. Data* **2002**, *47*, 367.

39 B. C. Lee, J. S. Lim, Y.-W. Lee, *J. Chem. Eng. Data* **2003**, *48*, 774.

40 J. M. Lee, B.-C. Lee, S.-J. Hwang, *J. Chem. Eng. Data* **2000**, *45*, 1162.

41 Y.-M. Kuk, B.-C. Lee, Y. W. Lee, J. S. Lim, *J. Chem. Eng. Data* **2001**, *46*, 1344.

42 Z. Bayraktar, E. Kiran, *J. Appl. Polym. Sci.* **2000**, *75*, 1397.

43 G. Dris, S. W. Barton, *Polym. Mat. Sci. Eng.* **1996**, *74*, 226.

44 C. A. Mertogan, T. P. DiNoia, M. A. McHugh, *Macromolecules* **1997**, *30*, 7511.

45 C. A. Mertogan, H.-S. Byun, M. A. McHugh, W. H. Tuminello, *Macromolecules* **1996**, *29*, 6548.

46 M. A. McHugh, C. A. Mertogan, T. P. DiNoia, C. Anolick, W. H. Tuminello, R. Wheland, *Macromolecules* **1998**, *31*, 2252.

47 Y.-L. Hsiao, E. E. Maury, J. M. DeSimone, S. Mawson, K. P. Johnston, *Macromolecules* **1995**, *28*, 8195.

48 S. Mawson, K. P. Johnston, J. R. Combes, J. M. DeSimone, *Macromolecules* **1995**, *28*, 3182.

49 M. Seiler, J. Rolker, W. Arlt, *Macromolecules* **2003**, *36*, 2085.

50 S. Beuermann, M. Buback, M. Jürgens, *Polym. Mat. Sci. Eng.* **2001**, *84*, 45.

51 Z. Shen, M. A. McHugh, J. Xu, J. Belardi, S. Kilic, A. Mesiano, S. Bane, C. Karnikas, E. Beckman, R. Enick, *Polymer* **2003**, *44*, 1491.

52 Y. Xiong, E. Kiran, *Polymer* **1995**, *36*, 4817.

53 I. C. Sanchez, R. H. Lacombe, *J. Phys. Chem.* **1976**, *80*, 2352.

54 L. A. Kleintjens, R. Koningsveld, *Colloid Polym. Sci.* **1980**, *258*, 711.

55 A. Garg, E. Gulari, C. W. Manke, *Macromolecules* **1994**, *27*, 5643.

56 E. J. Beckman, R. Koningsveld, R. S. Porter, *Macromolecules* **1994**, *23*, 2321.

57 N. F. Carnahan, K. E. Starling, *J. Chem. Phys.* **1969**, *51*, 635.

58 W. G. Chapman, K. E. Gubbins, G. Jackson, M. Radosz, *Fluid Phase Equilib.* **1989**, *52*, 31.

59 W. G. Chapman, K. E. Gubbins, G. Jackson, M. Radosz, *IEC Res.* **1990**, *29*, 1709.

60 S. H. Huang, M. Radosz, *IEC Res.* **1990**, *29*, 2284.

61 S. H. Huang, M. Radosz, *IEC Res.* **1991**, *30*, 1994.

62 S. S. Chen, A. Kreglewski, *Ber. Bunsen Ges.* **1977**, *81*, 1048.

63 M. S. Wertheim, *J. Chem. Phys.* **1987**, *87*, 7323.

64 Y. Song, S. M. Lambert, J. M. Prausnitz, *IEC Res.* **1994**, *33*, 1047.

65 J. Chang, S. I. Sandler, *Mol. Phys.* **1994**, *81*, 745.

66 A. Gil-Villegas, A. Galindo, P. J. Whitehead, S. J. Mills, G. Jackson, A. N. Burgess, *J. Chem. Phys.* **1997**, *106*, 4168.

67 T. Hino, J. M. Prausnitz, *Fluid Phase Equilib.* **1997**, *138*, 105.

68 C. M. Colina, C. K. Hall, K. E. Gubbins, *Fluid Phase Equilib.* **2002**, *194–197*, 553.

69 S. Behme, G. Sadowski, W. Arlt, *Fluid Phase Equilib.* **1999**, *158–160*, 869.

70 M. Banaszak, Y. C. Chiew, M. Radosz, *Phys. Rev. E* **1993**, *48*, 3760.

71 F. W. Tavares, J. Chang, S. I. Sandler, *Mol. Phys.* **1995**, *86*, 1451.

72 W. G. Chapman, *J. Chem. Phys.* **1990**, *93*, 4299.

73 E. A. Müller, L. F. Vega, K. E. Gubbins, *Mol. Phys.* **1994**, *83*, 1209.

74 F. J. Blas, L. F. Vega, *Mol. Phys.* **1997**, *92*, 135.

75 J. Gross, G. Sadowski, *IEC Res.* **2001**, *40*, 1244.

76 J. Gross, G. Sadowski, *IEC Res.* **2002**, *41*, 1084.

77 F. Tumakaka, J. Gross, G. Sadowski, *Fluid Phase Equilib.* **2002**, *194–197*, 541.

78 J. Gross, O. Spuhl, F. Tumakaka, G. Sadowski, *IEC Res.* **2003**, *42*, 1266.

79 F. Tumakaka, PhD thesis, University of Dortmund **2004**.

80 C. F. Kirby, M. A. McHugh, *Chem. Rev.* **1999**, *99*, 565.

81 M. A. van Schilt, W. J. van Meerendonk, M. F. Kemmere, J. T. F. Keurentjes, M. Kleiner, G. Sadowski, Th. W. de Loos, *IEC Res.* **2005**, submitted.

3
Transport Properties of Supercritical Carbon Dioxide *

Frederic Lavanchy, Eric Fourcade, Evert de Koeijer, Johan Wijers,
Thierry Meyer, and Jos Keurentjes

3.1
Introduction

For many processes performed in supercritical fluids, the transport properties of the medium will play an important role. For polymerizations, this includes mass transfer for mixing reactants and to allow proper contact between monomer and catalyst. Polymerization reactions are usually highly exothermic, so that the heat of reaction needs to be absorbed and transported through the supercritical fluid. Virtually all studies described in the literature have been performed on a relatively small scale, and scale-up aspects, for which mass and heat transfer are major issues, have generally been disregarded. This chapter will describe an experimental study of some aspects of mass and heat transfer in supercritical CO_2 (scCO_2), and a comparison will be made with the behavior of standard liquid systems.

Supercritical fluids (SCFs) generally exhibit particular properties in the vicinity of the critical point. SCFs are regarded as viscous Newtonian, heat conducting, and highly expandable fluids [1]. Motion of SCFs is described by the Navier-Stokes equations in combination with an adequate equation of state (EOS). For CO_2, thermodynamic parameters can be estimated by the Wagner and Span equation of state [2] and transport properties by the Vesovic et al. [3] equations. The heat capacity (c_p), the thermal conductivity (λ) and dynamic viscosity (μ) can then be calculated as shown in Fig. 1a, b, and c, respectively. It can be seen that the behavior of these parameters is very much in line with the behavior in the liquid and gaseous states; however, in the vicinity of the critical point (P_c=7.39 MPa and T_c=31 °C), variations are much more pronounced. The heat capacity at constant pressure diverges at the critical point, and its value can be several orders of magnitude higher than "normal" values. This effect is most pronounced for transitions from the liquid-gas curve to the supercritical region

---

* The symbols used in this chapter are listed at the end of the text, under "Notation".

Supercritical Carbon Dioxide: in Polymer Reaction Engineering
Edited by Maartje F. Kemmere and Thierry Meyer
Copyright © 2005 WILEY-VCH Verlag GmbH & Co. KGaA, Weinheim
ISBN: 3-527-31092-4

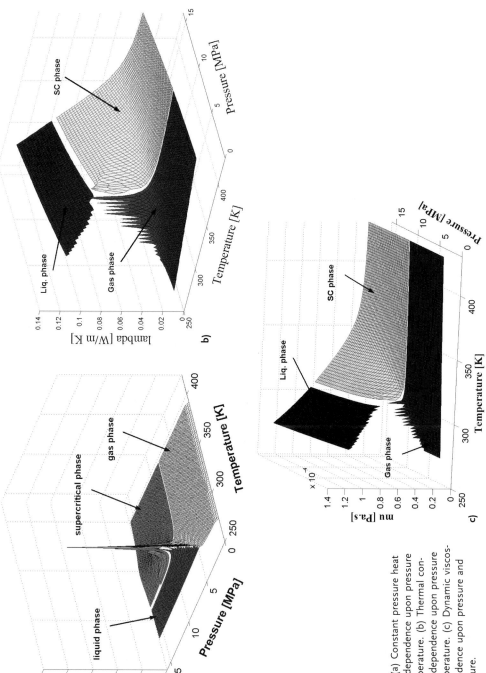

Fig. 3.1 (a) Constant pressure heat capacity dependence upon pressure and temperature. (b) Thermal conductivity dependence upon pressure and temperature. (c) Dynamic viscosity dependence upon pressure and temperature.

and for the resulting near-critical density values. When the transition occurs from pure gas or liquid phase to the supercritical phase, the increase is smoother and even disappears far from the critical point, which corresponds to high (liquid-like) or low (gas-like) densities. The thermal conductivity, like the isothermal compressibility, exhibits a similar critical enhancement in the vicinity of the critical pressure, temperature, and density. Also, the viscosity shows a similar tendency except that the critical enhancement is in a narrower region around the critical point and is less pronounced [4].

The influence of the variation in transport parameters on the design of process equipment is not well studied when compared to the case of relatively incompressible liquids. Therefore, this Chapter will focus on experiments describing the hydrodynamic behavior and the heat transfer properties of $scCO_2$. Although this analysis is crucial for the study and promotion of chemical reactions in SCFs, it is also important for the application of SCFs in refrigeration and cooling.

3.2
Hydrodynamics and Mixing

Mixing has a large influence on the yield and selectivity of a broad range of chemical processes, and the design and operation of mixing devices can determine the profitability of the whole plant. The interaction between mixing and chemical reaction has been investigated for stirred-tank reactors using water as the liquid medium, and rules have been obtained to predict the selectivity of a reaction as a function of the design of the mixing system [5]. In view of the specific behavior as mentioned in Section 4.1, it is not obvious whether the design rules obtained for common liquid solvents are also valid for supercritical fluids [1].

3.2.1
Laser-Doppler Velocimetry and Computational Fluid Dynamics

To compare the hydrodynamic behavior of supercritical CO_2 and water, laser-Doppler velocimetry (LDV) measurements have been performed in a specially designed high-pressure mixing vessel provided with glass windows. For the same geometry, Computational Fluid Dynamics (CFD) calculations have been made for both media.

The stainless steel high-pressure vessel used for the flow measurements, designed by ITTB Heerenveen (The Netherlands), is presented in Fig. 3.2. The vessel has an internal diameter of 6 cm and is designed for pressures up to 15 MPa. The temperature is controlled by pumping water from a thermostatic bath through channels in the vessel wall. Pitched-blade impellers with a diameter of half the vessel diameter are used to stir the vessel content. Two glass windows allow for measurement of velocity components in three directions with laser-Doppler velocimetry. The LDV equipment consists of a 2D fiber optics system

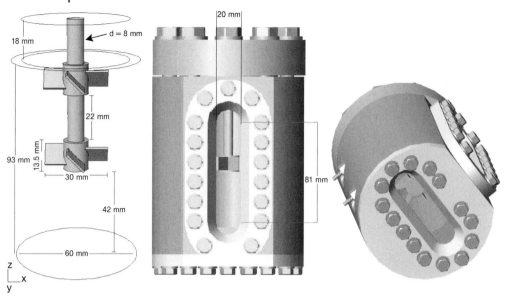

Fig. 3.2 High-pressure stirred vessel with glass windows.

with Burst Spectrum Analyzers supplied by Dantec. As seeding material, hollow glass particles with a diameter of 10 μm (Dantec) are used. The velocities were measured in water and in supercritical CO_2 with a density of 640 kg/m^3 and a temperature of 34 °C, in both a baffled (4 baffles with a blade width of 6 mm) and an unbaffled system at a stirrer speed of 500 rpm.

The velocities in the unbaffled vessel have also been calculated by means of the Computational Fluid Dynamics program CFX version 5.7 using the Shear Stress Transport (SST) model. A mesh independence study was carried out on three different meshes. Results presented here are based on the intermediate mesh, which contains 405 154 nodes and 1 626 400 elements (hexahedrons). All calculations are isothermal steady state and were performed in a rotating coordinate frame. To minimize numerical diffusion, calculations use a blended advection scheme with a blend factor of one (fully second order) for the momentum and continuity equations. A high-resolution advection scheme has been used for the turbulence equations for k and ε. The high-resolution scheme is not fully second order, but it is bounded and therefore more suitable for variables that are always positive like k and ε. Simulations are assumed to be converged when all the RMS residuals are below 10^{-6}. ScCO$_2$ is modeled using the real fluid model capabilities of CFX 5.7. ScCO$_2$ properties are entered as tables (RGP file), which represent, in a discrete manner, functions dependent on the pressure and temperature. Temperature- and pressure-dependent values for relevant physical parameters have been obtained from the NIST database [6].

3.2.2
Flow Characteristics

A selection of measured mean velocity data is presented in dimensionless form through dividing by the stirrer tip speed, V_{tip}, in Figs. 3 to 5. The mean velocities in Figs. 3 and 4, respectively in tangential and axial direction, are measured in the baffled vessel in a plane at the level of the lower stirrer (Figs. 3a and 4a) and in a plane 29 mm above the vessel bottom, which is around 10 mm below the lower stirrer (Figs. 3b and 4b). Fig. 3.5 shows the mean tangential velocities measured in the unbaffled system in a plane 29 mm above the vessel bottom both for water (Fig. 3.5a) and scCO$_2$ (Fig. 3.5b).

The pump number of a stirrer, N_F, correlates in a dimensionless way the total volume flow F produced by the stirrer with the stirrer speed N and the stirrer diameter d_s

$$N_F = \frac{F}{N \cdot d_s^3} \tag{1}$$

The value of N_F is constant in the turbulent regime and is characteristic for a defined vessel/stirrer set-up. Since the pump number of a stirrer is independent

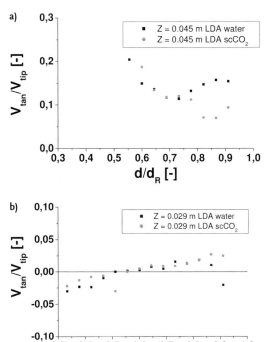

Fig. 3.3 V_{tan} as a function of the dimensionless vessel diameter for water and scCO$_2$ (640 kg m^{-3}) in the baffled system.

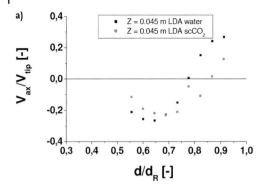

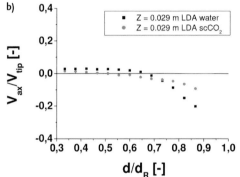

Fig. 3.4 V_{ax} as a function of the dimensionless vessel diameter for water and $scCO_2$ (640 kg m^{-3}) in the baffled system.

of the density, it is expected that, for the same stirrer speed, the mean velocities for $scCO_2$ are equal to those of water, which is in accordance with the measured data.

The velocities calculated with CFX 5, using the SST turbulence model, are also shown in Fig. 5 a and b for the unbaffled system. Especially near the centerline, the calculated values differ from the measured ones. In general, the values of the measured mean velocities are somewhat underpredicted by the calculations, which is in accordance with literature [7]. In general, however, the measured trends are relatively well predicted by the computations both for water and $scCO_2$.

About 10000 data points (N) are collected with LDV at a single point in the vessel with a data rate of around 200 Hz. From these data points the mean velocity is calculated (as given in Figs. 3 to 5) but also the fluctuating velocity V_i' as defined by

$$V_i' = \sqrt{\sum_{j=1}^{N} \frac{(V_i - \bar{V})_j^2}{N}} \tag{2}$$

From the fluctuating velocities in the three Cartesian directions the local turbulent kinetic energy k is calculated by Eq. (3) assuming that the influence of periodic fluctuations is negligible.

a)

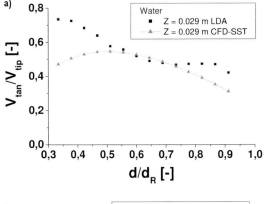

b)

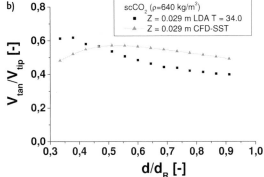

Fig. 3.5 V_{tan} measured (LDA) and calculated (CFD) as a function of the dimensionless vessel diameter for water and scCO$_2$ (640 kg m^{-3}) in the baffled system.

$$k = \frac{1}{2} \cdot \sum_{i=1}^{3} (V_i')^2 \qquad (3)$$

Since fluctuating velocities are not calculated in CFX, only the calculated turbulent kinetic energy k can be compared with the measured k values obtained through Eq. (3). Both measured and calculated turbulent kinetic energies are presented in Fig. 3.6a for water and in Fig. 3.6b for scCO$_2$ for the unbaffled system in a plane 29 mm above the vessel bottom.

The k values for water and scCO$_2$ are somewhat underpredicted by the calculations, which is also mentioned in literature [7]. The power consumption P of a stirrer in the turbulent regime is related to the stirrer speed N and the stirrer diameter d_s by the dimensionless power number N_p

$$N_p = \frac{P}{\rho \cdot N^3 \cdot d_s^5} \qquad (4)$$

The power number N_p is, as is the pump number, constant in the turbulent regime and has a characteristic value for a defined vessel/stirrer geometry. As-

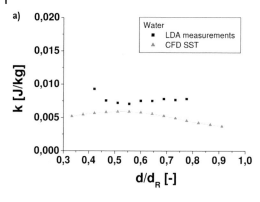

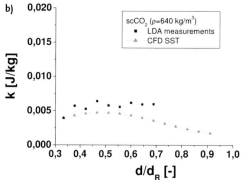

Fig. 3.6 Turbulent kinetic energy k measured (LDA) and calculated (CFD) as a function of the dimensionless vessel diameter for water and $scCO_2$ (640 kg m^{-3}) in the unbaffled system at $Z=0.029$ m.

suming that the local turbulent energy k scales with the stirrer power, which is linear with the fluid density ρ, the k values for $scCO_2$ should be 0.64 times those of water. As it can be seen from Fig. 3.6, this is fairly well in agreement with both measured and calculated data.

3.3
Heat Transfer

The transfer of heat to and from process fluids is an essential part of most processes. In general, heat flows from one location to another by three distinct mechanisms:

- *by conduction*, or the transfer of energy from matter to adjacent matter by direct contact without intermixing or flow of material;
- *by convection*, or the transfer of energy by the bulk mixing of material. In natural convection it is the density difference of hot and cold fluid that causes the mixing. In forced convection, usually a mechanical agitator is used;
- *by radiation* of light, infrared, ultraviolet, or radio waves emanated from a hot body, which can be absorbed by a cool body.

In this part we will concentrate on heat transfer in SCF reactors. For this, we will look at the mechanisms that govern heat transfer from the inside of the reactor (bulk) toward the coolant in the jacket. Many expressions and correlations have been developed for stirred vessels, depending on the vessel geometry, the stirrer type and geometry, and the liquid medium [8–10]. However, none of them have been specifically derived for supercritical fluids.

3.3.1
Specific for Near-Critical Fluids: the Piston Effect

In thermally non-homogeneous supercritical fluids, very intense convective motion can occur [1]. Moreover, thermal transport measurements report a very fast heat transport although the heat diffusivity is extremely small. In 1985, experiments were performed in a sounding rocket in which the bulk temperature followed the wall temperature with a very short time delay [11]. This implies that instead of a critical slowing down of heat transport, an adiabatic critical speeding up was observed, although this was not interpreted as such at that time. In 1990 the thermo-compressive nature of this phenomenon was explained in a pure thermodynamic approach in which the phenomenon has been called "adiabatic effect" [12]. Based on a semi-hydrodynamic method [13] and numerically solved Navier-Stokes equations for a Van der Waals fluid [14], the speeding effect is called the "piston effect". The piston effect can be observed in the very close vicinity of the critical point and has some remarkable properties [1, 15]:

1. *The piston effect is a thermoacoustic phenomenon.* Acoustic compression waves are emitted at heated boundaries, which provoke a homogeneous increase in the bulk temperature.

2. *The piston effect is a fourth heat transport mechanism.* Depending on the applied boundary conditions, the piston effect can transport heat from one side of a thermostat-controlled container to the other on a very short time scale. As the bulk phase is homogeneously heated by the piston effect, a boundary layer is formed at the thermostat-controlled wall.

3. *Temperature and density relaxations are uncoupled.* After the piston effect has equalized the temperature, the fluid of the boundary layer can no longer expand and the remaining temperature inhomogeneities can only disappear by diffusion. Although these very small inhomogeneities are associated with rather large density gradients due to the diverging compressibility, density relaxation seems to occur at a significantly lower rate than temperature relaxation.

4. *Temperature can propagate with the speed of sound.* When the critical point is approached more closely than a crossover value given by asymptote analysis, the characteristic time of the piston effect does not monotonically decrease to zero, but tends to reach a constant value, which is the characteristic acoustic time. For CO_2 contained in a 10 mm long container set at 1 K above its critical temperature, the crossover value is some mK. At these conditions, the

temperature wave emitted by the thermal boundary layer has the same amplitude as the temperature increase at the wall.

3.3.2
Reaction Calorimetry

To study heat transfer aspects of reactors operated under supercritical conditions, a special reaction calorimeter has been developed in collaboration with Mettler-Toledo GmbH (Switzerland) [4, 16]. The equipment is based on an RC1e thermostat coupled with a 1.3 L high-pressure stainless steel reactor as depicted in Fig. 3.7. The thermostat unit controls the reaction temperature by pumping silicone oil at high speed through the double jacket of the reactor. Process and control variables are monitored and controlled using WinRC-NT software. The maximum operating pressure and temperature are 35 MPa and 300 °C. The reactor is equipped with a magnetic stirrer drive, a 25 W calibration heater, a PT100 temperature sensor, and a pressure sensor. The stirrers used are a three-stage Ekato MIG®, a gassing stirrer, and a two-stage turbine (Table 3.1, see p. 52).

The calorimeter is able to work in four different operating modes: *adiabatic*, where the jacket temperature (T_j) is adjusted such that there is no heat transfer through the reactor wall; *isoperibolic*, where the jacket temperature is kept constant and the reaction temperature (T_r) follows the reaction profile; *isothermal*, where the desired reaction temperature is set to a constant value and T_j is changed automatically to maintain T_r at the specified value; and *distillation/reflux mode* or *crystallization*, where the respective (T_j–T_r) or (T_r–T_j) is maintained constant.

Fig. 3.7 Picture of the autoclave.

Classical reaction calorimetry

Supercritical reaction calorimetry

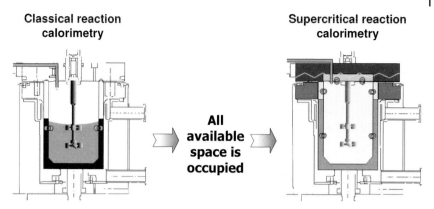

All available space is occupied

Fig. 3.8 Difference between classical liquid and supercritical reaction calorimetry.

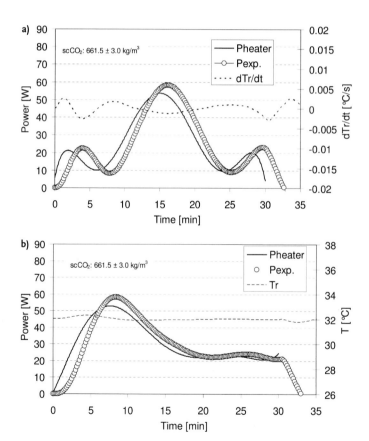

a)

scCO$_2$: 661.5 ± 3.0 kg/m^3

— Pheater
—o— Pexp.
···· dTr/dt

b)

scCO$_2$: 661.5 ± 3.0 kg/m^3

— Pheater
o Pexp.
---- Tr

Fig. 3.9 (a) Calorimeter response to a "Gaussian-like" signal input by the calibration heater in simulation mode at a density of scCO$_2$ of 661.5 kg m^{-3}. (b) Calorimeter response to a "peak" signal input by the calibration heater in simulation mode at a density of scCO$_2$ of 661.5 kg m^{-3}.

Since the supercritical phase occupies all space available, as illustrated in Fig. 3.8, not only has the jacket area to be perfectly controlled, but also the cover and the other parts have to be temperature controlled in order to avoid additional heat transfer interferences. In this case, all the reactor parts in contact with the reactor contents are adjusted to T_r.

The performance of the calorimeter has been validated by comparing measured properties with available literature data [16–18]. Also, heat flow calibration has been performed by creating a specific heat flow using a calibration probe inserted in the reactor. Such an introduction of heat, without any actual reaction occurring, shows the ability of the calorimeter to react to heat flows, separate from any reaction-related effects. In the ideal situation (no temperature gradient and a time constant of the equipment equal to zero) the two signals should be identical. It can be seen in Fig. 9a and b that the calorimetric signal correctly follows the input heat flow generated by the calibration probe, where the dynamics of the system induce a minor delay in the response.

3.3.3
Heat Transfer in Stirred Vessel with SCFs

The general equation for heat transfer across an interface is

$$dQ = dA \cdot U \cdot \Delta T \tag{5}$$

where dA is the element of surface area required to transfer an amount of heat dQ for a bulk temperature difference between the two streams ΔT at an overall heat transfer coefficient U. A general methodology to analyze heat transfer phenomena in stirred tanks is the Wilson plot analysis [19]. For this, the overall heat transfer resistance $1/U$ is expressed as the sum of three resistances in series (Eq. 6): one for the internal film $(1/h_r)$, one for the reactor wall (e/λ_w) and one for the external coolant film $(1/h_e)$, in which the last two resistances are often combined in a resistance $1/\phi$ being independent of the internal medium [20].

$$\frac{1}{U} = \frac{1}{h_r} + \frac{e}{\lambda_w} + \frac{1}{h_e} \tag{6}$$

To correlate internal mixing with heat transfer, the Nusselt number $(Nu=h_r d_R/\lambda)$ of the reactor is related to the Reynolds $(Re=\rho N d_s^2/\mu)$ and Prandtl $(Pr=\mu c_p/\lambda)$ number according to

$$Nu = C \cdot Re^a \cdot Pr^b \left(\frac{\mu}{\mu_w}\right)^c \tag{7}$$

in which it should be noted that Nu is related to the vessel diameter and Re to the impeller diameter. In liquid media the exponents a and b in the Nusselt correlation are usually found to be 2/3 and 1/3, respectively, for a stirred tank reactor equipped with a turbine impeller [21]. In isothermal and steady-state conditions, the viscosity ratio (with exponent $c=0.14$) can be included in the constant C. As all other properties remain constant at isothermal conditions, this allows the isolation of h_r:

$$h_r = C \cdot \left(\frac{d_s^2 \cdot \rho}{\mu} \right)^{2/3} \cdot \left(\frac{c_p \cdot \mu}{\lambda} \right)^{1/3} \cdot \frac{\lambda}{d_r} \cdot N^{2/3} = C' \cdot N^{2/3} \tag{8}$$

Combining Eqs. (6) and (8) then yields Eq. (9), in which the overall heat transfer resistance is coupled to the stirrer speed N:

$$\frac{1}{U} = \frac{1}{C' \cdot N^{2/3}} + \frac{1}{\phi} \tag{9}$$

In Eq. (9), C' and ϕ are system-specific constants.

An example of an experimentally determined Wilson plot is given in Fig. 3.10, in which the calorimeter reactor has been used in combination with two types of stirrers, i.e. a double-stage turbine and a gassing stirrer. In isothermal mode and using the calibration probe introducing a defined amount of heat, it is possible to estimate the overall heat transfer coefficient using Eq. (5). The linear plot confirms that the exponent of Re equals 2/3 in scCO$_2$, similar to "normal" liquids. Figure 3.10 clearly indicates that the overall heat transfer coefficient completely follows the Wilson plot regression, except for extremely low agitation speed (100 rpm). Obviously, this low agitation speed is not sufficient to homogenize the medium in terms of temperature and density [16]. Moreover, for the same temperature, the gassing stirrer is less efficient in terms of heat transfer than the double-stage turbine.

The slope of the regression lines allows calculation of the internal film transfer coefficient. From the intercept with the U axis the external part ϕ of the overall heat transfer coefficient can be determined. As the regression lines intercept be-

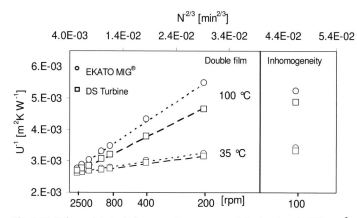

Fig. 3.10 Wilson plot study for two agitators at a scCO$_2$ density of 370 kg m^{-3}.

fore the U axis, it follows that the heat transfer behavior of the external film is the opposite of the internal film. The external heat transfer coefficient part, which includes conduction through the vessel wall and transfer through the coolant stagnant film, increases with temperature (Fig. 3.11). This effect results from a viscosity decrease of the coolant, which reduces the film thickness. It is clear that for the series at 305.15 K, where the slope is almost flat, almost all heat transfer resistance comes from the external part, independently of the stirrer rotation speed.

Fig. 3.12 compares the measured internal heat transfer coefficient (h_r) in classical liquids with supercritical carbon dioxide at 400 rpm [16, 17]. It indicates that the internal heat transfer coefficient increases asymptotically close to the critical point for scCO$_2$. In contrast to the normal liquids, in supercritical CO$_2$ a lower temperature (above the critical point) leads to a higher heat transfer coefficient. Although the behavior of the internal heat transfer coefficient differs from the trend observed for common liquids (water and methanol), the behavior is consistent with the changing thermodynamic properties around the critical point (Fig. 1 a–c), i.e. the constant-pressure heat capacity (c_p), thermal conductivity (λ) and bulk viscosity (μ). The internal heat transfer coefficient h_r is proportional to $\lambda^{2/3}$ and $c_p^{1/3}$ and is therefore subject to the same asymptotic divergence at the critical point. Although the viscosity is dependent on temperature to the power 1/3, which leads to some divergence around the critical point, μ has only a small range of densities where critical enhancement is significant (at less than 1 °C from the critical point) [22]. This explains that for rotation speeds leading to significant internal film resistances, the overall heat transfer coefficient is substantially improved for CO$_2$ when the temperature approaches the critical temperature. This effect has also been observed for a continuous flow system [23] for water and carbon dioxide and in a multi-loop system using carbon dioxide as a refrigerant [24].

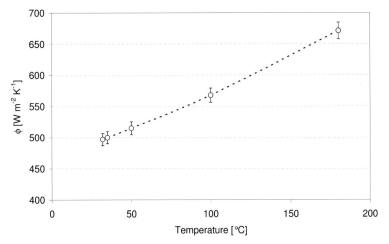

Fig. 3.11 Integrated wall and external resistance.

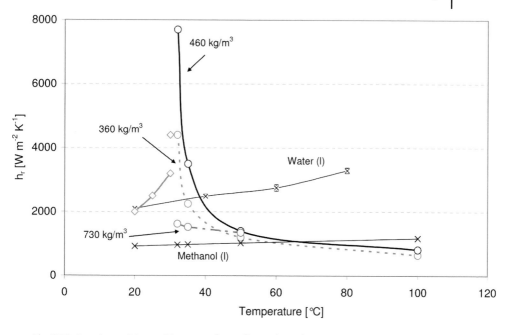

Fig. 3.12 Experimental internal heat transfer coefficient h_r in the reactor with double-stage turbine at 400 rpm.

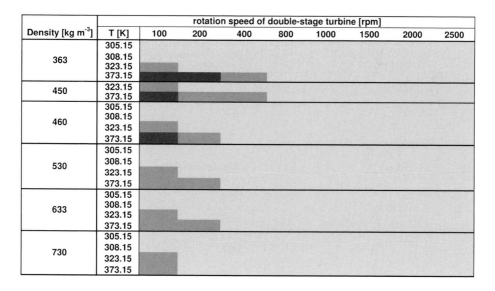

Density [kg m⁻³]	T [K]	rotation speed of double-stage turbine [rpm]							
		100	200	400	800	1000	1500	2000	2500
363	305.15								
	308.15								
	323.15								
	373.15								
450	323.15								
	373.15								
460	305.15								
	308.15								
	323.15								
	373.15								
530	305.15								
	308.15								
	323.15								
	373.15								
633	305.15								
	308.15								
	323.15								
	373.15								
730	305.15								
	308.15								
	323.15								
	373.15								

- internal film resistance
- external film resistance
- approx. same resistance film contribution [within 30 %]

Fig. 3.13 Limiting transfer resistance with temperature, CO_2 density, and rotation speed.

Parameters such as temperature, CO_2 density, and rotation speed directly affect the limiting heat transfer resistance. In this experimental set-up it appears that with $scCO_2$ the main resistance is located in the external film, except at extremely low stirrer speeds and high temperatures (low densities), where the limiting resistance is located inside the reactor (Fig. 3.13).

Theoretical values for c_p, λ and μ (obtained from EOS) and experimental h_r have been used to calculate the constant C of the Nusselt correlation (Eq. 7). An example of the results for the two stirrers and a $scCO_2$ density of 540 kg m^{-3} is presented in Fig. 3.14, leading to C values of 0.53 ± 0.03 for the turbine and 0.27 ± 0.03 for the Ekato MIG®, respectively.

C values are dependent on the liquid system, the vessel, the agitator type, and the applied hydrodynamic regime (turbulent, laminar, or transition). A comparison with standard values for typical stirrers (Table 3.1) [25–28] shows that the C values obtained for the turbine are very similar in liquids and $scCO_2$. For the Ekato MIG® stirrer, however, there appears to be a difference of a factor of two between the two types of system.

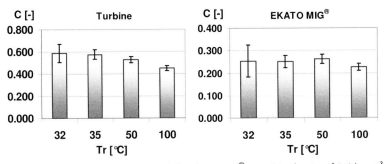

Fig. 3.14 C values for the turbine and the Ekato MIG® at $scCO_2$ density of 540 kg m^{-3}.

Table 3.1 Typical C values obtained in liquids for different stirrers [25–28].

Stirrer type	Characteristic length used in the Nusselt number	d_s/d_r [-]	C value [-]
Turbine [28]	Reactor diameter	0.5	0.535
Propeller 4 blades 45° [26]	Reactor diameter	0.05–0.4	0.54
Pitched-blade disk turbine [27]	Reactor diameter	0.33	0.53
Flat-blade disk turbine [27]	Reactor diameter	0.33	0.54
Ekato MIG®, two stages [25]	Reactor diameter	0.5–0.95	0.42
Ekato MIG®, three stages [25]	Reactor diameter	0.5–0.95	0.46

3.4
Conclusions

In this Chapter we have made an attempt to describe mixing and heat and mass transfer in supercritical fluids. Using Laser Doppler Anemometry, Computational Fluid Dynamics and High-Pressure Calorimetry, some basic guidelines have been derived. When compared to the behavior of ordinary liquids, the behavior of scCO$_2$ is quite different, especially near the critical point. However, when the thermodynamic behavior of CO$_2$ is taken into account (in terms of constant-pressure heat capacity, viscosity, and thermal conductivity), its behavior is consistent with that of other liquids.

Acknowledgments

The "Fonds National Suisse pour la Recherche" No. 21.61403.00 and 200020-101477 is gratefully acknowledged for its financial support.

Notation

Symbols

A	surface	[m^2]
c_p	constant-pressure heat capacity	[J kg^{-1} K^{-1}]
C	constant	[–]
C'	constant	[–]
d_R	reactor internal diameter	[m]
d_S	stirrer characteristic diameter	[m]
e	wall thickness	[m]
F	stirrer flow	[m^3 s^{-1}]
h_r	internal heat transfer coefficient	[W m^{-2} K^{-1}]
h_e	external heat transfer coefficient	[W m^{-2} K^{-1}]
k	local turbulent kinetic energy	[m^2 s^{-2}]
N	rotation speed of the stirrer	[s^{-1}]
N_p	stirrer power consumption	[W]
N_f	stirrer pump number	[–]
P	pressure	[MPa]
Q	heat	[J]
U	overall heat transfer coefficient	[W m^{-2} K^{-1}]
T	temperature	[K]
V_{tip}	stirrer tip speed	[m s^{-1}]
V_i'	fluctuating velocity	[m s^{-1}]
λ_w	thermal conductivity of the reactor wall	[W m^{-1} K^{-1}]
λ	bulk thermal conductivity	[W m^{-1} K^{-1}]
μ	dynamic viscosity	[kg m^{-1} s^{-1}]

ρ density   [kg m^{-3}]
ϕ constant   [W m^{-2} K^{-1}]
Re   Reynolds number
Pr   Prandtl number
Nu   Nusselt number

Acronyms
EOS   Equation of state
SCF   Supercritical fluid

References

1 Zappoli B., *C. R. Mecanique*, **2003**, 331, 713.
2 Span R., Wagner W., *J. Phys. Chem. Ref. Data*, **1996**, 25(6), 1509.
3 Vesovic V., Wakeham W. A., Olchowy G. A., Sengers J. V., Watson J. T. R., Millat J., *J. Phys. Chem. Ref. Data*, **1990**, 19(3), 763.
4 Lavanchy F., Fortini S., Meyer T., *Chimia*, **2002**, 56, 126.
5 Verschuren I. L. M., Wijers J. G., Keurentjes J. T. F., *AIChE J.* **2001**, 47, 1731.
6 NIST Standard Reference Database Number 69, March 2003, *http:// webbook.nist.gov/chemistry*
7 Montante G., Lee K. C., Brucato A., Yianneskis M., *Chem. Eng. Sci.* **2001**, 56, 3751.
8 VDI-Wärmeatlas, *Berechnungblätter für den Wärmeübergang*, Springer, Berlin Heidelberg, **2002**.
9 Sinnot R. K., *Coulson and Richardson's Chemical Engineering*, vol. 6, Butterworth Heinemann, Oxford, **2000**.
10 Wilke H. P., Weber C, Fries T., *Rührtechnik*, Dr. Alfred Hüthig, Heidelberg, **1988**.
11 Nitsche K. and Straub J., in *Proc. 6th Europ. Symp. on Mater. Sci. under Micro-g Conditions*, **1986**, ESA, SSP-256, 109.
12 Onuki A., Hao H., Ferrel R. A., *Phys. Rev.*, **1990**, A41, 2256.
13 Boukari H., Briggs M. E., Shaurneyer J. N., Gammon R. W., *Phys. Rev. Lett.*, **1990**, 65, 2654.
14 Zappoli B., Bailly D., Garrabos Y., Le Neindre B., Guenoun P., Beyseens D., *Phys. Rev.*, **1990**, A41, 2260.
15 T. Fröhlich, Nonlinear heat transfer mechanisms in supercritical fluids, PhD Thesis, **1997**, Munich.
16 Lavanchy F., Fortini S., Meyer Th., *Org. Proc. Res. Dev.*, **2004**, 8, 504.
17 Fortini S., Lavanchy F., Nising P., Meyer Th., *Macromol. Symp.*, **2004**, 206, 79.
18 Fortini S., Lavanchy F., Meyer Th., *Macromol. Mater. Eng.*, **2004**, 289, 757.
19 Wilson E. E., *Am. Soc. Chem. Eng.*, **1915**, 37, 47.
20 Prandtl L., *Physik Z.*, **1910**, 1072.
21 Bourne J. R., Buerli M., Regenass W., *Chem. Eng. Sci.*, **1981**, 36, 347.
22 Fenghour A., Wakeham W. A., Vesovic V., *J. Phys. Chem. Ref. Data*, **1998**, 27, 31.
23 Pioro I. L., Khartabil H. F., Duffey R. B., *Nucl. Eng. Des.*, **2004**, 230, 69.
24 Yoon S. H., Kim J. H., Hwang Y. W., Kim, M. S., Min K., Kim Y., *Int. J. Refrig.*, **2003**, 26, 857.
25 Ekato, Recherche et Développement dans la technologie de l'agitation, Ekato, Versailles, **1981**.
26 Bondy F., Lippa S., *Chem. Eng.*, **1983**, 90, 62.
27 Fletcher P., *Chem. Eng.*, **1987**, 435, 33.
28 Brooks G., Su G.-J., *Chem. Eng. Prog.*, **1959**, 55, 54.

4

Kinetics of Free-Radical Polymerization in Homogeneous Phase of Supercritical Carbon Dioxide *

Sabine Beuermann and Michael Buback

4.1
Introduction

Summarized in this contribution are the results of studies into initiation, propagation, termination, and chain-transfer rate coefficients for free-radical polymerization in homogeneous mixtures containing significant amounts of CO_2, mostly 40 wt%. Supercritical carbon dioxide ($scCO_2$) has been demonstrated to be a promising alternative reaction medium for free-radical polymerizations [1–3]. Particular advantages of CO_2 are associated with a pronounced lowering of viscosity, facilitating separation of this solvent from a polymeric product, and with the inertness of CO_2, which leads to the absence of chain-transfer-to-solvent reactions [4]. Moreover, CO_2 is an environmentally benign fluid with physical properties continuously tunable in the supercritical region. Although the literature mostly describes heterogeneous phase polymerizations in $scCO_2$, it was anticipated that, because of the cosolvent action of the monomer, reactions in homogeneous phase should also be possible. Detailed investigations into chemically initiated styrene polymerizations provided information on the dependence of the maximum monomer conversion (x_{max}) that may be reached in homogeneous phase on CO_2 content, temperature, pressure, and polymer molecular weight [5, 6]. For example, x_{max} of styrene homopolymerization is given as a function of CO_2 content at 30 and 50 MPa in Fig. 4.1. The data obtained at both pressures may be represented by a single line. x_{max} decreases from around 35% at a CO_2 content of 20 wt% to around 10% at a CO_2 content of 54 wt%.

The x_{max} data in Fig. 4.1 are remarkable in that polystyrene is considered to be the archetype of a polymer which is insoluble in $scCO_2$ [7]. The results are indicative of the significant solvent power of the monomer which is present in the reacting styrene–polystyrene–$scCO_2$ mixtures. It may be expected that polymerizations of other monomers or copolymerizations of styrene with other monomers may be carried out in homogeneous phase up to high monomer conversion. Styrene–

---

* The symbols used in this chapter are listed at the end of the text, under "Notation".

Supercritical Carbon Dioxide: in Polymer Reaction Engineering
Edited by Maartje F. Kemmere and Thierry Meyer
Copyright © 2005 WILEY-VCH Verlag GmbH & Co. KGaA, Weinheim
ISBN: 3-527-31092-4

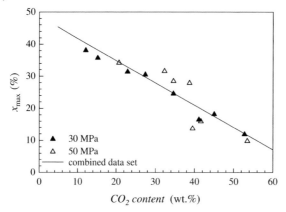

Fig. 4.1 Variation of maximum styrene conversion in homogeneous fluid phase, x_{max}, as a function of CO_2 content for chemically initiated polymerizations at 80 °C and two pressures, 30 and 50 MPa, in the presence of 1 mol% CBr_4 [5, 6]. The number average molecular weight of the resulting polymers is $M_n \approx 10^4$ g mol^{-1}.

methyl methacrylate–glycidyl methacrylate terpolymerizations to yield polymer with $M_n = 4000$ g mol^{-1} were carried out at 120 °C, 35 MPa in 20 wt% CO_2 up to almost complete conversion [2]. This reaction has been monitored via in-line FT-NIR spectroscopy in the region of the first overtone of olefinic CH-stretching vibrations [8]. An NIR spectral series recorded during a terpolymerization at 120 °C and 35 MPa is shown in Fig. 4.2. The arrow indicates the direction of change of the peak at around 6170 cm^{-1}, which is due to CH-stretching first overtone modes at the olefinic double bond of the monomers. The final spectrum corresponds to almost complete monomer conversion.

The examples in Figs. 4.1 and 4.2 show that high monomer conversions may be reached in homogeneous phase in the presence of significant amounts of CO_2. To identify suitable conditions for single-phase polymerization at supercritical conditions of CO_2, which are largely affected by the properties of the polymeric product, it is highly advantageous to simulate polymerization kinetics. Such simulations require reliable rate coefficients to be known, at least for the most important individual steps, i.e. for initiation, propagation, termination, and chain-transfer. With several such rate coefficients being already precisely known for bulk polymerizations, it appears interesting to see whether and to what extent the presence of CO_2 affects these coefficients. One may expect that chemically controlled reaction steps such as initiation, propagation, and chain-transfer will be only moderately affected by the presence of CO_2, whereas CO_2 may have a strong impact on the diffusion-controlled termination step. Aside from their relevance for modeling polymerization kinetics and polymer properties, rate coefficients measured in scCO_2 should also be useful for providing a better mechanistic understanding of the individual steps in free-radical polymerization, in particular of diffusion-controlled termination.

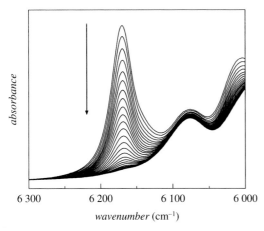

Fig. 4.2 FT-NIR spectral series recorded during an S-MMA-GMA terpolymerization in the presence of 20 wt% CO_2 at 120 °C and 35 MPa. The
initial monomer mole fractions are: $f_s = 0.07$, $f_{MMA} = 0.51$, and $f_{GMA} = 0.42$;
DDM concentration: 0.37 mol L^{-1}; DTBP concentration: 1.5×10^{-2} mol L^{-1};
optical path length: 2 mm [2].

4.2
Experimental

The majority of the kinetic data reported below is from experiments in the
authors' laboratory. Monomer-CO_2 mixtures were prepared prior to feeding and
pressurizing the reaction mixture in an experimental set-up described elsewhere
[9]. Pressure is varied at constant weight fractions of monomer, initiator, CO_2,
and, if applicable, chain-transfer agent (CTA) by means of a manually driven syringe pump which operates on the entire reaction mixture [9]. Thus, investigations into the pressure dependence of rate coefficients are not accompanied by
changes in CO_2 content such as those that occur in cases where pressure is applied by compressing the reaction system with CO_2.

The polymerization reactors are equipped with two sapphire windows [8],
which allow for initiation of the polymerization by UV laser pulses, for in-line
FT-NIR spectroscopic monitoring of the degree of monomer conversion, and for
visual and NIR-spectroscopic inspection of homogeneity of the polymerizing
system [10].

4.3
Initiation

Peroxides are the most important initiators for free-radical polymerization.
Among them, dialkyl or diaryl peroxides, peroxyesters, diacyl peroxides, and per-
oxycarbonates are of particular relevance. For a large number of such peroxides,
the temperature and pressure dependence of the decomposition rate coefficient,

k_d, has been experimentally determined in a dilute solution of *n*-heptane [11–15]. It is of interest to know whether and to what extent the large number of literature data may be used for estimating k_d for polymerizations with significant amounts of CO_2 present. Whereas scCO$_2$ and *n*-heptane are of similarly low polarity, the significantly lower viscosity (and thus higher diffusivity) in scCO$_2$ may affect initiator decomposition rate. k_d is defined (and is experimentally determined) as the overall first-order rate coefficient of peroxide decomposition according to the rate expression:

$$-dc_{PO}/dt = k_d \cdot c_{PO} \tag{1}$$

where c_{PO} is peroxide concentration. As will be shown further below, the decay in peroxide concentration, $- dc_{PO}/dt$, may be affected by several individual reaction steps. It is for this reason that k_d is referred to as the overall rate coefficient.

The first study of peroxide decomposition in scCO$_2$ has been carried out on diethyl peroxodicarbonate by the group of DeSimone [16]. An extended study into initiator decomposition rates in solution in scCO$_2$ has been performed by Barner [17] in which measurements up to 160 °C and 200 MPa were carried out for the peroxyesters *tert*-amyl peroxypivalate (TAPP), *tert*-butyl peroxypivalate (TBPP), and *tert*-butyl peroxyacetate (TBPA), and for the diacyl peroxides dioctanoyl peroxide (DOP) and bis(3,5,5-trimethylhexanoyl) peroxide (BTMHP). Compared in Fig. 4.3 are k_d data for TBPA and TBPP decomposition in solution of

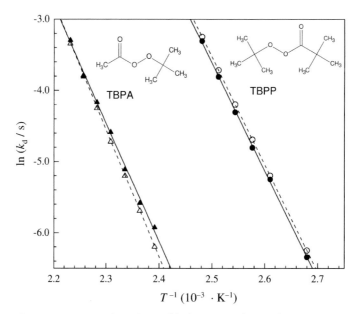

Fig. 4.3 Temperature dependence of k_d for TBPA and TBPP decomposition in solution of *n*-heptane (open symbols) [12, 13] and of scCO$_2$ [17] (filled symbols) at 50 MPa.

scCO$_2$ (full symbols) and of n-heptane (open symbols) at 50 MPa and various temperatures. The k_d data plotted for peroxyesters dissolved in scCO$_2$ are the experimental values provided by Barner [17]. The rate coefficients for peroxyester decomposition in n-heptane were calculated (for the same temperatures as those selected for the scCO$_2$ experiments) from the equations reported for k_d in n-heptane as a function of pressure and temperature [12, 13]. The dashed lines represent these literature expressions for 50 MPa, whereas the full lines are Arrhenius fits of the measured k_d data for decomposition in scCO$_2$.

Fig. 4.3 tells us that k_d values at otherwise identical conditions are only slightly affected, on average by about 10%, upon replacing the solvent n-heptane by scCO$_2$. It should, however, be noted that these minor changes occur into opposite directions. Whereas k_d of TBPP is slightly higher in n-heptane than in scCO$_2$, the rate coefficients for TBPA in scCO$_2$ are above the k_d values measured in n-heptane. This difference may be understood as resulting from the pronounced differences in decomposition behavior. TBPP decomposition is the archetype of very rapid successive two-bond scission of the peroxyester accompanied by an almost instantaneous production of CO$_2$, whereas TBPA undergoes single-bond scission to form a methylcarbonyloxy radical and a *tert*-butoxy radical.

A simplified illustration of the major kinetic steps is given in Scheme 4.1. Primary dissociation of R$_1$C(O)OOR$_2$ with rate coefficient k_1 yields the radicals R$_1$C(O)O· and ·OR$_2$ in a solvent-caged situation. This reaction may be followed by recombination of the two radicals, with rate coefficient k_2, by decarboxylation of R$_1$C(O)O· with rate coefficient k_3, or by out-of-cage diffusion of at least one of the two radicals with rate coefficient k_{diff}. Overall, k_d will be lower than k_1 in the case where recombination plays a major role, but approaches k_1 in situa-

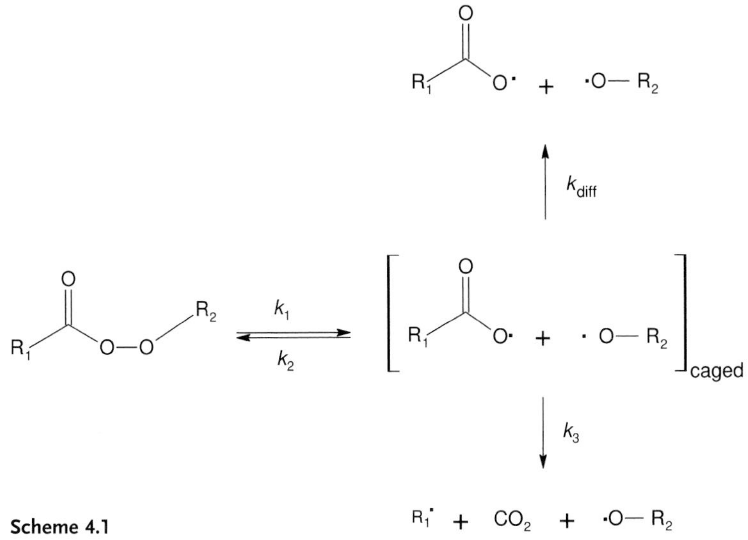

Scheme 4.1

tions where decarboxylation or out-of-cage reaction become very fast. If the decarboxylation reaction of the intermediate acyloxy radicals is very fast, e.g., occurs in an almost concerted fashion together with primary scission of the peroxy linkage, no reversible reaction will take place and overall k_d will be identical to k_1. Investigations into a large series of *tert*-butyl peroxyesters, with *n*-heptane being the solvent, demonstrated that components with the carbon atom of the R_1 moiety that sits in α-position to the carbonyl group being secondary or tertiary (such as TBPP) undergo close-to-concerted two-bond scission, whereas peroxyesters with this α-carbon atom being primary (as with TBPA) undergo single-bond scission [12]. According to these findings, the increase in k_d seen for TBPA upon replacing the solvent *n*-heptane by scCO$_2$ reflects the enhanced out-of-cage diffusivity of $R_1C(O)O\cdot$ and $\cdot OR_2$ radicals in a solvent environment of scCO$_2$ which is associated with a lowering of radical recombination and thus with an increase in overall k_d. As close-to-concerted two-bond scission occurs with TBPP, an enhancement of out-of-cage diffusivity would not result in an increase of k_d. The effect seen for TBPP thus seems to represent the weak effect of the solvent environment on the barrier to dissociation. Thus, interaction with an *n*-heptane environment appears to be associated with a higher decomposition rate coefficient than does interaction with CO$_2$. The TBPA data in Fig. 4.3 provide some indication that the slopes to the full and dashed lines are slightly different, which suggests different activation energies for TBPA decomposition in *n*-heptane and for that in scCO$_2$. The experimental material is, however, not extensive enough to allow for a subdivision of the overall activation energy resulting from the Arrhenius representation of k_d into activation energies of the contributing individual steps of primary bond scission, recombination, out-of-cage diffusion, and decarboxylation.

The results of the experimental k_d studies into alkyl peroxyester decomposition in CO$_2$ [17] may be summarized as follows: at least at pressures up to 100 MPa, which range encompasses the entire area of technical relevance, k_d values for peroxyester decomposition in scCO$_2$ differ by less than ±20%, and mostly by less than ±10%, from the associated values measured in *n*-heptane solution. A more detailed inspection of the data reveals that peroxyesters which undergo single-bond scission decompose faster in CO$_2$, whereas, in the case of close-to-concerted two-bond scission, decomposition of the peroxyester seems to occur at a slightly slower rate in CO$_2$ than in solution in *n*-heptane.

Investigations by Charpentier et al. into the decomposition of diethyl peroxodicarbonate in scCO$_2$ [16] provided no evidence for k_d being different from the values reported for decomposition in conventional solvents. The solvent effects seen for diacyl peroxides are somewhat larger, as is illustrated by the temperature dependence of BTMHP decomposition in Fig. 4.4. The k_d data with scCO$_2$ are the experimental points [17], whereas the *n*-heptane data were again calculated from fit functions to the underlying experiments [11]. With respect to reaction in *n*-heptane, BTMHP decomposition in scCO$_2$ is faster by about 40%. The activation energy, however, appears to be the same with both solvents. A similar situation is encountered for the decomposition of dioctanoyl peroxide, where k_d for reaction

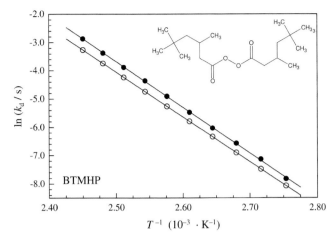

Fig. 4.4 Temperature dependence of k_d for BTMHP decomposition in n-heptane [14] (open symbols) and in scCO$_2$ [17] (filled symbols) at 50 MPa.

in scCO$_2$ is also enhanced by about 40% compared to using n-heptane as the solvent [11]. The diacyloxy radicals that are formed in the primary dissociation step are similar to the methylcarbonyloxy radical produced from TBPA (see Scheme 4.1) in that the carbon atom in the α-position to the carbonyl group is primary. Thus, delayed decarboxylation of the radicals from BTMHP and from DOP is expected to occur, which, as with TBPA, suggests that replacing the solvent n-heptane by CO$_2$ is accompanied by an enhancement of k_d. This is indeed what is observed. The effect is, however, larger than that expected from TBPA, where a change by about 10% was seen. Investigations into the influence on the decomposition of BTMHP of several organic solvents, including highly polar ones such as acetonitrile and dichloromethane, revealed that k_d is satisfactorily correlated with the Dimroth-Reichardt empirical parameter E_T^N, which is a measure of solvent polarity [14]. With increasing E_T^N parameter of the solvent, the decomposition rate is enhanced, e.g., by about a factor of 7, in going from n-heptane to acetonitrile. The E_T^N parameter of scCO$_2$ is slightly above the one for n-heptane, which is consistent with the experimental observation that k_d for BTMHP decomposition in scCO$_2$ is above the associated value for decomposition in n-heptane. Thus, two effects, an increase in diffusivity and an increase in polarity (in terms of the E_T^N parameter), appear to contribute to the fact that the enhancement of k_d, upon replacing n-heptane by CO$_2$, is larger for BTMHP than for TBPA.

In addition to common initiators, the group of DeSimone [18] studied the decomposition kinetics of bis(perfluoro-2-N-propoxypropionyl) peroxide (BPPP) in dense CO$_2$ and in a series of fluorinated solvents. At temperatures above T_c of CO$_2$, k_d is very close to the values measured in the fluorinated solvents.

Investigations into the decomposition of non-peroxide initiators in scCO$_2$ are rare. The decomposition of azo-bis-isobutyronitrile (AIBN), however, has been studied by two groups [19, 20]. Literature data from the two sources for decomposition of AIBN in scCO$_2$ differ by about a factor of two, which poses problems in clearly identifying the effects of changing the solvent environment. To allow for this comparison, the literature rate data have been shifted to identical pressure, via the activation volume of $\Delta V^{\#}(k_d) = 4.3$ cm^3 mol^{-1} [21]. Irrespective of the imprecise knowledge of k_d for decomposition in scCO$_2$, the replacement of the solvent environment of benzene [19] or styrene [21] by scCO$_2$ appears to result in a clear lowering of k_d. The direction of change induced by CO$_2$ thus is the same as that seen above for TBPP with the solvents scCO$_2$ and *n*-heptane, which peroxyester also decomposes in a close-to-concerted two-bond scission mode as does AIBN.

In summary, with the exception of AIBN, where the data from various sources allow for no clear decision about the size of solvent effects on k_d, the literature data on initiator decomposition suggest that k_d for decomposition in scCO$_2$ is very close to the associated values in the non-polar solvent *n*-heptane. Thus, the extended set of literature values reported for peroxide decomposition in *n*-heptane should be very suitable for estimating k_d in polymerizations of non-polar monomers even in the presence of significant amounts of CO$_2$.

4.4
Propagation

4.4.1
Propagation Rate Coefficients

Propagation rate coefficients for styrene, vinyl acetate, and various acrylates and methacrylates were determined by the so-called PLP-SEC technique, which combines pulsed-laser-initiated polymerization (PLP) with subsequent polymer analysis via size-exclusion chromatography (SEC). The PLP-SEC experiment measures the product of propagation rate coefficient, k_p, and monomer concentration, c_M:

$$L_1/t_0 = k_p \cdot c_M \tag{2}$$

The experimentally accessible quantities in Eq. (2) are the time between two successive laser pulses, t_0, and the number of propagation steps between two subsequent laser pulses, L_1, as obtained from SEC. In bulk polymerizations, c_M is identified with overall monomer concentration. In solution experiments, in particular in situations where the solvent power of the monomer is rather different from that of the solvent, the relevant monomer concentration at the site of the free-radical, $c_{M,loc}$, may differ from the analytically determined overall monomer concentration in the solution, $c_{M,a}$. One may think of the two limiting situations that either (1) the relevant monomer concentration is given by $c_{M,a}$ and differences in bulk and solution $k_p \cdot c_M$ are assigned to changes in propagation rate coefficient (to a solution quantity $k_{p,app}$) or (2) the propagation rate coefficient is the same for bulk and solution polymerization, $k_p = k_{p,kin}$, and variations in $k_p \cdot c_M$ are due to changes in effective monomer concentration, with the solution value being assigned as $c_{M,loc}$. The limiting situations are indicated in Eq. (3):

$$k_p \cdot c_M = k_{p,app} \cdot c_{M,a} = k_{p,kin} \cdot c_{M,loc} \qquad (3)$$

Variations in $k_p \cdot c_M$ observed upon changing the molecular environment of the free-radical chain end may also result from variations in both k_p and c_M. In what follows, experimental data will be presented which were derived according to the relation $k_p \cdot c_M = k_{p,app} \cdot c_{M,a}$. As will be seen below, there are good arguments that the observed changes in $k_p \cdot c_M$ actually are changes in monomer concentration. The apparent propagation rate coefficient, $k_{p,app}$, is calculated from L_1/t_0 with the analytical overall monomer concentration, $c_{M,a}$. For details on the PLP-SEC technique the reader is referred to, e.g., Beuermann et al. [22]. The influence of CO_2 concentration on propagation rate was determined for methyl methacrylate (MMA) and butyl acrylate (BA) [22]. Results for both monomers are presented as relative propagation rate coefficients, $k_{p,app}/k_{p,bulk}$, plotted vs relative monomer concentration, $c_{M,CO_2}/c_{M,bulk}$ in Fig. 4.5. Starting from bulk polymerization, $c_{M,CO_2}/c_{M,bulk} = 1$, $k_{p,app}/k_{p,bulk}$ decreases steadily to values of 0.6 at around $c_{M,CO_2}/c_{M,bulk} = 0.5$. A further reduction toward even lower $c_{M,CO_2}/c_{M,bulk}$ is not seen.

In addition to BA and MMA homopolymerizations, $k_{p,app}$ at a relative monomer concentration of $c_{M,CO_2}/c_{M,bulk} = 0.6$ corresponding to 40 wt% CO_2 was determined for various other monomers. The results are listed in Table 4.1. With the exception

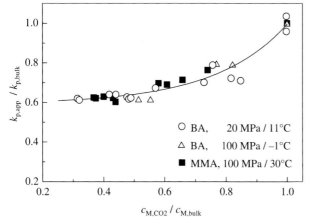

Fig. 4.5 Variation of the propagation rate coefficient $k_{p,app}$ with monomer concentration, c_{M,CO_2}, for MMA and BA homopolymerizations in CO_2. Propagation rate coefficient and monomer concentrations are given relative to the bulk polymerization values, $k_{p,bulk}$ and $c_{M,bulk}$, respectively. (Reprinted with permission from [22]).

Table 4.1 The ratio of $k_{p,app}/k_{p,bulk}$, where $k_{p,app}$ refers to polymerizations in 40 wt% CO_2. For further details see text.

Monomer		$k_{p,app}/k_{p,bulk}$	Ref.
Methyl acrylate	MA	0.60 ± 0.10 [a]	[23]
Butyl acrylate	BA	0.65 ± 0.05 [a]	[22]
Dodecyl acrylate	DA	0.75 ± 0.05	[23]
Methyl methacrylate	MMA	0.65 ± 0.05 [a]	[22]
Butyl methacrylate	BMA	0.78 ± 0.05 [a]	[23]
Dodecyl methacrylate	DMA	0.83 ± 0.05 [a]	[23]
Glycidyl methacrylate	GMA	0.87 ± 0.05 [a]	[23]
Cyclohexyl methacrylate	CHMA	0.75 ± 0.05	[23]
iso-Bornyl methacrylate	iBoMA	0.60 ± 0.05	[23]
Styrene	S	1 ± 0.05	[23]
Vinyl acetate	VAc	1 ± 0.05	[23]
Hydroxypropyl methacrylate	HPMA	1 ± 0.05	[24]

a) The number differs slightly from the previously published value [25], which changes results from additional experimental data that became available and from a modification of the data evaluation mode.

of styrene, vinyl acetate (VAc), and hydroxypropyl methacrylate (HPMA) polymerizations, for which no CO_2-induced variation in $k_{p,app}$ was seen, a reduction in $k_{p,app}$ in the presence of CO_2 was found. For alkyl acrylates and alkyl methacrylates, the CO_2 influence on $k_{p,app}$ is most pronounced for the methyl ester. The longer the alkyl ester group the smaller is the variation in $k_{p,app}$. On the other hand, for methacrylates with cyclic or bridged ester groups, such as glycidyl methacrylate (GMA), cyclohexyl methacrylate (CHMA), and iso-bornyl methacrylate (iBoMA), the CO_2 influence is largest for the most bulky iso-bornyl ester group (see Scheme 4.2), whereas for GMA with the comparatively small glycidyl ester group a rather minor reduction in $k_{p,app}$, by 13%, is found.

To gain a better understanding of the CO_2 influence on $k_{p,app}$, the pressure and temperature dependencies of $k_{p,app}$, with around 40 wt% CO_2 being present, were investigated. Fig. 4.6 presents the temperature dependence of $k_{p,app}$ for BA at 100 MPa and iBoMA at 30 MPa. The full lines were obtained from linear fitting of the experimental data, and the dashed and dotted lines represent the temperature dependence of $k_{p,bulk}$. For a given monomer, the slopes of the Arrhenius fits

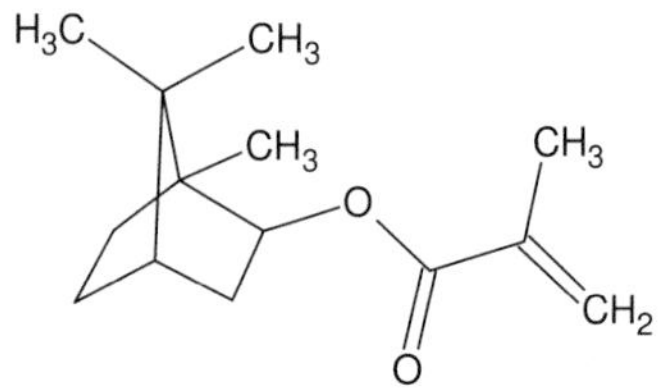

Scheme 4.2

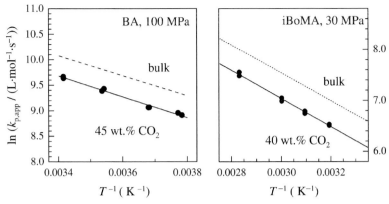

Fig. 4.6 Temperature dependence of $k_{p,app}$ for BA and iBoMA homo-polymerizations in CO_2 and in bulk. The dashed line for $k_{p,bulk}$ of BA is extrapolated from [27]. The dotted line for $k_{p,app}$ of iBoMA is extra-polated from [28]. For further details see text. (Reprinted with permission from [23], copyright Wiley-VCH).

for reactions in bulk and in the presence of 40 wt% CO_2 are identical within experimental accuracy. The same behavior was found for GMA, CHMA, and dodecyl acrylate (DA). Investigations into the pressure dependence of $k_{p,app}$ for BA [26], methyl acrylate (MA), DA, CHMA, and GMA [23] showed that the activation volume $\Delta V^{\#}(k_{p,app})$ is also not influenced by the presence of 40 wt% CO_2. The finding that both activation parameters, $\Delta V^{\#}(k_{p,app})$ and $E_A(k_{p,app})$, are not affected by CO_2, suggests that the CO_2-induced variation of propagation rate is not an intrinsic kinetic effect in the sense that the propagation rate coefficient varies. Thus, according to Eq. (2), local monomer concentration, $c_{M,loc}$, is assumed to be lower than the analytical overall monomer concentration, whereas $k_{p,kin}$ should be more or less identical to $k_{p,bulk}$.

The occurrence of differences between $c_{M,loc}$ and overall monomer concentration, $c_{M,a}$, is at least partly due to differences in solvation of polymer segments by CO_2–monomer mixtures and by the pure monomer. The segment-solvent interactions compete with intramolecular interactions of polymer segments. Along these lines, the observed effects of CO_2 on acrylate and methacrylate $k_{p,app}$ may be understood as an indication of effects on $c_{M,loc}$ resulting from enhanced interactions between polar polymeric segments and from a simultaneous weakening of segment-solvent interactions. As a consequence, $c_{M,loc}$ will be lower than $c_{M,a}$. This argument is consistent with the experimental finding that no reduction in $k_{p,app}$ occurs (1) for systems where polar intersegmental interactions are absent, as in the case of styrene polymerization, and (2) for situations where CO_2 has a solvent power for polymer segments which is not too far below that of the associated monomer. The latter situation seems to apply in the case of vinyl acetate polymerization.

It should further be noted that, even for monomers where $k_{p,app}$ is below $k_{p,bulk}$, this difference may become smaller or may disappear if low-molecular-weight polymer is produced in the PLP-SEC experiments, as is the case when high laser pulse repetition rates are used. For example, the MMA data contained in Fig. 4.5 and Table 4.1 refer to polymerizations where L_1 is at least 56 000 g mol^{-1} [22]. Van Herk et al. reported for MMA that $k_{p,app}$ is not affected for reaction conditions at which L_1 is around 10000 g mol^{-1} [29]. Such low-molecular-weight poly(methyl methacrylate) shows better solubility in the monomer-scCO$_2$ mixture than high-molecular-weight material. Moreover, the tendency of low-molecular-weight polymer to form coils is less pronounced, which may disfavor intramolecular segment-segment interactions. As a consequence, $c_{M,loc}$ and $c_{M,a}$ should differ to a smaller extent for low-molecular-weight material.

The differences in CO$_2$ influence on $k_{p,app}$ within one monomer family, e.g., within the acrylate or methacrylate families, as listed in Table 4.1, are assigned to the different shielding capabilities of the ester groups [23]. The long flexible alkyl ester groups may significantly shield carbonyl groups, thus reducing intra-polymer interactions of polar segments to a larger extent than is the case with small alkyl ester groups. As a consequence, the reduction in $k_{p,app}$ upon the addition of CO$_2$ should be less pronounced for (meth)acrylates with larger n-alkyl groups, which is indeed what is experimentally observed. Comparison of $k_{p,app}$/$k_{p,bulk}$ values of acrylates and methacrylates with identical alkyl ester group (Table 4.1) indicates that this reduction is smaller for methacrylates, which may be understood as being due to some additional shielding action of the a-methyl group at the poly(methacrylate) backbone.

As can be seen from Table 4.1, the correlation between the impact of CO$_2$ on $k_{p,app}$ and ester size may not be transferred to cyclic esters. For the two monomers with smaller ester size, GMA and CHMA, a lower CO$_2$-induced reduction of $k_{p,app}$ is seen than with iBoMA. The unexpectedly small difference between GMA polymerizations in bulk and those in solution in CO$_2$ suggests that solubility arguments play a major role with this monomer. Presumably as a consequence of the oxygen-containing ring, poly(glycidyl methacrylate) exhibits a significantly higher solubility in CO$_2$ than alkyl methacrylate polymers [30], which is indicative of favorable interactions between GMA segments and CO$_2$. Therefore, only small differences between $c_{M,loc}$ and $c_{M,a}$ occur for GMA polymerizations in solution in CO$_2$. The situation should be similar to the one observed for VAc-polymerization [23], where $k_{p,app}$ in bulk and in solution of CO$_2$ is the same because of the remarkable solubility of poly(vinyl acetate) (PVAc) in CO$_2$ [31]. The significant lowering of $k_{p,app}/k_{p,bulk}$ in the case of iBoMA is attributed to a relative enhancement of CO$_2$ concentration at the free-radical site, which occurs because the small linear CO$_2$ molecules fit more easily into the rigid and crowded structure around the macroradical chain. Based on the observed changes of propagation rate in the presence of CO$_2$, MMA and iBoMA homo-propagation has recently been studied for a wide variety of solvents [32]. The message from these studies is that the addition of solvents may vary $k_{p,app}$ to significant extents, but results in no change of $E_A(k_{p,app})$. The solvent-induced changes in $k_p \cdot c_M$ were assigned to local monomer concentration effects rather than to effects on the propagation rate coefficient. $k_{p,app}$ for solution polymerizations was found to be linearly correlated with the difference in molar volumes of monomer and solvent molecules.

4.4.2
Reactivity Ratios

As has been shown above, the CO_2-induced reduction in $k_{p,app}$ amounts up to about 40%. It is interesting to see whether reactivity ratios, which determine co-polymer composition, may also be affected by the addition of scCO_2. Special attention has been paid to copolymerization of such monomer pairs for which homopolymerization $k_{p,app}$ values are affected by CO_2 in different ways, such as styrene-(meth)acrylate systems. In addition, acrylate-(meth)acrylate copolymerization systems were studied.

According to the terminal model, the composition of binary copolymers is determined by the reactivity ratios r_i and by the composition of the monomer mixture via Eq. (4) [33]:

$$F_1 = \frac{r_1 f_1^2 + f_1 f_2}{r_1 f_1^2 + 2 f_1 f_2 + r_2 f_2^2} \quad \text{with} \quad r_i = k_{p,ii}/k_{p,ij}(i = 1, 2 \text{ and } i \neq j) \tag{4}$$

where F_1 is the mole fraction of monomer 1 in the copolymer, f_1 and f_2 are the mole fractions of monomers 1 and 2, respectively, in the monomer feed, $k_{p,ii}$ is the homopropagation rate coefficient of monomer i, and $k_{p,ij}$ the cross-propagation rate coefficient for adding monomer j to a propagating radical terminating in an i unit. Fig. 4.7 shows data for styrene (S)-butyl acrylate (BA) copolymerizations at 80 °C and 30 MPa carried out in both 40 wt% scCO_2 and in bulk. Copolymer composition of samples from reaction to low degrees of monomer conversion

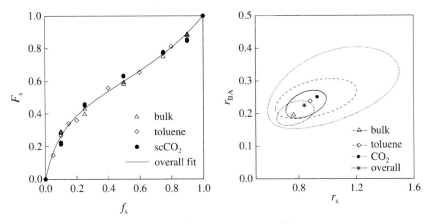

Fig. 4.7 Copolymer composition data (left) and 95% joint confidence intervals (right) obtained for styrene-BA polymerizations in two solvents and in bulk [34].

was determined by 300 MHz ^{1}H NMR spectroscopy. In addition, data for S-BA co-polymerizations in toluene (40 wt%) at 80 °C and ambient pressure are included [34]. The copolymerization diagram on the left hand side of Fig. 4.7 indicates that the styrene copolymer content, F_S, does not depend on whether the reaction is carried out in bulk or in solution in scCO$_2$ or in toluene. The entire set of composition data may be represented by a single fit to Eq. (4). The fitting procedure yields r_S and r_{BA} values. These reactivity ratios were determined for the individual data sets of each system and for the combined data set. The resulting 95% joint confidence intervals [35] for r_S and r_{BA} are plotted on the right hand side of Fig. 4.7. The r_i values for reactions in bulk, in toluene, and in scCO$_2$ are not significantly different. Similar observations have been made for S-MMA, S-BMA, S-DMA [34], S-GMA [30], MA-DA, and for DA-DMA copolymerizations [38]. This observation suggests that the reactivity ratios which are available for a wide variety of systems are well applicable for predicting copolymer composition for fluid systems containing significant amounts of CO$_2$.

The finding that reactivity ratios are not influenced by adding CO$_2$, even in cases where the associated homopropagation rate coefficients are affected to different extents by the presence of CO$_2$, is understood by assuming that CO$_2$ may affect $c_{M,loc}$, but leaves $k_{p,kin}$ unchanged. The sum of local concentrations of the two monomers may differ from total overall concentration in the presence of significant amounts of CO$_2$. In the case that both monomers are, however, of similar solvent quality for the copolymer, their relative amounts should be the same in both the local environment of the free-radical chain end and in the overall system irrespective of CO$_2$ content. As a consequence, the reactivity ratios should be the same in different solvent environments including the situation of the monomer mixture being considered as one particular type of solvent environment.

In summary, the propagation rate coefficients for bulk free-radical polymerization may be used to model polymerizations with significant amounts of CO$_2$ in cases where the monomer is non-polar, where the polymer is soluble in scCO$_2$, or where polymer molecular weight is not too high. Because of the limited solubility of most common high-molecular-weight polymers, solution polymerizations in fluid CO$_2$ target low-molecular-weight material, e.g., polymeric binders for coating applications at number average molecular weights of or even below 4000 g mol^{-1} [2]. As such low-molecular-weight polymer is mostly rather readily soluble, and as coiled structures (in which intersegmental interactions may give rise to $c_{M,loc}$ being different from $c_{M,a}$) will not develop, significant effects of adding CO$_2$ on either k_p or on c_M are not to be expected. Thus, modeling of polymerizations to low-molecular-weight material may be carried out using analytical monomer concentrations $c_{M,a}$ and the extended k_p data sets deduced from PLP-SEC studies of bulk homo- and copolymerizations [36]. For polymerization to high-molecular-weight material of polar monomers, the solvent power of scCO$_2$ for the polymer being poor, the analysis given above should be used and $c_{M,loc}$ and k_p values should be introduced into a detailed modeling. Reactivity ratios appear to be insensitive toward solvent environment. Thus bulk r_i values are suitable for modeling copolymer composition in fluid-phase polymerization with significant amounts of CO_2. It should, however, be noted that this insensitivity of copolymer composition toward the presence of CO_2 does not necessarily hold for copolymerization propagation rate. Even if the co-propagation rate coefficients (which should be described via the implicit penultimate unit effect (IPUE) model [39]) are identical for bulk and solution experiments, local monomer concentration of the monomers may deviate from the analytical overall concentrations and thus may affect propagation rate. First studies into such effects have already been carried out [5, 6, 38]. According to the above arguments for homopolymerization, these effects will be of minor importance in the case of polymerization to low molecular weight, which is probably the technically most relevant situation.

4.5
Termination

In contrast to chemically controlled propagation, the termination reaction is diffusion controlled throughout the entire range of monomer conversions. It is to be expected that termination rate coefficients may increase upon addition of CO_2, as the viscosity is lowered relative to bulk polymerization. In addition to the dependence on temperature, pressure, and degree of monomer conversion, termination rate coefficients, as is generally accepted, are affected by the chain lengths i and j of the two terminating radicals, which is represented by the notation $k_t(i,j)$ [39]. Powerful methods for studying chain-length-dependent termination have emerged during recent years [40–42], but have mostly been applied to bulk polymerizations. For estimating rates of polymerization and molecular weights, the chain-length-averaged termination rate coefficient, $\langle k_t \rangle$ has proven to be a useful quantity. In what follows, we will report on the influence of $scCO_2$ on $\langle k_t \rangle$ of acrylate, methacrylate, and styrene homopolymerizations.

Termination kinetics were studied using the SP–PLP technique, which combines single-pulse initiation of polymerization with in-line NIR-spectroscopic detection of monomer concentration, $c_M(t)$. Time-resolution of the measurement of monomer concentration may be as low as a few microseconds, and the time interval for analysis may be extended up to a few seconds after firing the laser pulse at $t=0$.

$$\frac{c_M(t)}{c_M^0} = \left(2 \cdot \langle k_t \rangle \cdot c_R^0 \cdot t + 1\right)^{-k_p/2\langle k_t \rangle} \tag{5}$$

Equation (5), in which c_M^0 and c_R^0 are the monomer concentration and the pulse-induced initial radical concentration at $t=0$, respectively, is used for fitting the experimental monomer concentration vs time traces to yield the parameters $\langle k_t \rangle / k_p$ and $\langle k_t \rangle \cdot c_R^0$. If k_p is known from independent PLP–SEC experiments, $\langle k_t \rangle$ is obtained as a function of monomer conversion from successive SP–PLP experiments, each covering a very small monomer conversion interval, typically far below one per cent. This allows for a point-wise probing of termination kinetics over extended ranges of monomer conversion. Details of the experimental procedure and of the evaluation method are given elsewhere [9, 43]. The method is particularly well suited for the analysis of termination of rapidly propagating and slowly terminating monomers. To study termination kinetics of methacrylates and styrene, which are low k_p-high k_t monomers, chemically initiated polymerizations were carried out in which in-line FT-NIR spectroscopy was used to monitor monomer conversion as a function of polymerization time [10]. For such chemically initiated polymerizations, $\langle k_t \rangle$ is obtained via the polymerization rate expression, Eq. (6), with k_p and the initiator decomposition rate coefficient, k_d, together with initiator efficiency f, being known from independent experiments [44]:

$$R_p = -\frac{dc_M}{dt} = c_M \cdot k_p \cdot \left(\frac{f \cdot k_d \cdot c_1}{\langle k_t \rangle} \right)^{0.5} \tag{6}$$

Shown in Fig. 4.8 is the conversion dependence of $\langle k_t \rangle$ as determined via SP–PLP for MA and DA homopolymerizations at 40 °C and 100 MPa both in bulk and in solution of 40 wt% CO_2. The pressure and temperature conditions were chosen such as to yield excellent signal-to-noise quality of the monomer concentration vs time (after firing the laser pulse) traces. For in-depth mechanistic studies, so-called mid-chain radicals have to be accounted for [45]. However, the

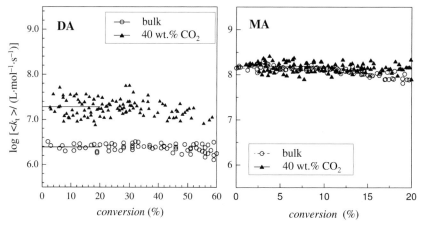

Fig. 4.8 Conversion dependence of chain-length-averaged termination rate coefficients, $\langle k_t \rangle$, for DA and MA homopolymerization in bulk and in solution containing 40 wt% CO_2 at 40 °C and 100 MPa. The $\langle k_t \rangle$ values are calculated from $\langle k_t \rangle / k_p$ by using the $k_{p,app}$ values presented in the previous section. (Reprinted from S. Beuermann, M. Buback, *Progr. Polym. Sci.* **2002**, *27*, 191–254, with permission from Elsevier).

influence of mid-chain radical propagation on the effective propagation rate coefficient is minor at significant monomer concentrations. A detailed discussion is provided by Asua et al. [45].

The data on the left hand side of Fig. 4.8 demonstrate that $\langle k_t \rangle$ of DA is constant over the extended conversion range under investigation. It is primarily because of the cosolvent action of the monomer that polymerization in the presence of a significant amount of CO_2 may be carried out up to fairly high conversion, at least up to 60%, and thus up to high polymer contents. Bulk DA polymerization in homogeneous phase may be run to even higher conversion. The remarkable point to note from Fig. 4.8 is that $\langle k_t \rangle$ is invariant toward monomer conversion, although viscosity increases significantly with polymer content. On the other hand, $\langle k_t \rangle$ greatly increases (by a factor of eight) upon replacing part of the DA monomer by CO_2.

In clear contrast to the situation with DA, no difference between the bulk and the solution data (with 40 wt% CO_2) is seen for methyl acrylate (MA) homopolymerization at monomer conversions up to 20% (right hand side of Fig. 4.8). The difference in $\langle k_t \rangle$ behavior between DA and MA homopolymerizations is considered to be due to the shielding action of the dodecyl group in DA, which is much more effective than that of the methyl group in MA. Under the conditions where termination is under segmental control [46], the process of segment reorientation of two entangled macroradicals determines termination rate (rather than the process of two radicals getting into this entangled situation by the so-called translational diffusion process). The relevant viscosity for segmental diffusion is determined by the intra-coil medium, i.e. by monomer viscosity in the case of bulk polymerizations and by the viscosity of the monomer/solvent mixture in the case of solution polymerizations. Translational diffusion, on the other hand, is essentially determined by the concentration and type of polymer which is contained in the polymerizing mixture. In addition to intra-coil viscosity, the shielding effects of units that sit close to the radical functionality may affect segmental diffusion. Because of the significant shielding action by the dodecyl group, entanglement of two DA macroradicals does not necessarily result in termination, and the two radicals may become separated before the two radical functionalities have approached each other. With CO_2 added, the intra-coil viscosity is significantly reduced, which enormously enhances the potential of the entangled radicals to explore their immediate environment and bring the radical functionalities into close contact. The enhanced mobility is thus associated with an increase in $\langle k_t \rangle$. A different situation applies with MA. Because of the poor shielding potential of the methyl group, entanglement of two MA macroradicals mostly results in a successful termination event. As a consequence, lowering intra-coil viscosity by adding CO_2 will not result in any further enhancement of termination probability.

Studies into $\langle k_t \rangle$ values for methacrylate homopolymerizations were carried out on MMA and DMA at 80 °C and 30 MPa in both bulk and solution with 40 wt% CO_2 present [38]. For both monomers, the measured conversion dependence of $\langle k_t \rangle$ is depicted in Fig. 4.9. In the conversion range under investigation, which however is not very extensive for MMA, $\langle k_t \rangle$ is independent of the degree of monomer conversion. Other than with the associated members of the acrylate family (Fig. 4.8), both methacrylates exhibit almost the same increase in $\langle k_t \rangle$ (by about a factor of four) in going from bulk polymerization to the solution polymerization with 40 wt% CO_2 present. This finding suggests that the shielding effect is controlled by the a-methyl group located in close proximity to the radical functionality rather than by the size of the alkyl ester group.

For styrene polymerizations carried out at 80 °C and 30 MPa with CO_2 contents up to 40 wt%, even larger enhancements in $\langle k_t \rangle$, by up to one order of magnitude, have been reported to occur upon the addition of CO_2 [10]. However, the activation parameters for 40 wt%, $E_A(\langle k_t \rangle)$ and $\Delta V^{\#}(\langle k_t \rangle)$, are more or less identical to the corresponding bulk polymerization data [5, 10]. The increase in $\langle k_t \rangle$ essentially occurs in the region up to 10 wt% CO_2. Further increase in CO_2 content does not result in any significant change of $\langle k_t \rangle$. Changes in solvent quality may be responsible for the pronounced effects on $\langle k_t \rangle$. Lower solvent quality may be associated with the formation of more tightly coiled polymeric species, including more tightly coiled macroradicals which terminate at a higher rate [47]. Similar arguments have been used in the discussion of free-radical polymerizations in conventional solvents [47]. The quantitative analysis of the effects of CO_2 addition on $\langle k_t \rangle$ would certainly benefit from further independent experiments into polystyrene coil size and mobility at various CO_2 contents.

In summary, the influence of CO_2 on the termination kinetics strongly depends on the monomer and may not be generalized, with the exception of methacrylates, for which an enhancement of $\langle k_t \rangle$ by a factor of 4 was found.

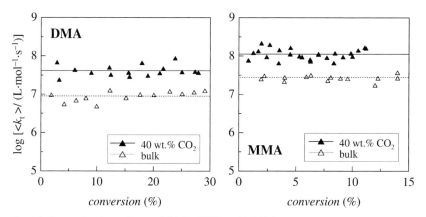

Fig. 4.9 Conversion dependence of $\langle k_t \rangle$ for DMA and MMA homopolymerizations in bulk and with 40 wt% CO_2 at 80 °C and 30 MPa. The data are derived from chemically initiated polymerizations.

4.6
Chain Transfer

For polymerizations in $scCO_2$ it is particularly important to be able to limit polymer molecular weight, as the range for polymerization in homogeneous fluid phase is significantly enhanced for reactions that yield lower molecular weights. An efficient way of controlling molecular weight is by adding chain-transfer agents (CTAs). To do this, it is important to know whether and to which extent chain-transfer rate coefficients, k_{tr}, are influenced by the presence of CO_2. The influence of CO_2 on k_{tr} values for the conventional CTA dodecyl mercaptan (DDM), of the MMA trimer, and of the catalytic CTA bis(borondifluoro diphenylglyoximato) cobaltate (CoPhBF) (see Scheme 4.3) were studied in MMA and in styrene homo- and copolymerizations [48].

k_{tr} may be measured by the well-known Mayo method [49], in which k_{tr}/k_p is derived from M_n values. For relatively low-molecular-weight polymers it may be difficult to determine accurate M_n values because of significant uncertainties of SEC associated with end-group effects [50] due to the loss of low-molecular-weight material during the polymer isolation procedure, residual monomer, or CTA with low volatility, e.g., DDM remaining with the polymer sample. Thus, it was decided to use the chain-length distribution (CLD) method for determination of the chain-transfer constant C_T, where $C_T = k_{tr}/k_p$. Details of the CLD method have been presented in references [51–53]. In contrast to the Mayo method, only sections of the MWD at moderate and high molecular weights are used for analysis, which reduces or even eliminates the above-mentioned problems. Within the CLD method, the MWDs obtained as $w(\log M)$ from SEC analysis were transformed into the natural logarithm of the number distribution, $\ln f(M)$. The slope of the linear $\ln f(M)$ vs M correlation in the region of the peak molecular weight is plotted as a function of the ratio of CTA and mono-

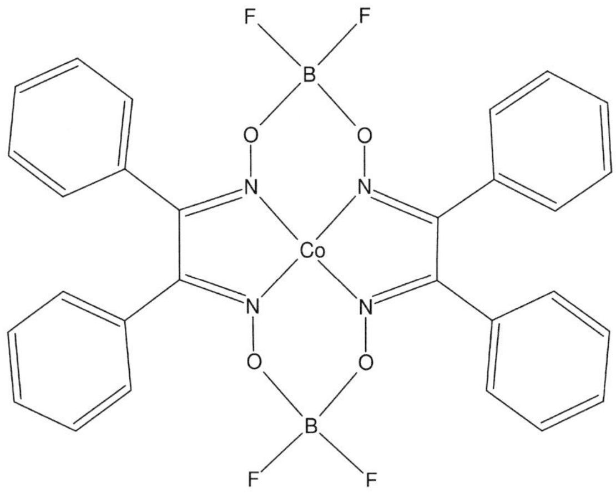

Scheme 4.3

mer concentrations, c_{CTA}/c_M. After the discussion of local monomer concentrations in the Section 4.4, the question arises which monomer concentration should be used. According to the above arguments, the relative amounts of CTA and of monomer should not be changed by the presence of CO_2. Therefore, it is assumed that the overall concentrations of CTA and monomer should provide good estimates for the data evaluation.

As anticipated for a chemically controlled reaction, CO_2 has only a minor influence on the rate coefficient for chain-transfer to DDM and to the MMA trimer in MMA and styrene homo- and copolymerizations. Going from bulk polymerization to solution polymerization with 40 wt% CO_2 present enhances C_T by about 10%, but leaves the associated activation volume, $\Delta V^{\#}(C_T)$, unchanged [48]. As pointed out in the previous section, the observed lowering of $k_{p,app}$ upon increasing CO_2 content is no true kinetic effect, and the propagation rate coefficient $k_{p,kin}$ most likely remains unaffected by the presence of CO_2. Thus, k_{tr} for DDM and for the MMA trimer should not be significantly varied by the presence of CO_2.

Chain transfer with CoPhBF acting as the CTA is a catalytic process. The reaction mechanism is represented by Scheme 4.4 below [50, 54].

$$R_n \;+\; Co(II) \;\longrightarrow\; P_n \;+\; Co(III)-H$$

$$Co(III)-H \;+\; M \;\longrightarrow\; Co(II) \;+\; R_1$$

Scheme 4.4

First, an H-atom is transferred from the macroradical R_n to the Co(II)-complex forming a polymer molecule P_n with a terminal double bond and a Co(III)-complex. In the subsequent step, re-initiation occurs via hydrogen transfer from the Co(III)-complex to a monomer molecule, thus generating a radical of chain length 1. Studies into MMA homopolymerization revealed that the C_T of CoPhBF is significantly enhanced by the presence of CO_2. Fig. 4.10 shows the pressure dependence of C_T measured for MMA polymerizations at 80 °C in bulk and in the presence of 40 wt% CO_2. The data point for ambient pressure is in very satisfactory agreement with the literature values [55]. C_T data for polymerization in CO_2 are higher than those for bulk, e.g., by 50% at 30 MPa. As pointed out above, the true propagation rate coefficient should not be affected by the presence of CO_2. The observed enhancement of C_T thus suggests an increase by about 50% for the k_{tr} of CoPhBF upon adding 40% CO_2 to the reaction mixture.

In addition to the data summarized above, the CO_2 influence on C_T for CoPhBF in MMA polymerizations has recently been studied by Davis and co-workers [57] at, however, quite different reaction conditions, i.e. 50 °C, 150 MPa, with a CO_2 content of 80%. An enhancement of C_T by approximately one order of magnitude compared to polymerizations in solution in toluene was reported.

Fig. 4.10 indicates that the fitted straight lines for the pressure dependence of CoPhBF C_T in bulk and in solution with 40 wt% CO_2 have slightly different

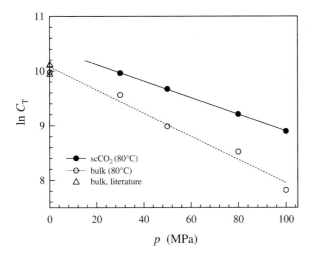

Fig. 4.10 Pressure dependence of the chain-transfer constant for MMA homopolymerizations in bulk and in 40 wt% CO_2 at 80 °C. Literature data were taken from [55, 56].

slopes. The bulk data were fitted without consideration of the data point obtained for bulk polymerizations at 100 MPa, because under these conditions polymerizations may proceed very quickly, and gel-effect conditions may have been reached under which polymer molecular weight may be influenced by effects other than the action of the CTA. As has been pointed out in the section on propagation rate, $\Delta V^{\#}(k_p)$ is not affected by the presence of CO_2. Thus, with $\Delta V^{\#}(k_p) = -16 \ cm^3 \ mol^{-1}$, as measured for MMA polymerizations [36], an activation volume of $\Delta V^{\#}(k_{tr}) = (29 \pm 2) \ cm^3 \ mol^{-1}$ is obtained for polymerizations in 40 wt% CO_2. This value is close to the activation volume of k_t observed for MMA bulk polymerizations [58, 59].

The results in this section indicate that chemically controlled transfer reactions, such as with DDM or with the MMA trimer, are adequately represented by the rate coefficients reported for bulk polymerization. Catalytic chain transfer processes, as with CoPhBF, are speeded up by the presence of CO_2.

4.7
Conclusions

Propagation rate is moderately influenced by CO_2. Reductions by up to about 40% compared to the propagation rate of the respective bulk polymerizations have been found. Such effects have been considered to be due to local monomer concentration at the free-radical site being different from the overall monomer concentration determined by analysis in the case of solution polymerizations with significant amounts of CO_2. Studies into reactivity ratios provide no evidence for any effect of CO_2 content. The rate coefficients for chemically controlled chain transfer are also insensitive to CO_2 content, whereas the catalytic chain transfer with the Co(II) complex is faster by about 50% in the presence of 40 wt% CO_2 than in bulk MMA polymerization. The CO_2 content may largely influence the diffusion-controlled termination reaction, this effect being strongly dependent on the particular type of monomer and on the mode of termination rate control [46]. A large effect is seen for DA at monomer conversions, at least up to 60%, in which range the presence of 40 wt% CO_2 gives rise to an enhancement of chain-length averaged k_t by a factor of eight. The reported rate coefficients are useful for modeling polymerizations of conventional monomers in the presence of CO_2. They have also been successfully used in modeling MMA dispersion polymerization experiments in CO_2 [60].

Acknowledgments

Financial support by the *Deutsche Forschungsgemeinschaft* (Schwerpunktprogramm: "Überkritische Fluide als Lösungs- und Reaktionsmittel" and European Graduate School: "Microstructural Control in Radical Polymerization") and by the *EU* within the TMR network "Superclean Chemistry 2" is gratefully acknowledged, as is support by the *Fonds der Chemischen Industrie*. We are grateful to Akzo Nobel and DuPont Performance Coatings for financial support and the cooperation enjoyed.

Notation

Chemicals

AIBN   azo-*bis*-isobutyronitrile
BA   butyl acrylate
BMA   butyl methacrylate
BPPP   bis(perfluoro-2-*N*-propoxypropionyl) peroxide
BTMHP   bis(3,5,5-trimethylhexanoyl) peroxide
BzMA   benzyl methacrylate
CBr_4 carbon tetrabromide
CHMA   cyclohexyl methacrylate
CO_2 carbon dioxide
CoPhBF   bis(borondifluoro diphenylglyoximato) cobaltate
CLD   chain-length dependence
CTA   chain-transfer agent
DA   dodecyl acrylate
DDM   *n*-dodecyl mercaptan
DMA   dodecyl methacrylate
DOP   dioctanoyl peroxide
DTBP   di-*tert*-butyl peroxide

FT-NIR	Fourier transform near infrared
GMA	glycidyl methacrylate
HPMA	2-hydroxypropyl methacrylate
iBoMA	*iso*-bornyl methacrylate
IPUE	implicit penultimate unit model
M	monomer
MA	methyl acrylate
MMA	methyl methacrylate
NMR	nuclear magnetic resonance
PLP	pulsed laser polymerization
PVAc	poly(vinyl acetate)
S	styrene
sc	supercritical
SEC	size-exclusion chromatography
SP-PLP	single-pulse pulsed laser initiation
TAPP	*tert*-amyl peroxypivalate
TBPA	*tert*-butyl peroxyacetate
TBPP	*tert*-butyl peroxypivalate
TBPO	*tert*-butyl peroxy-2-ethylhexanoate
VAc	vinyl acetate

Symbols

c_{CTA}	chain-transfer agent concentration	$(mol\ L^{-1})$
c_I	initiator concentration	$(mol\ L^{-1})$
c_M	monomer concentration	$(mol\ L^{-1})$
c_M^0	initial monomer concentration	$(mol\ L^{-1})$
$c_{M,a}$	analytical (overall) monomer concentration	$(mol\ L^{-1})$
c_{M,CO_2}	monomer concentration in solution of CO_2	$(mol\ L^{-1})$
$c_{M,loc}$	local monomer concentration	$(mol\ L^{-1})$
c_{PO}	peroxide concentration	$(mol\ L^{-1})$
c_R	radical concentration	$(mol\ L^{-1})$
c_R^0	radical concentration generated by a single laser pulse	$(mol\ L^{-1})$
C_T	chain-transfer constant	
E_A	activation energy	$(kJ\ mol^{-1})$
E_T^N	Dimroth-Reichardt parameter	
f	initiator efficiency	
f_i	molar ratio of comonomer i in the mixture	
F_i	molar fraction of monomer i in the copolymer	
$f(M)$	number molecular weight distribution	
i	chain length	
k_1	primary dissociation rate coefficient	(s^{-1})
k_2	recombination rate coefficient	(s^{-1})
k_3	decarboxylation rate coefficient	(s^{-1})
k_d	initiator decomposition rate coefficient	(s^{-1})
k_{diff}	out-of-cage diffusion rate coefficient	(s^{-1})

k_p	propagation rate coefficient	$(\text{L mol}^{-1} \text{ s}^{-1})$
$k_{p,ii}$	homopropagation rate coefficient	$(\text{L mol}^{-1} \text{ s}^{-1})$
$k_{p,ij}$	crosspropagation rate coefficient	$(\text{L mol}^{-1} \text{ s}^{-1})$
$k_{p,app}$	apparent propagation rate coefficient	$(\text{L mol}^{-1} \text{ s}^{-1})$
$k_{p,bulk}$	propagation rate coefficient of bulk polymerization	$(\text{L mol}^{-1} \text{ s}^{-1})$
$k_{p,kin}$	true kinetic propagation rate coefficient	$(\text{L mol}^{-1} \text{ s}^{-1})$
k_t	termination rate coefficient	$(\text{L mol}^{-1} \text{ s}^{-1})$
$\langle k_t \rangle$	chain-length averaged termination rate coefficient	$(\text{L mol}^{-1} \text{ s}^{-1})$
$k_t(i,j)$	termination rate coefficient as a function of chain lengths i and j	$(\text{L mol}^{-1} \text{ s}^{-1})$
k_{tr}	chain-transfer rate coefficient	$(\text{L mol}^{-1} \text{ s}^{-1})$
L_1	number of propagation steps between two successive laser pulses	
M_n	number average molecular weight	(g mol^{-1})
MW	molecular weight	(g mol^{-1})
MWD	molecular weight distribution	
p	pressure	(MPa)
P_n	polymer molecule of chain length n	
r_i	reactivity ratio of monomer i	
R_n	radical of chain length n	
R_p	rate of polymerization	$(\text{mol L}^{-1} \text{ s}^{-1})$
s_i	radical reactivity of monomer i	
t	time	(s)
T	temperature	(K)
T_c	critical temperature	(K)
t_0	time between two laser pulses	(s)
$\Delta V^{\#}$	activation volume	$(\text{cm}^3 \text{ mol}^{-1})$
$w(\log M)$	weight fraction of polymer with molecular weight M	
wt%	weight percentage	(%)
x	monomer conversion	
x_{max}	maximum monomer conversion accessible in homogeneous phase	

References

1 J.L. Kendall, D.A. Canelas, J.L. Young, J.M. DeSimone, *Chem. Rev.* **1999**, *99*, 543.

2 S. Beuermann, M. Buback, M. Jürgens, *Ind. Eng. Chem. Res.* **2003**, *42*, 6338.

3 T.A. Davidson, J.M. DeSimone, *Polymerizations in Dense Carbon Dioxide*, in P.G. Jessop, W. Leitner (eds.), *Chemical Synthesis Using Supercritical Fluids*, Wiley-VCH, Weinheim, p. 297, **1999**.

4 T.J. Romack, E.E. Maury, J.M. DeSimone, *Macromolecules* **1995**, *28*, 912.

5 A. Wahl, *PhD. Thesis*, Göttingen, **2000**.

6 C. Isemer, *PhD. Thesis*, Göttingen, **2000**.

7 M.A. McHugh, V. Krukonis, *Supercritical Fluid Extraction: Principles and Practice*, 2nd edn., Butterworths Publishers, Stoneham, **1993**.

8 M. Buback, *Angew. Chem. Int. Ed. Engl.* **1991**, *30*, 1654.

9 S. Beuermann, M. Buback, C. Schmaltz, *Ind. Eng. Chem. Res.* **1999**, *38*, 3338.

10 S. Beuermann, M. Buback, C. Isemer, A. Wahl, *Macromol. Rapid Commun.* **1999**, *20*, 26.

11 M. Buback, C. Hinton, *Z. Phys. Chem. (Munich)*, **1996**, *193*, 61.

12 M. Buback, S. Klingbeil, J. Sandmann, M.-B. Sderra, H.-P. Vögele, H. Wackerbarth, L. Wittkowski, *Z. Phys. Chem.* **1999**, *210*, 199.

13 M. Buback, J. Sandmann, *Z. Phys. Chem.* **2000**, *214*, 583.

14 M. Buback, C. Hinton, *Z. Phys. Chem.* **1997**, *199*, 229.

15 M. Buback, D. Nelke, H.P. Vögele, *Z. Phys. Chem.* **2003**, *217*, 1169.

16 P.A. Charpentier, J.M. DeSimone, G.W. Roberts, *Chem. Eng. Sci.* **2000**, *55*, 5341.

17 Y.L. Barner, *PhD. Thesis*, Göttingen, **1997**.

18 W.C. Bunyard, J.F. Kadla, J.M. DeSimone, *J. Am. Chem. Soc.* **2001**, *123*, 7109.

19 Z. Guan, J.R. Combes, Y.Z. Menceloglu, J.M. DeSimone, *Macromolecules* **1993**, *26*, 2663.

20 R.E. Morris, A.E. Mera, R.F. Brady, Jr., *Fuel* **2000**, *79*, 1101.

21 F.-D. Kuchta, *PhD. Thesis*, Göttingen, **1995**.

22 S. Beuermann, M. Buback, F.-D. Kuchta, C. Schmaltz, *Macromol. Chem. Phys.* **1998**, *199*, 1209.

23 S. Beuermann, M. Buback, V. El Rezzi, M. Jürgens, D. Nelke, *Macromol. Chem. Phys.* **2004**, *205*, 876.

24 S. Beuermann, D. Nelke, *Macromol. Chem. Phys.* **2003**, *204*, 460.

25 S. Beuermann, M. Buback, in: *Supercritical Fluids as Solvents and Reaction Media*, Ed.: G. Brunner, VCH, **2004**.

26 S. Beuermann, M. Buback, C. Schmaltz, *Macromolecules* **1998**, *31*, 8069.

27 S. Beuermann, D.A. Paquet, Jr., J.H. McMinn, R.A. Hutchinson, *Macromolecules* **1996**, *29*, 4206.

28 S. Beuermann, M. Buback, T.P. Davis, N. García, R.G. Gilbert, R.A. Hutchinson, A. Kajiwara, M. Kamachi, I. Lacík, G.T. Russell, *Macromol. Chem. Phys.* **2003**, *204*, 1338.

29 A.M. van Herk, B.G. Manders, D.A. Canelas, M. Quadir, J.M. DeSimone, *Macromolecules* **1997**, *30*, 4780.

30 M. Jürgens, *PhD. Thesis*, Göttingen, **2001**.

31 F. Rindfleisch, T.P. DiNoia, M.A. McHugh, *J. Phys. Chem.* **1996**, *100*, 15581.

32 S. Beuermann, N. García, *Macromolecules* **2004**, *37*, 3018.

33 F.R. Mayo, F.M. Lewis, *J. Am. Chem. Soc.* **1944**, *66*, 1954.

34 S. Beuermann, M. Buback, C. Isemer, A. Wahl, *Proceedings of the 6th Meeting on Supercritical Fluids, Chemistry and Materials*, p. 331, Nottingham, **1999**.

35 A.M. v. Herk, *J. Chem. Educ.* **1995**, *72*, 138.

36 S. Beuermann, M. Buback, *Prog. Polym. Sci.* **2002**, *27*, 191.

37 T. Fukuda, Y.-D. Ma, H. Inagaki, *Macromolecules* **1985**, *18*, 17.

38 D. Nelke, *PhD. Thesis*, Göttingen, **2002**.

39 M. Buback, M. Egorov, R.G. Gilbert, V. Kaminsky, O.F. Olaj, G.T. Russell, P. Vana, G. Zifferer, *Macromol. Chem. Phys.* **2002**, *203*, 2570.

40 M. Buback, M. Egorov, T. Junkers, E. Panchenko, *Macromol. Rapid Commun.* **2004**, *25*, 1004.

41 M. Buback, M. Egorov, A. Feldermann, *Macromolecules* **2004**, *37*, 1768.

42 P. Vana, T. P. Davis, C. Barner-Kowollik, *Macromol. Rapid Commun.* **2002**, *23*, 952.

43 M. Buback, H. Hippler, J. Schweer, H.-P. Vögele, *Makromol. Chem., Rapid Commun.* **1986**, *7*, 261.

44 G. Odian, *Principles of Polymerization*, Wiley, New York, **1991**.

45 J. M. Asua, S. Beuermann, M. Buback, P. Castignolles, B. Charleux, R. G. Gilbert, R. A. Hutchinson, J. R. Leiza, A. N. Nikitin, J.-P. Vairon, A. M. van Herk, *Macromol. Chem. Phys.* **2004**, *205*, 2151.

46 M. Buback, *Makromol. Chem.* **1990**, *191*, 1575.

47 J. Dionisio, H. K. Mahabadi, K. F. O'Driscoll, E. Abuin, E. A. Lissi, *J. Polym. Sci., Polym. Chem. Ed.* **1979**, *17*, 1891.

48 V. El Rezzi, *PhD. Thesis*, Göttingen, **2001**.

49 F. R. Mayo, *J. Am. Chem. Soc.* **1943**, *65*, 2324.

50 A. A. Gridnev, S. D. Ittel, *Chem. Rev.* **2001**, *101*, 3611.

51 P. A. Clay, R. G. Gilbert, *Macromolecules* **1995**, *28*, 552.

52 R. A. Hutchinson, D. A. Paquet, Jr., J. H. McMinn, *Macromolecules* **1995**, *28*, 5655.

53 R. G. Gilbert, *Trends. Polym. Sci.* **1995**, *3*, 222.

54 J. P. A. Heuts, G. E. Roberts, J. D. Biasutti, *Aust. J. Chem.* **2002**, *55*, 381.

55 J. P. A. Heuts, D. J. Forster, T. P. Davis, *Macromolecules* **1999**, *32*, 3907; J. P. A. Heuts, D. J. Forster, T. P. Davis, *Macromol. Rapid Commun.* **1999**, *20*, 299.

56 J. P. A. Heuts, D. J. Forster, T. P. Davis, B. Yamada, H. Yamazoe, M. Azukizawa, *Macromolecules* **1999**, *32*, 2511.

57 D. J. Forster, J. P. A. Heuts, F. P. Lucien, T. P. Davis, *Macromolecules* **1999**, *32*, 5514.

58 S. Beuermann, *PhD Thesis*, Göttingen, **1993**.

59 M. Yokawa, Y. Ogo, T. Imoto, *Macromol. Chem.* **1974**, *175*, 179.

60 U. Fehrenbacher, O. Muth, T. Hirth, M. Ballauff, *Macromol. Chem. Phys.* **2000**, *201*, 1532; U. Fehrenbacher, M. Ballauff, *Macromolecules* **2002**, *35*, 3653.

5
Monitoring Reactions in Supercritical Media*

Thierry Meyer, Sophie Fortini, and Charalampos Mantelis

5.1
Introduction

Chemical reactions in supercritical fluids (SCF) have been extensively studied during the past 30 years. Although many of these studies have been performed on a small scale (<60 mL), recent developments tend to attain the liter scale in order to gain engineering as well as chemical and physical information. To carry out chemical reactions on an industrial scale requires a detailed and comprehensive understanding of the energetics of exothermic reactions. The development of an intrinsically safe process requires data on kinetics, physicochemical properties, thermicity, and safety aspects [1].

The monitoring of chemical reactions is necessary to obtain reliable chemical information on the one hand and to control process parameters on the other. It is often observed that measured quantities are strongly dependent on the methodology and the operating conditions used. Thus, it is mandatory to master the operating conditions in such a manner that measured parameters can be determined in controlled and desired conditions. The monitoring is generally not a single stand-alone observation, but is composed of several individual techniques coupled or linked together.

On-line monitoring of chemical reactions encompasses "on-stream" and "on-reactor" applications of analytical methods to

• monitor the evolution of the chemical composition of a reaction mixture,
• identify process-related chemical species,
• quantify the concentration of reaction ingredients, products, and by-products.

On-line analysis of physical parameters, such as temperature, pressure, volume, density, etc., may also reflect the extent of a chemical reaction in addition to revealing the state of the reactor.

---

* The symbols used in this chapter are listed at the end of the text, under "Notation".

Supercritical Carbon Dioxide: in Polymer Reaction Engineering
Edited by Maartje F. Kemmere and Thierry Meyer
Copyright © 2005 WILEY-VCH Verlag GmbH & Co. KGaA, Weinheim
ISBN: 3-527-31092-4

On-line monitoring and control of chemical reactions helps to

- guarantee and improve product quality and consistency,
- ensure safe reactor operation by monitoring the process and reactor parameters,
- increase the efficiency of the process,
- save time for analysis and sampling,
- understand the fundamentals of the chemical reaction itself.

This chapter will concentrate on monitoring techniques applied to polymerization reactions in supercritical fluids. Different available techniques will be discussed, ending with the coupling of analytical and calorimetric measurements. This kind of coupling could be one solution to the problem of simultaneous evaluation of physicochemical properties, kinetic data, and engineering information such as heat transfer and thermicity.

5.2
On-line Analytical Methods Used in SCF

Because of the rapidly growing number of reactions which can be carried out in supercritical fluids, there is an increasing demand for *in situ* techniques to monitor the course of chemical syntheses in these reaction media. There is a growing need to have efficient analytical techniques in order to determine chemical properties (like concentration and chemical species), physicochemical parameters (like heat capacities, conductivity, density, refractive index, and solubility), thermodynamical information (like phase behavior and boundaries, partitioning, and critical points) and/or engineering information (like transfer phenomena, mixing, and scale-up).

5.2.1
Spectroscopic Methods

Spectroscopic methods are often used as in-line or off-line analytical tools to identify chemical species or determine chemical concentrations. Optical spectroscopy may cover the entire spectral range of wavelengths from the ultraviolet (UV, $\lambda > 10$ nm) to the infrared (IR, $\lambda < 1$ mm). Spectra can be recorded in either absorption (UV, Vis, NIR, IR) or emission (IR, Raman, fluorescence). These methods are used in supercritical media and especially, but not exclusively, in supercritical carbon dioxide ($scCO_2$).

5.2.1.1 **FTIR**
Fourier Transform Infrared Spectroscopy (FTIR) is often used in high-pressure cells to obtain time-resolved absorption bands evolving during the course of the reaction. The spectral information is recorded in the time domain as an interfer-

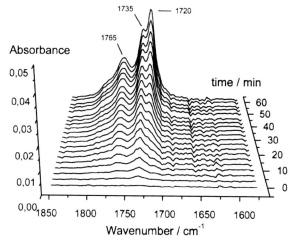

Fig. 5.1 Time-resolved ATR-IR spectra of cyclooctane oxidation in compressed carbon dioxide showing the region of carbonyl stretching frequencies. (Reproduced by permission of The Royal Society of Chemistry [4]).

ogram first, and the Fourier transformed into the frequency domain. The specific advantages are excellent sensitivity at fast spectral acquisition and the capability to simultaneously monitor the variation of absorption bands. One of the first systems, developed in the 1980s, was connected to a capillary flow injection analysis as described by Olesik et al. [2]. A larger-scale system was studied by Kainz et al. [3] in a 200 mL high-pressure reactor equipped with special windows. Their techniques avoids some of the problems frequently encountered with conventional high-pressure IR cells, such as difficulties in pumping the reaction medium and inefficient mass transport due to the small path lengths. Their probe design made the whole set-up more flexible than reactors integrated with conventional ATR sensors (Attenuation Total Reflection). A typical picture of time-resolved IR spectra is shown in Fig. 5.1 [4].

Some recent work, by Schneider at al. [5], presents a new type of high-pressure spectroscopy view cell, with a maximal volume of 67 mL, for the investigation of multiphase reactions. Their set-up allows the quasi-simultaneous measurements of the reactor cell's upper part by transmission spectroscopy with variable path length and of the cell's bottom part by attenuated ATR spectroscopy. Phase behavior and spectroscopic measurements are simultaneously possible, gaining deeper insight into fundamental aspects of heterogeneous or multiphase chemical reactions in supercritical fluids. The relatively small size of the cell may induce some undesired "wall effects" and mixing effects. A comparison of the FTIR spectra in the gas phase and the ATR measurements in the liquid phases for the formylation of morpholine with $scCO_2$ are depicted in Fig. 5.2.

It should be noted that FTIR or ATR could give reliable results in terms of chemical species, concentration, and phase behavior even in CO_2-rich media.

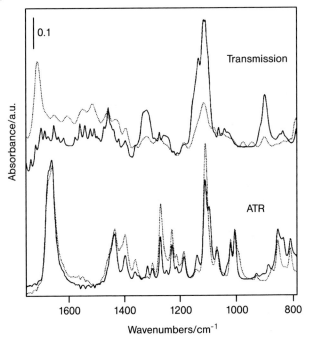

Fig. 5.2 Comparison of FTIR spectra between the gaseous (transmission measurements) and the liquid phase (ATR measurements) for the formylation of morpholine with hydrogen and CO_2. In each measuring mode, spectra were taken before (solid line) and after (dotted lines) the addition of CO_2. The reaction mixture contained morpholine, *N*-formylmorpholine, and water corresponding to a conversion of 50% at 80 °C; hydrogen pressure was 5 Mpa. The background spectrum was measured in nitrogen atmosphere. (Reproduced by permission of The Royal Society of Chemistry [5]).

5.2.1.2 Raman Spectroscopy

Vibrational spectroscopy like Raman or IR is usually quite sensitive to the environment of the molecule. Thus, vibrational spectra are excellent probes of conditions within the fluid. Raman experiments usually involve excitation of the sample by a focused and relatively intense laser beam. The scattered light is then detected in a direction perpendicular to the laser beam. Raman scattering is a relatively weak effect, and so this scattered light has to be collected efficiently (this being one of the main challenges). The intensity of the Raman line is directly proportional to the number of corresponding oscillators in the scattering volume and the intensity of the illuminating radiation.

Performing IR spectroscopy at elevated pressures is often difficult, since the most commonly used high-pressure spectroscopic windows, quartz and glass, strongly absorb IR radiation. This consideration makes Raman spectroscopy an attractive alternative technique for examining solutions in gaseous and supercritical fluids. In Raman spectroscopy, both the excitation and the scattered radiation are generally in the visible region, allowing the use of more economical

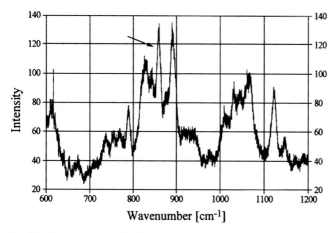

Fig. 5.3 Raman spectra of terbutyl peroxypivalate in n-heptane [7].

glass, quartz, and sapphire windows. Like IR spectroscopy, Raman provides information concerning the vibrational and reorientational dynamics in a system.

An excellent review on vibrational spectroscopy in supercritical fluids was published in 1995 by Poliakoff et al. [6]. In the late 1990s, Kessler et al. [7] developed IR and Raman spectroscopy for the investigation of rapid high-pressure reactions in optical cells. Raman was preferred to IR for the determination of the decomposition rate of peroxides under high pressure. They studied the decomposition of *tert*-butyl peroxypivalate at pressure up to 180 MPa and temperatures of 90–160 °C. A typical Raman spectrum is presented in Fig. 5.3.

More recently, Blatchford and Wallen [8] examined the utility of Raman and NMR spectroscopy as a means of following changes in density of SCF solutions. Their investigation, comparing the area of a Raman band at an arbitrary reference pressure, showed that any change in the area at subsequent pressures should reflect a change in the molar amount of material present in the excitation volume, assuming that the other factors that could affect the peak remain constant. They observed that the Raman integration method provides an excellent means of following the density changes in the system.

5.2.1.3 UV/Vis
UV-Vis spectra are generally highly sensitive but less informative, because they typically consist of a few broad absorption peaks. Chemical reaction monitoring using UV-Vis spectroscopy is less common than using other spectroscopic techniques. Two major devices have been developed for supercritical fluids: the fiber-optic and the cell device. Hunt et al. [9] reported the development of a fiber-optic-based reactor connected directly to a CCD array UV-Vis spectrometer for *in situ* determination of reaction rates in scCO$_2$. The cell can be configured either to study the kinetics of chemical reactions or to determine the rate of dis-

solution and the solubility of compounds in SCF. The domain applicability range for chemical processes occurring is probably in the order of tens of seconds to several minutes.

Fehrenbacher et al. [10] use a UV-Vis cell to measure directly the turbidity in the reactor. It allows monitoring the earliest stage of polymerization with sufficient time resolution. They demonstrated that turbidity is practically insensitive toward multiple scattering of light. Hence, measurements of turbidity as a function of the incident wavelength may be used to study turbid systems such as polymerization in scCO$_2$. Conventional measurements by static and dynamic light scattering would be profoundly disturbed by multiple scattering.

Fehrenbacher and Ballauff [11] demonstrated that turbidimetric measurements can be used to monitor the early stage of the dispersion polymerization of methyl methacrylate (MMA) in scCO$_2$. Turbidimetry is the method of choice for following the growth of particles out of a homogeneous medium. Using this methodology, they demonstrated unambiguously that the early stage of polymerization takes place in the continuous phase. Polymerization within the particles can play no significant role. Hence, the particles can only grow through precipitation of polymer onto their surface. If growing radicals are adsorbed onto a particle, the polymerization may go on for some time, but this process cannot contribute significantly to the observed rate. In the later stage, where much larger particles are present and where the concentration of monomer in the homogeneous phase has decreased considerably, polymerization can take place within the particles [12]. They also stated that the stabilization by the macromonomer stabilizer, polydimethylsiloxane methacrylate (PDMS-MA), is the decisive factor that determines the critical diameter of the particles and hence their final size.

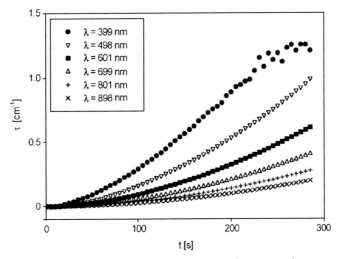

Fig. 5.4 Analysis of the kinetics of the dispersion polymerization by turbidimetric measurements. The turbidimetry measured at different wavelengths given in the insets is plotted against time [10].

5.2.1.4 **NMR**

The study of supercritical fluid systems by Nuclear Magnetic Resonance (NMR) has been accomplished in two ways: the high-pressure probe method and the high-pressure cell method. At this present stage, no paper on the study of polymerization reactions in SCF by NMR could be found. Nevertheless, other reactions have been studied in supercritical media and at extreme conditions [13], and these can be compared with polymerization reactions.

Hoffmann and Conradi [14] monitored hydrogen exchange reactions in supercritical media by *in situ* NMR. They found that higher pressure at constant temperature increases the exchange rate, indicating that hydrogen exchange at supercritical conditions proceeds via charged intermediates. Wallen et al. [15] and Blatchford and Wallen [8] used a polymer NMR cell for monitoring density changes in supercritical fluids. The understanding of the fundamental interactions between surfactants, CO_2-philic molecules, and CO_2, such as the solvatation structure and dynamics, are crucial to the design and development of CO_2-based processes. This method could be used to determine and compare the density increase of either a neat gaseous or supercritical fluid system or a solution in an SCF. Gaemers et al. [16] developed a titanium-sapphire high-pressure cell for multinuclear magnetic resonance under high pressure or supercritical conditions. A schematic drawing of the cell is presented in Fig. 5.5. They could monitor the chemical shift changes of different species upon reaching the critical state. This set-up has two important characteristics: simplicity and flexibility of operation.

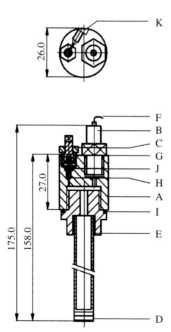

Fig. 5.5 Schematic diagram of the high-pressure cell [16].

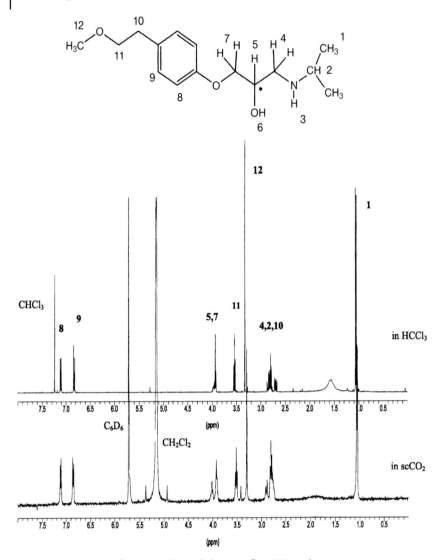

Fig. 5.6 NMR spectra of metoprolol recorded in stop-flow SFC mode and under liquid conditions. (Reprinted with permission from [17], Copyright (2003) American Chemical Society).

Fisher at al. [17] reported the use of on-line NMR spectroscopy with a flow probe for supercritical fluid chromatography (SFC) used for reaction monitoring purposes. They monitored aliphatic amines in $scCO_2$. A typical NMR spectrum realized both in classical media and in $scCO_2$ is presented in Fig. 5.6.

They concluded that the use of on-line NMR spectroscopy employing an SFC probe enables the investigation of reactions in supercritical medium.

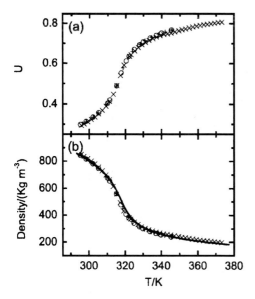

Fig. 5.7 (a) Signal from three different fiber probes showing the variation with temperature at constant pressure of 9.9 MPa of pure CO_2. (b) Densities calculated from **a** compared with NIST values (solid line). (Reproduced with permission from American Institute of Physics [18]).

5.2.2
Reflectometry

The operating principle is the measurement of the intensity of light from a light-emitting diode (He-Ne laser), which is reflected from the end of a fiber immersed in the medium. The amount of reflected light depends on the difference between the refractive index of the fiber and that of the medium, which is related to its density. Avdeev et al. [18] developed this approach for quantitative determination of refractive index in compressed gases/liquids and for monitoring its evolution in supercritical reactors. This approach allows the determination of phase boundaries of a mixture in a heterogeneous supercritical state. As an example, the on-line density measurement of CO_2 at 9.9 MPa is depicted in Fig. 5.7.

5.2.3
Acoustic Methods

The propagation of small pressure pulses or sound waves is a physical effect which can be a source of information for single and multiphase systems. Ultrasonic methods (between 20 kHz and 100 MHz) are easy to use, safe, non-destructive and non-invasive. Morbidelli et al. [19] used this acoustic method, with success, for the determination of the evolution of conversion in various copolymeric systems in the dispersed phase (emulsion polymerization). The same methodology has been applied to bulk and solution systems by Cavin et al. [20, 21] and Zeilmann et al. [22] for monitoring high-solids content polymerization of styrene and MMA.

More recently, Oag et al. [23] developed an optical view cell (volume 80–260 mL) equipped with an ultrasonic sensor to determine phase boundaries and vapor/liquid critical points in supercritical fluids. They observed that acoustic techniques offer a rapid search algorithm for locating the critical points of single-component fluids. Even for temperatures several degrees above a critical point, the acoustic velocity exhibits a minimum when measured as a function of pressure along an isotherm. They coupled this acoustic method with a shear-mode piezoelectric sensor to determine phase equilibrium and a phase diagram. They indicate the need to use simultaneous and complementary techniques in determining the phase diagram and critical points of the mixture relevant to supercritical studies.

5.3
Calorimetric Methods

Most chemical processes are accompanied by temperature changes of the reaction mixture due to the release or consumption of heat during the course of the reaction. Heat evolution is a definite, reproducible, and directly measurable characteristic of a chemical reaction. Thermometric and enthalpimetric methods have been applied to process control and process optimization for many decades [24]. Only recently has this field been open to supercritical conditions [25].

The heat generated by the reaction is directly proportional to the reaction rate for simple systems. The interpretation of the thermogram is more complicated in the case of multiple reactions or simultaneous enthalpic processes such as mixing, dissolution, phase transition, crystallization, etc. Two different calorimetric methods will be discussed: power compensation and heat flow calorimetry.

5.3.1
Power Compensation Calorimetry

Wang et al. [26] were the first to develop a power compensation calorimeter. They assumed that heat losses are constant throughout the reaction in order to link the measured and the generated heat. This can be achieved by maintaining the outside of the autoclave at a constant temperature that is slightly lower than that of the reaction. This temperature differential is maintained by the use of a smaller internal heater whose power can be externally controlled. If an exotherm occurs, then the system automatically reduces the power supplied to the internal heater to compensate and to maintain a constant temperature. The opposite happens when an endotherm occurs.

Their preliminary experimental results demonstrate the feasibility of the technique [26]. They studied, as an example, the polymerization of MMA in $scCO_2$. Their results are in general agreement with previously reported data. The reactor volume used (60 mL) has some advantages and disadvantages. The advantages concern the small amount of reactants that are necessary and the fairly low price of the equipment. The disadvantages are that little engineering information could be obtained and that "wall effects" are still important. It should be noted that the inertia of the system is rather large because of the small ratio between the mass of the reactant and that of the reactor. Nevertheless the preliminary results obtained are promising and encouraging for the future.

5.3.2
Heat Flow Calorimetry

The basis for reaction calorimetry is the energy balance around the reactor. Reaction calorimetry is an efficient tool used to obtain kinetic, thermodynamic, and safety data. A recently developed reaction calorimeter, the RC1e-HP350, allows chemical reactions under supercritical conditions to be investigated [25]. The main technical difference, compared with a classical liquid system, is that the whole reactor volume is occupied by the media. The reaction calorimeter for supercritical thermal analysis has been developed in collaboration with Mettler-Toledo GmbH. The reaction vessel is a 1.3 L high-pressure autoclave operating at up to 35 MPa and 300 °C (Fig. 5.8). The reactor is coupled to a thermostat unit (RC1e) which controls the reaction temperature, T_r. Moreover, the reactor is equipped with a magnetic drive, a 25 W calibration heater, a Pt100 reactor

Fig. 5.8 Picture of the autoclave.

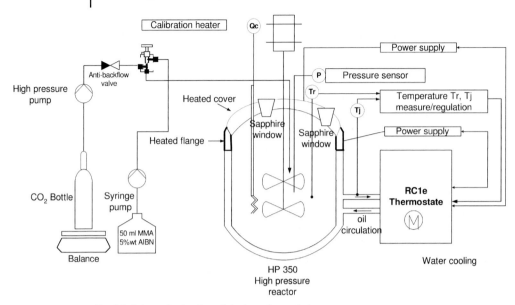

Fig. 5.9 Schematic drawing of the heat flow calorimeter.

temperature sensor, and a pressure sensor (Fig. 5.9). The calorimeter can be operated in three different modes: *adiabatic, isoperibolic,* and *isothermal*.

The set-up gives the opportunity to work at larger scale with reaction conditions close to those of industrial reactors. In fact, most of the studies related to SCF chemical reaction applications are realized in small-scale batch or tubular reactors (1–60 mL). So far, very few publications deal with calorimetry applied to the supercritical phase.

Generally, in "classical" reaction calorimetry only the liquid phase is taken into account in the heat balance. This means that the gas phase in equilibrium with it is neglected because of its small contribution in terms of heat transfer and heat capacity. The situation with supercritical fluids becomes complicated as soon as they occupy all the available volume. This implies that the whole inner reactor surface has to be thermally perfectly controlled when working with supercritical fluids. In this case, the cover and the flange temperature are adjusted on-line to the reaction temperature in order to neglect the heat accumulation term.

5.3.2.1 **Heat Balance Equations** [27]

The most fundamental assumptions are that reactor and jacket temperatures are homogeneous and that the reaction system is perfectly mixed. The heat balance for a semi-batch process is given by

$$\dot{Q}_r + \dot{Q}_c + \dot{Q}_{stir} = \dot{Q}_{acc} + \dot{Q}_{dos} + \dot{Q}_{flow} + \dot{Q}_{loss} \tag{1}$$

where $\dot{Q}_r$ is the heat generation of the reaction, $\dot{Q}_{acc}$ is the accumulation term, $\dot{Q}_c$ is the heat delivered by the calibration probe, $\dot{Q}_{dos}$ corresponds to the amount of heat due to addition of reactants, $\dot{Q}_{flow}$ is the heat flow through the reactor wall, $\dot{Q}_{loss}$ is the heat losses to the surroundings and $\dot{Q}_{stir}$ is the heat dissipated by the stirrer. All heat flows are expressed in watts [W].

The heat released by a polymerization reaction is defined by Eq. (2), where $V_{continuous\ phase}$ is the total volume of the reactor, in this case V_r [m^3]; ΔH_r is the heat of polymerization of the monomer used [J mol^{-1}], and R_p is the global rate of polymerization [mol m^{-3} s^{-1}].

$$\dot{Q}_r = V_{continuous\ phase} \cdot (-\Delta H_r) \cdot R_p \qquad (2)$$

The accumulation term $\dot{Q}_{acc}$ defined by Eq. (3) corresponds to the heat needed to change the internal temperature T_r [K] of the media by a certain amount dT_r/dt [K s^{-1}]. m_r is the reaction mass [kg], c_{pr} is the specific heat capacity of the reaction mass [J kg^{-1} K^{-1}], and c_{pi} is the heat capacity of each insert [J K^{-1}]. Equation (3) does not include the accumulation term for the reactor itself as it is compensated in the WinRC® software by the use of a corrected jacked temperature.

$$\dot{Q}_{acc} = \left(m_r \cdot c_{pr} + \sum_i c_{pi} \right) \cdot \frac{dT_r}{dt} \qquad (3)$$

The calibration probe delivers an amount of heat $\dot{Q}_c$ that is measured on-line and has a value around 25 W. It is used to measure the overall heat transfer coefficient, UA, either during Wilson plot experiments or before and after reactions according to Eq. (4). Calibration runs (10 min) are at constant internal temperature T_r.

$$UA = \frac{\int_{t1}^{t2} (\dot{Q}_c - \dot{Q}_b) \cdot dt}{\int_{t1}^{t2} (T_r - T_j) \cdot dt} \qquad (4)$$

$\dot{Q}_b$ is the baseline term, which will be discussed later, and T_j is the jacket temperature [K].

The dosing term, defined in Eq. (5), is the amount of heat absorbed or released due to addition of reactant at a different temperature from that of the reaction mass:

$$\dot{Q}_{dos} = \frac{dm_d}{dt} \cdot c_{pd} \cdot (T_r - T_{dos}) \qquad (5)$$

where dm_d/dt is the mass flow of dosing [kg s^{-1}], c_{pd} is the specific heat of the added substance [J kg^{-1} K^{-1}], T_r is the current temperature of the reactor contents [K], and T_{dos} is the temperature of the added substance.

The term $\dot{Q}_{flow}$ in Eq. (1) is the heat flow through the reactor wall:

$$\dot{Q}_{flow} = U \cdot A \cdot (T_r - T_j) \tag{6}$$

where U is the overall heat transfer coefficient [W m^{-2} K^{-1}], and A is the heat exchange surface [m^2]. As soon as the supercritical fluid occupies all the volume available, the heat exchange area corresponds to its maximum geometrical value.

The terms $\dot{Q}_{loss}$ and $\dot{Q}_{stir}$ in Eq. (1) are normally constant during an isothermal run and can be combined in the baseline term $\dot{Q}_b$. This means that the integration of a heat production or a calibration peak already takes into account heat losses and the heat dissipated by stirring as soon as stirrer speed and temperature do not change. For reactions with only small changes of physico-chemical properties, the measurement of $U \cdot A$ before and after the reaction taking into account a constant baseline ($\dot{Q}_b$) is sufficient. For polymerizations, this assumption is limited, as the viscosity can change significantly during the process. Thus, the two terms $\dot{Q}_{stir}$ and $U \cdot A$, which are directly dependent on the media viscosity, can evolve during the polymerization process. The influence of $\dot{Q}_{stir}$ could be taken into account by choosing an appropriate baseline. A mean value of U measured at the start and the end of the reaction can be assumed as a first approximation.

From Eqs. (1) and (2) it can be seen that the reaction rate is related to the evolution of the measured heat. For single reactions or one dominant reaction, like the propagation reaction in free-radical or chain polymerization processes, $\dot{Q}_r$ is directly proportional to the measured heat generated by the reaction. For complex systems, where consecutive or parallel reactions with similar thermal contributions occur, the signal corresponds to the addition of all heat contributions, i.e. the macrokinetic. The final and simplified heat balance equation used for the polymerization part is given in Eq. (7):

$$\dot{Q}_r = \dot{Q}_{flow} + \dot{Q}_{acc} + \dot{Q}_{dos} \tag{7}$$

The accumulation term can be neglected as soon as the system is working in a pure isothermal mode. The dosing term can be neglected because of the small quantity of added reactant. Another possibility is to set the dosing temperature the same as the reaction temperature. A more simplified equation is obtained when $\dot{Q}_{acc} + \dot{Q}_{dos}$ are relatively small compared to $\dot{Q}_{flow}$.

5.3.2.2 **Determination of Physico-Chemical Parameters**

Heat Capacity

Heat capacity (c_v) values of CO_2 and $scCO_2$ have been measured and compared to calculated ones derived from Span and Wagner's equation of state [28] in order to evaluate the accuracy of the system's heat flow model. Theoretical c_p and c_v values for $scCO_2$ and liquid-vapor CO_2 at equilibrium are compared with experimental c_v values obtained through the T_r–T_j signal integration (Fig. 5.10). A good agreement between experimental and calculated values is observed with a deviation less than 10%. This indicates that calorimetric evaluations are correct and allows reaction calorimetry to be performed in reliable conditions.

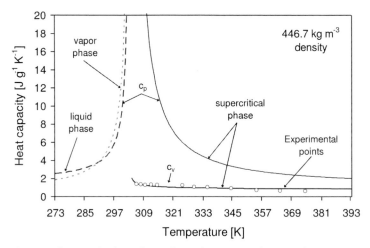

Fig. 5.10 Theoretical values of c_p and c_v and experimental points of c_v.

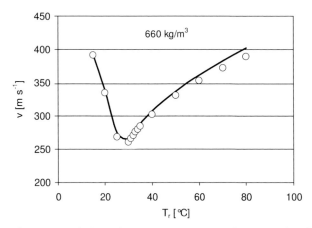

Fig. 5.11 Speed of sound measurements compared to NIST values for $scCO_2$ at a density of 660 kg m^{-3}.

Density

Medium density can be measured on-line by acoustic methods. An ultrasonic sensor was installed in the autoclave and allowed on-line measuring of the speed of sound in the medium. An example is given in Fig. 5.11 for pure $scCO_2$. These values are compared with the NIST database and reveal an excellent agreement. Knowing the speed of sound and pressure/temperature, it is possible to determine the medium density in the case of single or even multiple components.

5.3.2.3 Calorimeter Validation by Heat Generation Simulation

The calorimeter has been validated by measuring thermodynamical parameters and physical properties (Fig. 5.10). Another validation method used is the reaction heat generation simulated by the mean of the calibration probe. A specific heat flow input was created using a Labview interface and the calibration probe serving as internal heater. Simultaneously the response of the calorimeter was measured. The heat generation profile was promoted without an actual chemical reaction happening, so that the calorimetric signal measured represents the ability of the system to measure heat flows. In an ideal theoretical situation, the two signals must be identical. It can be seen in Fig. 5.12 that the calorimetric signal correctly follows the heat flow input given by the calibration probe used as an internal heater except at the very beginning, where the dynamics of the system tend to delay the response. This delay is quite short, less than two minutes, and unavoidable, and is due to the time constant of the calorimeter (in order to have a negligible time constant, the system must have a mass tending to zero). It could be mathematically suppressed by a deconvolution method, which will not be discussed here. By integrating the two signals we have access to the heat generated (51.3 kJ) and measured (49.2 kJ), indicating that the error is below 4%.

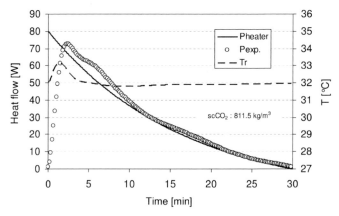

Fig. 5.12 Calorimeter response to a signal input by the calibration heater in simulation mode at a density of $scCO_2$ of 811.5 kg m^{-3}.

This result, together with the accuracy of thermodynamical parameters previously measured, confirms that the supercritical calorimeter is operating correctly and validates the equipment for calorimetric evaluation of reactions in supercritical media.

5.4
MMA Polymerization as an Example

Several "analytical" techniques that could be used to monitor reactions in supercritical fluids have been presented in the preceding sections. In this section, the dispersion polymerization of MMA in $scCO_2$ will be discussed in the light of on-line monitoring.

The dispersion polymerization of methyl methacrylate in supercritical carbon dioxide, using the commercially available poly(dimethylsiloxane) monomethacrylate [29, 30] (PDMS macromonomer) as dispersant, is used as a model reaction. Macromonomers are polymers with a polymerizable terminal functional group commonly used for the formation of graft copolymers. The disadvantage of such molecules is that they are chemically incorporated in the final polymer. Moreover, stirring speed and stabilizer concentration will have an impact on the product quality. However, the polymerization has to be in agreement with the basic requirements of reaction calorimetry, which means having a detectable thermal signal and efficient stirring. Since efficient stirring can destabilize the dispersion, the first approach is consequently to try to find a compromise between operating and synthesis conditions.

5.4.1
Calorimetric Results

A typical reaction procedure is the following. The reaction vessel is charged with monomer (200 g) and the required amount of stabilizer, and is then closed and filled with CO_2 (around 800 g). The temperature is raised to 80 °C. At this temperature, a solution of 50 mL MMA containing 5.1 wt% AIBN is introduced into the reactor under pressure using a syringe pump. The final ratio of AIBN/ MMA is 1 wt% for all the experiments. At the end of the reaction, the reactor content is quenched by cooling and CO_2 venting. Global product yield and the quantity of PDMS monomethacrylate chemically incorporated in the final product are evaluated gravimetrically.

The experiments conducted with 10 wt% PDMS macromonomer at stirring speeds of 200, 400, and 600 rpm yield a fine white powder with high molecular weight ($M_w \approx 120$ kg mol^{-1}) at monomer conversions of about 95% and a well-defined spherical morphology observable by SEM micrograph (Fig. 5.13) [31]. Molecular weights are lower than the ones published by Giles et al. [30], where the experiments were realized in the absence of stirring and at lower temperature. The quantity of PDMS macromonomer chemically incorporated is of the same order of magnitude.

Fig. 5.13 SEM micrograph (5000×) of PMMA produced at 400 rpm.

It can be seen in Fig. 5.14 that the heat flow measured with respect to the total weight of monomer does not depend on stirring speed. Moreover, the good reproducibility of the reaction calorimeter is confirmed. By integration of the thermal curves, the enthalpy of polymerization can be calculated as -56.9 ± 2.2 kJ mol^{-1} and is in good agreement with the literature's value of -57.8 kJ mol^{-1} obtained in conventional solvent [32]. The error of 2.2 kJ mol^{-1} is a standard deviation calculated from a series of seven measurements [33].

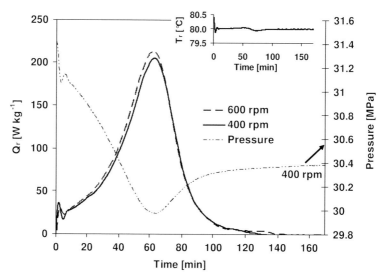

Fig. 5.14 Evolution of the heat released by MMA polymerization at 80 °C, at 400 and 600 rpm and the pressure evolution at 400 rpm.

The pressure can also be used to monitor the polymerization, because when the pressure reaches a plateau this means that the conversion is higher than 90%, as also reported by Lepilleur and Beckman [12] and Wang et al. [26]. It should be noted that the thermal signal obtained by calorimetry is much more sensitive than the pressure and gives more information. This result reveals all the potential of reaction calorimetry for supercritical fluid investigations and polymerization monitoring.

In order to validate the system, external sampling under pressure during the reaction was performed. The reactant contents were analyzed by dynamic head space gas chromatography and the polymer molecular weight distribution by size exclusion chromatography with triple detection.

The thermal monomer conversion is obtained by integration of the heat generation rate expressed in Fig. 5.14. The chemical monomer conversion is obtained by gas chromatography off-line analysis. A very good agreement between the two different measures, as expressed in Fig. 5.15, indicates that calorimetric information excellently describes what is happening during the reaction and also confirms that the propagation step is the main contributor to heat generation [34]. The calorimetric information enables MMA polymerization in supercritical fluids to be correctly monitored.

Wang et al. [26] have discussed some preliminary results using alternative techniques by monitoring polymerizations in $scCO_2$ using a compensation calorimeter with a total volume of 60 mL. They concluded that, at the present moment, this technique was not of very high precision, mainly because of calorimetric problems resulting from the large thermal inertia of the reactor and the high ratio of reactor mass to reaction mass. This can be seen in Fig. 5.16, where the scattering has a significant effect on calorimetric evaluations. They obtained

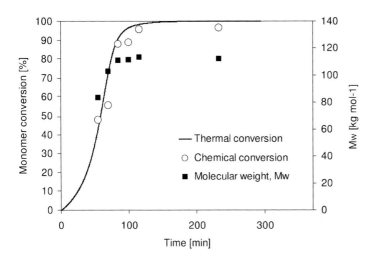

Fig. 5.15 Experimental thermal and chemical monomer conversions and mean molecular weight at 400 rpm.

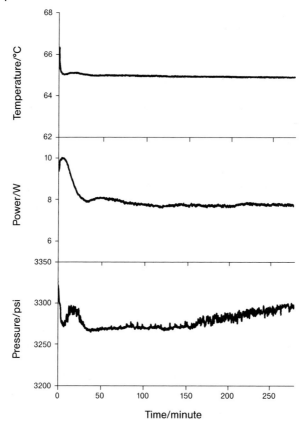

Fig. 5.16 Data of reaction temperature, heater power, and reaction pressure for the polymerization of MMA in scCO$_2$ (AIBN 1.5%, without stabilizer). (Reprinted from [26], Copyright (2003), with permission from Elsevier).

an enthalpy of polymerization from –52.6 to –59.7 kJ mol^{-1}, leading to an accuracy of ca. 10%.

5.4.2
The Coupling of Calorimetry and On-Line Analysis

One of the main advantages of reaction calorimetry on the larger scale is the possibility of inserting into the reactor special analytical probes for on-line measurements. Some preliminary results obtained by coupling an ultrasonic sensor with calorimetry are presented in Fig. 5.17. The sensor is directly inserted into the reactor, its contribution in terms of heat accumulation having been previously determined so that the calorimetric signal is only related to the chemical reaction and process. At the moment, only the sound wave measurement is compared to the

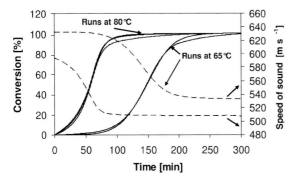

Fig. 5.17 Calorimetric and speed of sound measurements for MMA polymerization at 65 and 80 °C.

thermal conversion obtained by the integration of the calorimetric signal. It can be seen that good agreement exists, indicating that sound speed can be related to monomer conversion. It can also be seen that at high solid content or monomer conversion, the speed of sound remains unchanged, indicating that the system has reached its sensitivity limit due to the high solid content present.

The use of heat flow calorimetry on a larger scale (1.3 L) was efficient and gave reproducible results. The development of supercritical calorimetry required a very sensitive apparatus and good analysis of the heat transfer parameters. The newly developed supercritical reaction calorimeter allows us to measure thermodynamical properties and kinetic and engineering parameters such as reaction enthalpy and the overall heat transfer coefficient. These are also used for process safety and scale-up and optimization studies.

5.5
Conclusions

Different methods for on-line monitoring reactions and polymerization in supercritical fluids have been presented. The most promising of these are based on the coupling of calorimetry with spectroscopic or acoustic devices. This allows us to monitor thermal, engineering, physical, physico-chemical, kinetic, and reactant changes during the course of the reaction. The selection of the most suitable devices to be coupled will depend on the system studied. The final goal of monitoring reactions in supercritical media remains to

- guarantee and improve product quality and consistency,
- ensure safe reactor operation by monitoring the process and reactor parameters,
- increase the efficiency of the process,
- save time in analysis and sampling,
- understand the fundamentals of the chemical reaction itself,
- access all necessary data for a reliable scale-up.

Notation

Symbols

A	heat exchange surface	$[m^2]$
C_{pi}	heat capacity of each insert	$[J\ K^{-1}]$
c_{pd}	specific heat capacity of the dosing mass	$[J\ kg^{-1}\ K^{-1}]$
c_{pr}	specific heat capacity of the reaction mass	$[J\ kg^{-1}\ K^{-1}]$
m_r	reaction mass	$[kg]$
$\dot{Q}_{acc}$	accumulation term	$[w]$
$\dot{Q}_c$	heat flow delivered by the calibration probe	$[W]$
$\dot{Q}_{dos}$	amount of heat flow due to addition of reactants	$[W]$
$\dot{Q}_{flow}$	heat flow through the reactor wall	$[W]$
$\dot{Q}_{loss}$	heat losses to the surroundings	$[W]$
$\dot{Q}_r$	heat generated by the reaction	$[W]$
$\dot{Q}_{stir}$	heat flow dissipated by the stirrer	$[W]$
R_p	global rate of polymerization	$[mol\ m^{-3}\ s^{-1}]$
T_{dos}	temperature of the added substance	$[K]$
T_r	internal temperature	$[K]$
T_j	jacket temperature	$[K]$
U	overall heat transfer coefficient	$[W\ m^{-2}\ K^{-1}]$
$V_{continuous\ phase}$	total volume of the reactor V_r	$[m^3]$
$(-\Delta H_r)$	heat of polymerization	$[J\ mol^{-1}]$

Acronyms

AIBN	2,2′-Azobisisobutyronitrile
ATR	Attenuation total reflection
CO_2	Carbon dioxide
FTIR	Fourier transform infrared spectroscopy
IR	Infrared
MMA	Methyl methacrylate
NIR	Near infrared
NIST	National Institute of Standards and Technology (http://csrc.nist.gov/)
NMR	Nuclear magnetic resonance
PDMS	Poly dimethyl siloxane methacrylate
SEM	Scanning electron microscope
$scCO_2$	Supercritical CO_2
SCF	Supercritical fluids
SFC	Supercritical fluid chromatography
UV	Ultraviolet
Vis	Visible

References

1 Meyer Th. and Keurentjes J.T.F., *Handbook of Polymer Reaction Engineering*, Wiley-VCH, Weinheim, **2005**.

2 Olesik S.V., French S.B. and Novotny M., *Anal. Chem*, **1986**, 58, 2256.

3 Kainz S., Brinkmann A., Leitner W. and Rfaltz A., *J. Am. Chem. Soc.*, **1999**, 121, 6421.

4 Theyssen N. and Leitner W., *Chem. Commun.*, **2002**, 410.

5 Schneider M.S., Grunwaldt J.D., Bürgi T. and Baiker A., *Rev. Sci. Instrum.*, **2003**, 74(9), 4121.

6 Poliakoff M., Howdle, S.M. and Kazarian S.G., *Angew. Chem. Int. Ed. Engl.*, **1995**, 34, 1275.

7 Kessler W., Luft G. and Zeiss W., *Ber. Bunsenges. Phys. Chem.*, **1997**, 101, 698.

8 Blatchford M.A. and Wallen S. L., *Anal. Chem.*, **2002**, 74, 1922.

9 Hunt F., Ohde H. and Wai C.M., *Rev. Sci. Instrum.*, **1999**, 70(12), 4661.

10 Fehrenbacher U., Muth O., Hirth T. and Ballauff M., *Macromol. Chem. Phys.*, **2000**, 201(13), 1532.

11 Fehrenbacher U and Ballauff M., *Macromolecules*, **2002**, 35, 3653.

12 Lepilleur C. amd Beckmann E. J., *Macromolecules*, **1997**, 30, 745.

13 Akitt J.W. and Merbach A.E. in Diehl P., Fluck E., Günther H., Kosfeld R., Seelig J. and Jonas J., High pressure NMR, Vol. 24, Springer-Verlag, New York, **1991**.

14 Hoffmann M.M. and Conradi M.S., *J. Supercrit. Fluids*, **1998**, 14, 31.

15 Wallen S.L., Schoenbachler L.K., Dawson E.D. and Blatchford M.A., *Anal. Chem.*, **2000**, 72, 4230.

16 Gaemers S., Luyten H., Ernsting J.M. and Elsevier C.J., *Magn. Reson. Chem.*, **1999**, 37, 25.

17 Fisher O., Gyllenhaal O., Vessman and Albert K., *Anal. Chem.*, **2003**, 75, 622.

18 Avdeev M.V., Konovalov A.N., Bagratashvili V.N., Popov V.K., Tsypina A.I., Sokolova M., Ke J and Poliakoff M., *Phys. Chem. Chem. Phys.*, **2004**, 6, 1258.

19 Morbidelli M., Storti G. and Siani A. in Asua J.M., *Polymeric dispersions: principles and applications*, Kluwer Academic Publishers, **1997**.

20 Cavin L., Meyer Th. and Renken A., *Polym. React. Eng.*, **2000**, 8(3), 201.

21 Cavin L., Meyer Th. and Renken A., *Polym. React. Eng.*, **2000**, 8(3), 225.

22 Zeilmann T., Lavanchy F. and Meyer Th., *Chimia*, **2001**, 55, 249.

23 Oag R.M., King P.J., Mellor C.J., George M.W., Ke J., Poliakoff M., Popov V.K. and Bagrarashvili V.N., *J. Supercrit. Fluids*, **2004**, 30, 259.

24 Regenass W., *Themochim. Acta*, **1985**, 95, 351.

25 Lavanchy F., Fortini S. and Meyer Th, *Chimia*, **2002**, 56, 126.

26 Wang W., Griffiths R.M.T., Giles M.R., Williams P. and Howdle S.M., *Eur. Polym. J.*, **2003**, 39, 423.

27 Lavanchy F., Fortini S. and Meyer Th., *Org. Proc. Res. Dev.*, **2004**, 8, 504.

28 Span R. and Wagner W., *J. Phys. Chem. Ref. Data*, **1996**, 25 (6), 1509.

29 Shaffer K.A., Jones T.A., Canelas D.A., DeSimone J.M. and Wilkinson S.P., *Macromolecules*, **1996**, 29, 2704.

30 Giles M.R., Hay J.N., Howdle S.M. and Winder R.J., *Polymer*, **2000**, 41, 6715.

31 Fortini S., Lavanchy F., Nising P. and Meyer Th., *Macromol. Symp.*, **2004**, 206, 79.

32 Brandrup J. and Immergut E.H, *Polymer Handbook*, 3rd Edition, John Wiley & Sons, **1989**.

33 Fortini S., Lavanchy F. and Meyer Th., *Macromol. Mater. Eng.*, **2004**, 289, 757.

34 Fortini S. and Meyer Th., *Dechema Monograph.*, **2004**, 138, 361.

6

Heterogeneous Polymerization in Supercritical Carbon Dioxide *

Philipp A. Mueller, Barbara Bonavoglia, Giuseppe Storti, and Massimo Morbidelli

6.1
Introduction

Particle-forming or, more simply, heterogeneous polymerization processes are usually two-phase systems in which the starting monomer(s) and/or the resulting polymer is/are in the form of a fine dispersion in a continuous phase (the dispersant or polymerization medium). Depending upon (1) the initial state of the reacting mixture, (2) the mechanism of particle formation, and (3) the shape and size of the final particles, the different heterogeneous processes are assigned to the categories "suspension", "emulsion", "dispersion" and "precipitation". In the last two cases, the monomer and the initiator are both soluble in the polymerization medium, which is nevertheless a poor solvent for the resulting polymer. Since supercritical media are generally good solvents for most of the typical industrial monomers, heterogeneous polymerizations carried out in supercritical media are almost invariably of the dispersion or precipitation type.

While irregularly shaped particles or even a bulky phase are typical of precipitation polymerizations, in dispersion polymerizations surface-active molecules are added to the system recipe in order to prevent particle aggregation and coagulation. Depending upon the stabilizing power of the selected compound, spherical particles with average diameter from <1 up to about 10 μm are easily obtained. In both processes, interphase surface area and partitioning of reactants between continuous and dispersed phase play a decisive role in determining the reaction extent in each phase and, therefore, the properties of the final product.

In this chapter, the main features of heterogeneous polymerization in scCO$_2$ are analyzed. After a critical review of the main contributions in the literature, a comprehensive mathematical model is proposed. Finally, two case studies, selected as representative of dispersion and precipitation process, respectively, are discussed with the aim of elucidating the main mechanisms of the two heterogeneous processes.

---

* The symbols used in this chapter are listed at the end of the text, under "Notation".

Supercritical Carbon Dioxide: in Polymer Reaction Engineering
Edited by Maartje F. Kemmere and Thierry Meyer
Copyright © 2005 WILEY-VCH Verlag GmbH & Co. KGaA, Weinheim
ISBN: 3-527-31092-4

6.2
Literature Review

In 1994, DeSimone et al. [1] reported the first successful free-radical dispersion polymerization in $scCO_2$. They made detailed studies of the use of poly(1,1-dihydroperfluorooctyl acrylate) (PFOA) as a stabilizer for the dispersion polymerization of methyl methacrylate (MMA) and, more recently, vinyl acetate and styrene. Hsiao et al. [2] optimized the dispersion polymerization of MMA in $scCO_2$ using PFOA, and found that very uniform PMMA particles could be formed with narrow particle size distributions (polydispersity index, $PDI = 1.01–1.21$). The observation of some kind of auto-acceleration made them believe that the polymerization takes place primarily in the swollen particles. In another work, Hsiao and DeSimone [3] investigated the influence of helium concentration on particle size and particle size distribution in MMA dispersion polymerization in $scCO_2$. They showed that the average particle size decreases with increasing helium concentration, which is a nonsolvent for PMMA. However, reactions conducted in helium/CO_2 mixtures exhibited significant differences in particle size characteristics compared with reactions in neat CO_2. In 1996, Schaffer et al. [4] reported the dispersion polymerization of MMA in CO_2 using a commercially available methacrylate-terminated poly(dimethylsiloxane) (PDMS) macromonomer. High molecular weight PMMA was formed in good yields, whereas the best polymer was produced in $scCO_2$ using rather high stabilizer concentrations (>5 wt% with respect to monomer). Lepilleur and Beckman [5] synthesized a series of effective graft surfactants. Block copolymers have been used extensively as stabilizers for dispersion polymerization in $scCO_2$. The most important property for this type of stabilizer was identified as the anchor-to-soluble balance, which corresponds to the relative block lengths of the CO_2-philic segment and the CO_2-phobic anchoring segment. O'Neill et al. [6, 7] have also investigated the dispersion polymerization of MMA in $scCO_2$ using PDMS macromonomer stabilizers. They concluded that, since pure CO_2 is a moderate solvent for PDMS, the dispersions showed a tendency to aggregate before polymerization was complete because of a loss of solvation power as the monomer was depleted. In addition, after a comprehensive analysis of the time evolution of molecular weight, particle size, product morphology, and polymerization rate, they confirmed the main polymerization locus to be the particle phase. Yates et al. [8] have reported the synthesis of a CO_2-soluble surfactant (PDMS-b-polymethacrylate acid) consisting of a CO_2-philic PDMS block and a hydrophilic PMA block. Uniform PMMA particles could be synthesized in $scCO_2$ by dispersion polymerization using this stabilizer. It is worth noting that this stabilizer is particularly interesting because, in contrast to other surfactants, it allows the formation of water-dispersible PMMA powders. Giles et al. [9] showed that PDMS monomethacrylates with a variety of molar masses are successful surfactants for the polymerization of MMA in $scCO_2$, producing particulate polymers above a critical concentration which is dependent on the molar mass of the macromonomer. Recently, Christian and Howdle [10] explored the use of single-point anchoring stabilizers in the free-radical dispersion polymerization of MMA in scCO$_2$. They reported high yields and high molecular weights in the presence of Krytox 157FSL, a commercially available carboxylic acid-terminated perfluoro-polyether. Novel graft stabilizers were used by Giles et al. [11] and found to be very effective at low concentrations for the same polymerization system. Li et al. [12] investigated the effect of various known stabilizers on the particle formation stage in MMA dispersion polymerization in scCO$_2$ by *in situ* turbidimetry and compared their efficiency. Similar investigations were done by Fehrenbacher et al. [13, 14] using PDMS monomethylacrylate as stabilizer. In addition, they elucidated the kinetics of the early stage of the polymerization, confirming that the continuous phase is the only polymerization locus at the very beginning of the reaction. Finally, Okubo et al. [15] reported the successful production of submicron-sized PMMA particles by dispersion polymerization with a PDMS-based azoinitiator in scCO$_2$. The interesting aspect in their work is that the initiator operated not only as a radical-producing species but also as a colloidal stabilizer ("inistab"), leading to particularly small particles ($r_p \approx 20$ nm).

About precipitation polymerization in scCO$_2$, many of the early investigations have been reported on industrially important vinyl monomers and are mostly described in the patent literature [16–18]. More recently, Romack et al. [19] reported the successful precipitation polymerization of acrylic acid in scCO$_2$. Investigating the same system, Hu et al. [20] explored the effect of temperature, pressure, monomer concentration, and initiator concentration. Xu et al. [21] found that cosolvents have a pronounced effect on the product molecular weight and glass transition temperature, again dealing with the same polymerization system. In particular, they reported that the average molecular weight decreases with increasing cosolvent concentration when ethanol is used as cosolvent. On the other hand, the average molecular weight increases with the increase of the cosolvent concentration when acetic acid is used. Cooper et al. [22] reported the synthesis of cross-linked copolymer microspheres in scCO$_2$ made of ethylvinyl-benzene/divinylbenzene. Teng et al. [23] studied the stereoregularity of polyacrylonitrile (PAN) produced by precipitation polymerization in scCO$_2$. They found that the sequence distribution is completely random, in contrast to the case where the same polymer is produced either by aqueous suspension polymerization or by aqueous precipitation polymerization. Such a lower polymer isotacticity has been attributed to the non-polar nature of CO$_2$. In another work [24], the same authors studied the effect of monomer concentration, initiator concentration, pressure, and the total reaction time on the molecular weight distribution (MWD) of PAN produced in scCO$_2$. They concluded that the MWD became broad as the monomer concentration increased, whereas no appreciable effect of initiator concentration was observed. Changing the density of the continuous phase by manipulating the CO$_2$ pressure had a dramatic effect on the reaction evolution. The molecular weight could be enhanced by high pressure, whereas the MWD became narrow. Okubo et al. reported the successful production of polydivinylbiphenyl particles [25] and PAN particles [26], both with clean surfaces, by precipitation polymerization in scCO$_2$.

Finally, special mention should be made of the application of the precipitation process in supercritical media to the production of an industrially very important class of products, the fluorinated polymers. Since 1995 DeSimone has reported the homopolymerization [27] and copolymerization of tetrafluoroethylene (TFE) with perfluoro(propylvinylether) and with hexafluoropropylene [28], resulting in high yields of high-molecular-weight polymers. This specific subject is treated in detail in Chapter 9 of this book. However, as an example of the very specific behavior of the process, we should mention that Charpentier et al. [29, 30] recently investigated the continuous polymerization of vinylidene fluoride (VDF), in scCO$_2$. For this same system, peculiar behavior has been reported by Saraf et al. [31, 32], who found bimodal molecular-weight distributions without arriving at a conclusive understanding of the underlying mechanism.

Summarizing, most of the above-mentioned investigations were aimed at proving the stabilizing ability of different compounds for dispersion polymerizations and at identifying the best process condition for both dispersion and precipitation reactions. Discussion of the results was almost invariably focused on the analysis of the resulting particle morphology and molecular weight, with much less emphasis on elucidation and quantification of the reaction kinetics. The model presented below is believed to be a useful tool toward a deeper understanding of the process mechanisms, a key prerequisite to the successful scale-up of the process.

6.3
Modeling of the Process

To represent dispersion polymerization in conventional liquid media, several models have been reported in the literature, mainly focused on the particle formation and growth [33, 34] or on the reaction kinetics. Since our first aim is the reliable description of the reaction kinetics, we focus on the second type of models only. The model developed by Ahmed and Poehlein [35, 36], applied to the dispersion polymerization of styrene in ethanol, was probably the first one from which the polymerization rates in the two reaction loci have been calculated. A more comprehensive model was later reported by Saenz and Asua [37] for the dispersion copolymerization of styrene and butyl acrylate in ethanol-water medium. The particle growth as well as the entire MWD were predicted, once more evaluating the reaction rates in both phases and accounting for an irreversible radical mass transport from the continuous to the dispersed phase. Finally, a further model predicting conversion, particle number, and particle size distribution was proposed by Araujo and Pinto [38] for the dispersion polymerization of styrene in ethanol.

With specific reference to heterogeneous polymerizations in supercritical media, several oversimplified pseudo-homogeneous descriptions have been occasionally applied aimed at estimating kinetic parameters or elucidating dominant mechanisms [5, 6, 13, 14, 31]. The first (and probably only) comprehensive model of dispersion polymerization in supercritical media was reported by Chat-zidoukas et al. [39]. This model is conceptually identical to that proposed by the same group for the suspension polymerization of polyvinyl chloride [40], and it accounts for all reaction steps typical of the free-radical polymerization taking place in both phases. Results in terms of monomer conversion as well as average particle size and MWD are provided. It is worth noting that in the proposed model low-molecular-weight species are at interphase equilibrium at all times during the reaction whereas the growing radical chains remain segregated in the phase where they were formed (i.e. no radical interphase transport was assumed to take place) [39, 40]. More recently, we published a similar type of model, with some significant differences detailed in the following [41, 42].

Following the previously published modeling work and exploiting the body of experimental work analyzed in the previous section, the key features of the process under examination can be summarized as follows:

- The initial stage of the reaction is homogeneous and characterized by polymerization in the continuous phase.
- In the presence of a stabilizing agent (dispersion process), the nucleation of the second phase, i.e. the particle formation step, is usually finished at a very early stage (below one percent of conversion, see [7, 14]). The stabilizing molecules are adsorbed or grafted on the surface and their concentration determines the total particle surface (i.e. particle number and size).
- From this point on, two reaction loci are present, the relative importance of which is the result of a complex interplay of different factors such as interphase partitioning and transport rates of the reactants and reaction rates.
- The interphase mass transport is usually fast for low-molecular-weight species (such as solvent, initiator, and monomer). However, the same transport is expected to be significantly reduced for species at higher molecular weight and, in particular, for the growing polymer chains.
- The polymerization in the continuous phase can be regarded as a solution polymerization without any diffusion limitations. On the other hand, a remarkable reduction of diffusivity can be expected within the polymer particles where the conditions resemble those of a bulk polymerization at high polymer content.

Based on this process schematization, a comprehensive kinetic model has been proposed [41, 42]. Its main features are:

- Two reaction loci are considered, the polymer-rich dispersed phase and the CO_2-rich continuous phase. A kinetic scheme typical of free-radical reactions and including initiation, propagation, terminations, and chain transfer to monomer and to polymer is applied to each phase.
- Low-molecular-weight species (solvent, initiator, and monomer) undergo very fast transport between the phases, and they are assumed to be at interphase thermodynamic equilibrium at all times. The Sanchez-Lacombe equation of state [43, 44] was used for monomer and solvent, while an oversimplified partition coefficient was assumed for the initiator.

- A chain length-dependent partition coefficient for polymer chains between continuous and dispersed phase is considered. The same functional form proposed and experimentally validated by Kumar et al. for polystyrene is adopted [45]. A chain length dependence of the corresponding interphase mass transport coefficient is also accounted for.
- Particle nucleation is not accounted for, i.e. a number of particles evaluated to reproduce the final value of the total specific surface area is assumed to be present from the beginning. As a consequence, the role of the stabilizer in particle formation and stabilization is not explicitly considered.

The considered kinetic scheme includes the following kinetic steps typical for the free-radical polymerization of vinyl monomers:

- *Initiation*

$$I_j \xrightarrow{k_{dj}} 2I_j^\bullet$$
$$I_j^\bullet + M_j \xrightarrow{k_{Ij}} R_{1,j}$$

- *Propagation*

$$R_{x,j} + M_j \xrightarrow{k_{pj}} R_{x+1,j}$$

- *Chain transfer*

$$R_{x,j} + M_j \xrightarrow{k_{fmi}} R_{x,j} + R_{1,j}$$
$$R_{x,j} + P_{y,j} \xrightarrow{k_{fpi}} R_{y,j} + P_{x,j}$$

- *Termination*

$$R_{x,j} + R_{y,j} \xrightarrow{k_{tdj}} P_{x,j} + P_{y,j}$$
$$R_{x,j} + R_{y,j} \xrightarrow{k_{tcj}} P_{x+y,j}$$

Note that when two subscripts are given, the first one (x or y; $x,y=[1,\infty]$) indicates the chain length and the second one the phase ($j=1$ for the continuous phase and $j=2$ for the dispersed phase). The above kinetic scheme is applied to both phases.

For the thermodynamic description of low-molecular-weight species, i.e. monomer, solvent, and initiator, the Sanchez-Lacombe model [46, 47] was used for the first two species, while a simple partition coefficient was assumed for the initiator. Since all equations related to these species are reported in detail in [41, 42, 48], here we summarize only the main equations, which are the mass balances of solvent, initiator and monomer:

$$\frac{dS}{dt} = \sum_{j=1}^{2} \frac{dS_j}{dt} = 0 \tag{1}$$

$$\frac{dI}{dt} = \sum_{j=1}^{2} \frac{dI_j}{dt} = -k_{d1}[I_1]V_1 + k_{d2}[I_2]V_2 \tag{2}$$

$$\frac{dM}{dt} = \sum_{j=1}^{2} \frac{dM_j}{dt} = -2f_1 k_{d1}[I_1]V_1 - 2f_2 k_{d2}[I_2]V_2$$
$$- (k_{p1}+k_{fm1})[M_1]V_1 \sum_{x=1}^{\infty}[R_{x,1}] - (k_{p2} + k_{fm2})[M_2]V_2 \sum_{x=1}^{\infty}[R_{x,2}] \tag{3}$$

with

$$S = [S_1]V_1 + [S_2]V_2 \tag{4}$$

$$I = [I_1]V_1 + [I_2]V_2 \tag{5}$$

$$M = [M_1]V_1 + [M_2]V_2 \tag{6}$$

and the equilibrium interphase partition conditions:

$$\mu_S^1 = \mu_S^2 \tag{7}$$

$$\mu_M^1 = \mu_M^2 \tag{8}$$

$$\frac{[I_1]}{[I_2]} = K_1 \tag{9}$$

where μ_i^j is the chemical potential of component i in phase j. The Sanchez-Lacombe equation of state is applied to compute the density of each phase and this, combined with the constraint of constant reactor volume, allows the volumes of the two phases, V_j, and the overall pressure, p, to be computed.

The high-molecular-weight species (i.e. active and terminated polymer chains) are described in terms of population balance equations. These equations are written for the active and the terminated chains for each chain length, $x = [1, \infty]$, and each phase, $j = 1, 2$, separately:

$$
\begin{aligned}
\frac{dR_{x,j}}{dt} = & k_{pj}[M_j][R_{x-1,j}]V_j(1 - \delta(x-1)) \\
& - \left((k_{pj} + k_{fmj})[M_j] + (k_{tcj} + k_{tdj})\sum_{y=1}^{\infty}[R_{y,j}]\right)[R_{x,j}]V_j \\
& + \left(2f_j k_{dj}[I_j]V_j + k_{fmj}[M_j]\sum_{y=1}^{\infty}[R_{y,j}]V_j\right)\delta(x-1) \\
& - k_{fpj}[R_{x,j}]V_j\sum_{y=1}^{\infty}y[P_{y,j}] + k_{fpj}x[P_{x,j}]V_j\sum_{y=1}^{\infty}[R_{y,j}] \\
& - K_{x,j}4\pi r_p^2 N_p([R_{x,j}] - [R_{x,j}^*])
\end{aligned} \tag{10}
$$

$$
\begin{aligned}
\frac{dP_{x,j}}{dt} = & \frac{1}{2}k_{tcj}\sum_{y=1}^{x-1}[R_{y,j}][R_{x-y,j}]V_j \\
& + \left(k_{tdj}\sum_{y=1}^{\infty}[R_{y,j}] + k_{fmj}[M_j]\right)[R_{x,j}]V_j \\
& + k_{fpj}[R_{x,j}]V_j\sum_{y=1}^{\infty}y[P_{y,j}] - k_{fpj}x[P_{x,j}]V_j\sum_{y=1}^{\infty}[R_{y,j}] \\
& - K_{x,j}4\pi r_p^2 N_p([P_{x,j}] - [P_{x,j}^*])
\end{aligned} \tag{11}
$$

where $\delta(x - x_0)$ indicates the Kronecker delta function, defined as equal to 1 for $x = x_0$ and 0 otherwise. The last term in the equations above describes the rate of interphase mass transport according to the two-film theory, where $[R^*_{x,j}]$ and $[P^*_{x,j}]$ are defined as the hypothetical concentrations in phase j in equilibrium with the bulk concentrations in the other phase j. This can be computed from the following chain length-dependent equilibrium partitioning condition:

$$\frac{[R^*_{x,1}]}{[R_{x,2}]} = \frac{[R_{x,1}]}{[R^*_{x,2}]} = \frac{[P^*_{x,1}]}{[P_{x,2}]} = \frac{[P_{x,1}]}{[P^*_{x,2}]} = m_x \tag{12}$$

where the functional form of the interphase partition coefficient, m_x, has been assumed equal to that found experimentally by Kumar et al. for polystyrene [45]:

$$\log(m_x) = \log(m_1) + a(x - 1) \tag{13}$$

where m_1 is the monomer partition coefficient and a is a constant, which has to be evaluated for each specific system. It is worth noting that, in the frame of this model, the only required information about the polymer phase morphology is the overall interphase surface area, which, assuming equal spherical polymer particles, is given by $A_p = 4\pi r_p^2 N_p$. In the following we use the final value of the polymer particle specific surface area, a_p^f as an adjustable parameter and estimate the number of particles under the assumption of spherical geometry as follows:

$$N_p = \frac{X^f m_M^0 \rho_p^2 a_p^{f3}}{36\pi} \tag{14}$$

where X^f is the final conversion (at which a_p^f has been obtained), m_M^0 the total initial monomer mass, and ρ_p the density of the polymer. During the reaction N_p is assumed constant in time, and, since V_2 obviously increases, we can back-compute the increase in time of r_p and, more importantly, of A_p, which is the only morphology parameter in the model.

The numerical solution of the resulting system of mixed algebraic-differential equations has been achieved using the discretization method by Kumar and Ramkrishna [49] as detailed elsewhere [41, 42].

The reliable evaluation of the large number of model parameters is a critical issue when developing a detailed kinetic model. It is very important to estimate as many parameter values as possible from independent sources in order to minimize direct fitting to the available kinetic data and to ensure genuine model reliability. Therefore, a summary of the adopted parameter evaluation procedure (and of our unavoidable arbitrary choices) is briefly reported in the following.

Numerical values of the kinetic rate constants at low conversion (i.e. without diffusion limitations) are reported in the literature for many polymerization systems. Since these constants are usually weakly dependent on the solvent, in the case where no values can be found for the specific solvent under examination, their values can be approximated to those measured in other solvents or in bulk at low conversion. In particular, Beuermann and Buback [50] found for various polymerization systems that the values of propagation and termination rate constants in bulk and scCO$_2$, respectively, do not differ by more than 50%. Moreover, Morrison et al. [51] reported negligible solvent effects on the propagation rate constant of styrene and MMA. Thus, summarizing, the following Arrhenius-type expressions were used to evaluate the intrinsic kinetic rate constants as a function of temperature and pressure:

$$k(T, p_0) = A \exp\left(-\frac{E}{RT}\right) \tag{15}$$

$$k(T, p) = k(t, p_0) \exp\left(-\frac{\Delta V^{\#}}{RT}(p - p_0)\right) \tag{16}$$

where A is a pre-exponential factor, E the activation energy, $\Delta V^{\#}$ the activation volume, and p_0 the reference pressure, usually equal to the pressure at which values of k are available.

The situation becomes more complicated in the polymer-rich phase, where diffusion limitations must be accounted for. In this model, the following expressions for cage, glass, and gel effect in the dispersed phase (subscript 2) have been adopted [41]:

$$f_2 = \left[1 - \frac{D_{M,0}}{D_M}\left(1 - \frac{1}{f_{2,0}}\right)\right]^{-1} \tag{17}$$

$$k_{p2} = \left(\frac{1}{k_{p2,0}} + \frac{1}{4\pi\sigma_M D_M N_A}\right)^{-1} \tag{18}$$

$$k_{t2}(x, y) = \left[\frac{1}{k_{t2,0}} + \frac{1}{8\pi a j_c^{1/2} N_A (D_{x,com} + D_{y,com} + k_{p2}[M_2]a^2/3)}\right]^{-1} \tag{19}$$

where D_M is the monomer diffusion coefficient evaluated in the frame of the free-volume theory of Vrentas and Duda [52, 53], σ_M the Lennard-Jones diameter of the monomer, a the root-mean-square end-to-end distance divided by the square root of the number of monomer units in the chain, j_c the entanglement spacing, and $D_{x,com}$ the center-of-mass diffusion coefficient for a chain of length x evaluated from the following universal scaling law [54]:

$$D_{x,com} = D_M x^{-(0.664 + 2.02\omega_p)} \tag{20}$$

ω_p being the polymer weight fraction. As anticipated, the monomer diffusion coefficient is evaluated from the free-volume theory through the following expression:

$$D_M = D_0 \exp\left(-\frac{E}{RT}\right) \exp\left[-\frac{\gamma(\omega_M V_M^* + \xi_{MS}\omega_S V_S^* + \xi_{MP}\omega_P V_P^*)}{V_{FH}}\right] \qquad (21)$$

The specific critical hole free volume, V_i^*, is estimated as the specific volume at 0 K, which in turn is obtained from group contribution methods. The ratio of the molar volumes of the jumping units of components i and j, ξ_{ij}, is computed using the values at 0 K. The pre-exponential factor, D_0, and the critical energy needed by a molecule to overcome the attractive force, E, are obtained by fitting the Dullien equation for the self-diffusion coefficient to viscosity versus temperature data. Finally, the average hole-free volume per gram of the mixture, V_{FH}/γ, can be estimated from those of the individual species:

$$\frac{V_{FH}}{\gamma} = \sum_i \omega_i \frac{V_{FH,i}}{\gamma} \qquad (22)$$

where $V_{FH,i}/\gamma$ is evaluated from the free volume parameters:

$$\frac{V_{FH,i}}{\gamma} = \frac{K_{1,i}}{\gamma}(K_{2,i} + T - T_{g,i}) \qquad (23)$$

The free-volume parameters are again obtained by fitting viscosity versus temperature data using either the adopted Doolittle expression (low-molecular-weight species) or the Williams-Landel-Ferry equation (polymers). The glass transition temperature, $T_{g,i}$, is as reported in the literature or can be estimated from the melting temperature.

Finally, the pure-component parameters in the Sanchez-Lacombe equation of state can be evaluated to reproduce pure-component properties (for example, density versus temperature data), whereas the mixture parameters are evaluated through selected mixing rules. These in turn involve one or two binary interaction parameters per component pair, which are usually obtained by fitting binary data (for example sorption data for the pair CO_2-polymer).

To conclude, a few residual model parameters usually remain undefined, and these have to be evaluated by direct fitting to the available experimental data. The specific parameters are system dependent, and detailed examples are presented in the following sections. Note that, since the model is not accounting for particle formation, the total interphase surface area is often an adjustable quantity. In fact, this quantity could be estimated from experimental data of particle size and number assuming spherical, non-porous particles. However, this is not the case in precipitation polymerization or unstable dispersions where the polymer-rich phase is recovered in the form of irregular fragments, porous structures, or even a bulky phase, thus preventing a reliable estimation of the actual surface area at reaction conditions.

Model validation, together with specific examples of parameter evaluation, is presented in the next sections with reference to two selected systems: the dispersion polymerization of a conventional monomer (MMA) and the precipitation polymerization of a fluorinated monomer (VDF). They are representative of largely different situations, thus representing a quite convincing test of the model reliability.

6.4
Case Study I: MMA Dispersion Polymerization

For this specific system, all parameter values have been estimated *a priori* (numerical values and literature sources are in Table 6.1), and experimental values of the particle size (spherical and non-porous) are available, thus allowing a reliable estimation of the particle surface to be made. Therefore, the model results can be considered fully predictive. Three different sets of experimental data have been considered [2, 6, 41] (a summary of the corresponding recipes and operating conditions is given in Table 6.2) and the same model parameter values reported in Table 6.1 have been used in all cases, without any parameter tuning. Moreover, an additional comparison has been done using the experimental data obtained by reaction calorimetry as described in Chapter 5 and Ref. [55]. Note that since these experiments were carried out at higher temperatures, different values of the binary interaction parameters of the Sanchez-Lacombe EOS have been used. The corresponding values are given in brackets in Table 6.1.

The comparison between experimental data and model predictions in terms of conversion versus reaction time as well as average molecular weights versus conversion is shown in Figs. 6.1 to 6.4 for each of the four experimental sets. It can be concluded that the agreement between experiment and prediction is quite satisfactory in all cases.

In all considered reactions, some kind of auto-acceleration in the reaction rate is quite evident. In order to elucidate the relevance of this phenomenon, the same simulation shown in Fig. 6.1 has been repeated neglecting cage, glass, and gel effects (i.e. using Eqs. 17–19 with $\omega_P = 0$). Fig. 6.5 shows the result: it can be seen that diffusion limitations play a major role in the system, and they need to be properly accounted for.

In Fig. 6.6, the calculated MWD is compared to the experimental one (run 1, Table 6.2). A satisfactory agreement is once more observed, and the diffusion limitations (mainly gel effect) are responsible for the slight shift of the maxima of the distributions toward higher molecular weights as the conversion increases. Moreover, the monomodal nature of the distribution indicates that the reaction is taking place predominantly in one phase, the polymer particles, as reported by different researchers [2, 6, 7].

The last result deserves more comment. Even though two reaction loci are potentially operative, only one dominates the system kinetics, and it would be useful to better understand why. Let us compare the characteristic times of the kinetic steps involved in the polymerization process under examination. Focusing on a generic active chain with length x and growing in phase j, at any given time it can undergo one of the following events: propagation, termination, and interphase transport. By introducing the corresponding overall mass transfer coefficient, K_j, the characteristic time of diffusion out of phase j is given by the following expression:

$$\tau_{\text{diff},j} = \frac{1}{K_j a_{p,j}} \tag{24}$$

where $a_{p,j} = A_p/V_j$ is the total particle surface area per volume of phase j. On the other hand, the time needed for the addition of one monomer unit in phase j is given by:

$$\tau_{p,j} = \frac{1}{k_{pj}[M_j]} \tag{25}$$

Table 6.1 Model parameter values and corresponding sources used in case study I.

Parameter	Value	Unit	Source
Kinetic parameters – dispersed phase			
$f_{2,0}$	0.50		[57, 58]
A_{d2}	1.97×10^{14}	s^{-1}	[59]
E_{d2}	123.5	kJ mol^{-1}	[59]
$\Delta V_{d2}^{\#}$	5.0	cm^3 mol^{-1}	[59]
$p_{0,d2}$	0.1	MPa	[59]
A_{p2}	4.92×10^5	L mol^{-1} s^{-1}	[60]
E_{p2}	18.2	kJ mol^{-1}	[60]
$\Delta V_{p2}^{\#}$	−16.7	cm^3 mol^{-1}	[61]
$p_{0,p2}$	0.1	MPa	[60]
A_{t2}	9.80×10^7	L mol^{-1} s^{-1}	[60]
E_{t2}	2.9	kJ mol^{-1}	[60]
$\Delta V_{t2}^{\#}$	15.0	kJ mol^{-1}	[62]
$p_{0,t2}$	0.1	Mpa	[60]
k_{fm2}/k_{p2}	5.15×10^{-5}		[63]
k_{td2}/k_{tc2}	4.37		[64]
Kinetic parameters – continuous phase			
$f_{1,0}$	0.83		[65]
A_{d1}	4.19×10^{15}	s^{-1}	[65]
E_{d1}	134.6	kJ mol^{-1}	[65]
$\Delta V_{d1}^{\#}$	0	cm^3 mol^{-1}	[65]
$p_{0,d1}$	24.8	MPa	[65]
A_{p1}	5.20×10^6	L mol^{-1} s^{-1}	[66]
E_{p1}	25.4	kJ mol^{-1}	[66]
$\Delta V_{p1}^{\#}$	−16.7	cm^3 mol^{-1}	[61]
$p_{0,p1}$	18.0	MPa	[66]
A_{t1}	9.80×10^7	L mol^{-1} s^{-1}	[50, 60]
E_{t1}	2.9	kJ mol^{-1}	[50, 60]
$\Delta V_{t1}^{\#}$	15.0	cm^3 mol^{-1}	[62]

Table 6.1 (continued)

Parameter	Value	Unit	Source
$p_{0,t1}$	0.1	MPa	[60]
k_{fm1}/k_{p1}	5.15×10^{-5}		[63]
k_{td1}/k_{tc1}	4.37		[64]

Free-volume parameters

Parameter	Value	Unit	Source
D_0	1.61×10^{-3}	$cm^2\ s^{-1}$	[67]
E	3.26×10^3	$J\ mol^{-1}$	[67]
$K_{1,MMA}/\gamma$	8.15×10^{-4}	$cm^3\ g^{-1}\ K^{-1}$	[67]
$K_{1,PMMA}/\gamma$	4.77×10^{-4}	$cm^3\ g^{-1}\ K^{-1}$	[67]
$K_{2,MMA}$	143	K	[67]
$K_{2,PMMA}$	52.4	K	[67]
$T_{g,MMA}$	143	K	[67]
$T_{g,PMMA}$	392	K	[67]
V^*_{PMMA}	0.870	$cm^3\ g^{-1}$	[67]
$V^*_{CO_2}$	0.589	$cm^3\ g^{-1}$	[68, 69]
V^*_{PMMA}	0.757	$cm^3\ g^{-1}$	[67]
V^{ref}_{FH,CO_2}	0.231	$cm^3\ g^{-1}$	[68, 69]
a_{CO_2}	8.76×10^{-4}	K^{-1}	[68, 69]
$\tilde{a}_{PMMA}$	0.44		[67]
$\xi_{MMA/PMMA}$	0.60		[67]
ξ_{MMA/CO_2}	0.18		[70]

Thermodynmic parameters

Parameter	Value	Unit	Source
ε^*_{MMA}	3850	$J\ mol^{-1}$	by fitting data from [71–73]
ε^*_{PMMA}	5787	$J\ mol^{-1}$	[74]
$\varepsilon^*_{CO_2}$	2536	$J\ mol^{-1}$	[74]
υ^*_{MMA}	7.66	$cm^3\ mol^{-1}$	by fitting data from [71–73]
υ^*_{PMMA}	11.51	$cm^3\ mol^{-1}$	[74]
$\upsilon^*_{CO_2}$	4.41	$cm^3\ mol^{-1}$	[74]
r_{MMA}	11.49		by fitting data from [71–73]
r_{CO_2}	6.60		[74]
ρ^*_{PMMA}	1.269	$g\ cm^{-3}$	[74]
k_{MMA/CO_2}	0.938 (0.945)		[42]
$k_{MMA/PMMA}$	0.970 (0.920)		[42]
$k_{CO_2/PMMA}$	1.144 (1.148)		by fitting data from [75]
K_{AIBN}	1.0		[41]
a	−0.3		by fitting data from [45, 76]
a_p^f	2.5	$m^2\ g^{-1}$	[42]

Table 6.2 Recipes and operation conditions of the experimental runs used in case study I.

Data set	Source	Initial amounts			Operating conditions			Particle conc.
		MMA [g]	CO_2 [g]	AIBN [g]	T [°C]	P_0 [MPa]	V [cm³]	Np/V [cm⁻³]
1	[41]	30	320	0.33	65	14.0	580	1×10^{10}
2	[2]	1.49	7.09	8.1×10^{-3}	65	34.5	10	3×10^{10}
3	[6]	0.38	1.38	5.7×10^{-3}	65	20.7	2.25	3×10^{10}

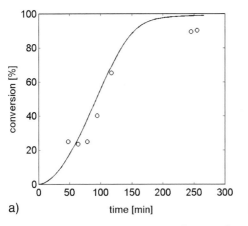

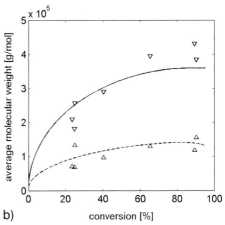

a) b)

Fig. 6.1 (a) Conversion as a function of time predicted by the model (line) and measured experimentally (○) under the operating conditions corresponding to the experimental run 1 in Table 6.2. (b) Weight average (solid line) and number average molecular weight (dashed line) as a function of conversion predicted by the model under the operating conditions corresponding to the experimental run 1 in Table 6.2. (Experimental data from Mueller et al. [41]: M_w (▽) and M_n (△) from [42]).

Thus, the maximum average length that chains produced in phase j can achieve before diffusing out of the phase can be estimated as the ratio between these two characteristic times:

$$x_{max,j} = \frac{\tau_{diff,j}}{\tau_{p,j}} = \frac{k_{pj}[M_j]}{K_j a_{p,j}} \tag{26}$$

However, the actual average length of the chains produced in phase j is given by the ratio between the characteristic times of termination and propagation, leading to:

$$x_{act,j} = \frac{k_{pj}[M_j]}{k_{tj}[R_j]} \tag{27}$$

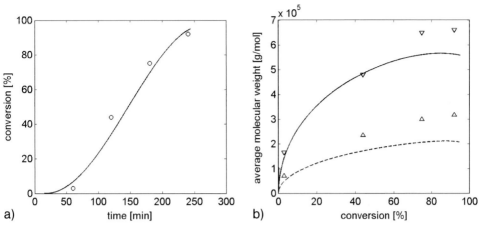

Fig. 6.2 (a) Conversion as a function of time predicted by the model (line) and measured experimentally (○) under the operating conditions corresponding to the experimental run 2 in Table 6.2. (b) Weight average (solid line) and number average molecular weight (dashed line) as a function of conversion predicted by the model under the operating conditions corresponding to the experimental run 2 in Table 6.2; experimental data from Hsiao et al. [2]: M_w (▽) and M_n (△) (from [42]).

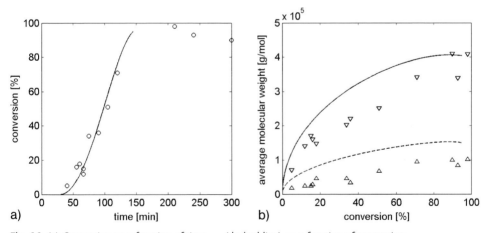

Fig. 6.3 (a) Conversion as a function of time predicted by the model (line) and measured experimentally (○) under the operating conditions corresponding to the experimental run 3 in Table 6.2. (b) Weight average (solid line) and number average molecular weight (dashed line) as a function of conversion predicted by the model under the operating conditions corresponding to the experimental run 3 in Table 6.2; experimental data from O'Neill et al. [6]: M_w (▽) and M_n (△), from [42]).

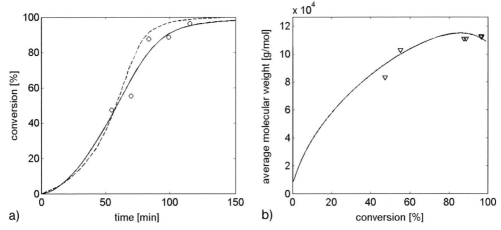

a)

b)

Fig. 6.4 (a) Conversion as a function of time predicted by the model (solid line) and measured experimentally by reaction calorimetry (dashed line) and gravimetrically (○).
(b) Weight average molecular weight as a function of conversion predicted by the model (line) and measured experimentally (▽).
Operating conditions described in Chapter 5 and [55].

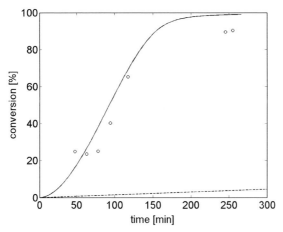

Fig. 6.5 Calculated conversion as a function of reaction time under the operating conditions 1 in Table 6.2 with (solid line, same as in Fig. 6.1) and without diffusion limitations (dashed line); circles are experimental data (from [42]).

   In the case where $x_{act} \ll x_{max}$, the chains are terminated before they can diffuse out of the phase of origin and we can therefore regard this phase as fully segregated. On the other hand, when $x_{act} \gg x_{max}$, the active chains are able to diffuse out of the phase before terminating, and interphase equilibrium prevails at any time. Extending the same arguments to the system under consideration

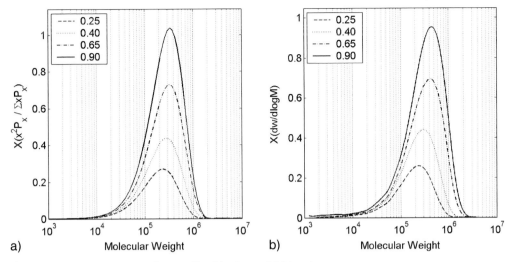

Fig. 6.6 MWD at various conversions predicted by the model (a) and measured experimentally (b) under the operating conditions corresponding to the experimental run 1 in Table 6.2; areas under the curves are proportional to the conversion (from [42]).

and including the chain length dependence of all the kinetic coefficients, we can introduce the following characteristic quantity for each phase and each chain length:

$$\Omega_j(x) = \frac{K_{x,j}A_p}{k_{tj}[R_j]V_j} \tag{28}$$

where $[R_j]$ is the total active chain concentration in phase j. Using the two-film theory [56], the overall mass transfer coefficients can be evaluated as follows (subscripts 1 and 2 indicate continuous and dispersed phase, respectively):

$$K_{x,1} = \left(\frac{1}{k_{x,1}} + \frac{m_x}{k_{x,2}}\right)^{-1} \quad \text{and} \quad K_{x,2} = \left(\frac{1}{m_x k_{x,1}} + \frac{1}{k_{x,2}}\right)^{-1} \tag{29}$$

where $k_{x,j} = D_{x,j}/r_p$ is the local mass transfer coefficient in phase j. Depending upon the actual values of the parameters $\Omega(x)$, four different conditions or operating regions can be identified for a generic heterogeneous polymerization process:

1. $\Omega_1(x)$ and $\Omega_2(x) < 1$: The probability of termination is larger than that of diffusion out of the phase for both continuous and dispersed phase. The chains terminate in the phase where they were initiated, without any significant chance of being transported to the other phase while active (region I).

2. $\Omega_1(x)$ and $\Omega_2(x) > 1$: The probability of termination is smaller than that of diffusion out of the phase for both continuous and dispersed phase. The chain mobility is so large that termination cannot prevent the achievement of complete equilibration of the chain concentration in the two phases (region III).

3. $\Omega_1(x) > 1$ and $\Omega_2(x) < 1$: The chains initiated in phase 1 (continuous) can diffuse to phase 2, but the opposite transport cannot occur because of termination in phase 2 (which is faster than diffusion). The net result is that the active chains initiated in phase 2 are segregated there while those initiated in phase 1 are transported to phase 2 where they terminate. This situation can be regarded as a sort of "irreversible transport'" of the radicals from the continuous to the dispersed phase (region IV).

4. $\Omega_1(x) < 1$ and $\Omega_2(x) > 1$: The situation is opposite to that in the previous case. The chains initiated in phase 2 (dispersed) can diffuse to phase 1 but the opposite transport does not occur because of termination in phase 1 (which is faster than diffusion). Therefore, active chains initiated in phase 1 are segregated there and those initiated in phase 2 are transported to phase 1 where they terminate. This corresponds to an "irreversible transport'" of the radicals from the dispersed to the continuous phase (region II).

All four operating regions are sketched in the plane $\Omega_1 - \Omega_2$ in Fig. 6.7, thus giving a kind of "master plot" for radical partitioning, which is of general validity since in principle it could be applied to any two-phase polymerization process (from this point on, the chain length dependence is omitted for brevity). The four quadrants obtained when drawing the two lines at $\Omega_1 = 1$ and $\Omega_2 = 1$

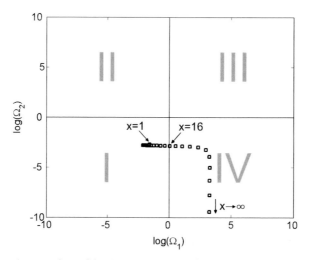

Fig. 6.7 Values of the $\Omega_1 - \Omega_2$ pairs as a function of chain length evaluated under the operating conditions corresponding to the experimental run 1 in Table 6.2.

have been enumerated clockwise from I to IV starting from the quadrant in the origin of the plane. As an example, a typical emulsion polymerization would be located in quadrant IV since, because of the practical insolubility of the polymer chains in the continuous (aqueous) phase, an irreversible mechanism of radical transport prevails. On the other hand, a typical precipitation polymerization should be located in quadrant I, since, because of the high reactivity of the growing chains and the usually small interphase surface area, the former are typically able to grow and terminate in the continuous phase before moving to the other phase.

Evaluating the $\Omega_1 - \Omega_2$ pairs for each chain length in each phase for our specific polymerization process, the trace shown in Fig. 6.7 is obtained at intermediate conversion under the experimental conditions of run 1. It is clear that the calculated path covers two regions only in the Ω plane: regions I and IV. In particular, we see that the points representative of very short chains ($x < 16$) lie in region I, indicating that these radicals are segregated because of their relatively large solubility in the continuous phase. However, at increasing chain length, the chain solubility decreases and the corresponding Ω pairs enter region IV, thus indicating that an irreversible transport from the continuous to the dispersed phase is dominant. In addition, it can be noted that above a certain chain length, Ω_1 remains practically constant as the chain length increases further, while Ω_2 decreases strongly. The maximum value of chain length shown in Fig. 6.7 is $x = 51$. If we consider longer chains, negligible solubility in the continuous phase, i.e. $m_x \to 0$, can be assumed, and, from Eqs. (28) and (29), it is readily seen that $\Omega_2(x) \to 0$. This result indicates that for systems where the polymer chains exhibit low solubility in the continuous phase, which is the case for most of the two-phase polymerization systems of practical interest, Ω_2 is usually smaller than 1 and the radical partitioning regime is always determined by the value of Ω_1 only. Therefore, the previous analysis could be effectively carried out in terms of one single Ω value (that of the continuous phase), and the two accessible regions correspond to either segregation (region I, $\Omega_1 < 1$) or irreversible transport from the continuous to the dispersed phase (region IV, $\Omega_1 > 1$). Moreover, the calculated trace shown in Fig. 6.7 confirms that monomodal MWD is expected for the specific system under examination. As a result of the irreversible transport of growing chains from the continuous to the dispersed phase, the latter is the dominant reaction locus: very short chains only can be segregated in the continuous phase, thus leading to the almost negligible tailing in the low-molecular-weight region of the experimental and calculated MWD shown in Fig. 6.6. To enhance the relevance of the low-MW mode of this distribution, the value of Ω_1 should be reduced for much larger values of the chain length. By inspection of Eq. (28), this could be done by reducing the stability of the polymer particles (i.e. the interphase surface area) or by increasing the radical concentration (i.e. the amount of initiator).

6.5
Case Study II: VDF Precipitation Polymerization

In contrast to the case of MMA, very little information on VDF polymerization in scCO$_2$ is reported in the literature, thus making the model parameter evaluation difficult. However, the selected values are summarized in Table 6.3 together with the corresponding sources (for more details, see [48]). Three parameters have been considered to be adjustable quantities: the VDF/PVDF binary interaction parameter (needed in the Sanchez-Lacombe equation of state), the chain transfer to polymer rate constant, and the final value of the total interphase surface area (i.e. the particle morphology). Since the role of each parameter is quite specific (the first and the last having a significant impact on the whole system behavior, while the second only affects the MWD broadening), it was possible to evaluate them with reference to a first specific experiment (so-called base case) while simulating in a truly predictive way all remaining experiments. It is worth noting that the crystalline part of the polymer is assumed to be impermeable to all species and is therefore simply not included as "effective" volume of polymer phase when evaluating interphase equilibrium. The batch experimental data have been provided by Solvay-Solexis, and the different recipes as well as the reaction conditions are presented in Table 6.4. In particular, the effect of changing initial monomer concentration on the one hand and density (i.e. pressure) on the other has been analyzed experimentally. Three different levels of initial monomer concentrations and densities were explored. As a further validation, the model equations were adapted to a continuous reactor (CSTR), and the results were compared to the experimental data reported by Saraf et al. [31].

The base case corresponds to an initial monomer concentration of 3.1 mol L^{-1} and a pressure of 20.4 MPa, respectively. The fitting to the experimental data in order to estimate the three adjustable parameters was done in terms of conversion versus time (Fig. 6.8) and MWD versus conversion (Fig. 6.9). Inspection of the results depicted in the two figures shows that the agreement is remarkably good, and the final values of the adjustable parameters are given in Table 6.3. Note that a value of 0.25 m^2 g^{-1} was used for the specific surface area. This value corresponds to a final particle diameter of around 10 μm, which is reasonable for the system under examination.

The conversion curve in Fig. 6.8 reveals an acceleration of the polymerization rate at increasing conversion, thus indicating once more that diffusion limitations are operative. In Fig. 6.9, it can be seen that the MWD exhibits two clearly defined modes. The relevance of these two modes changes during the reaction in such a way that the shorter chains are dominant at low conversion. In the course of the polymerization, the higher-molecular-weight chains increasingly get the upper hand. Moreover, looking at the shape of the two modes, it can be seen that the one at higher molecular weights is significantly broader than the other. All these features are well described by the model, thus enabling us to propose the following reaction mechanism.

Table 6.3 Model parameter values and corresponding sources used in case study II.

Parameter	Value	Unit	Source
Kinetic parameters – dispersed phase			
$f_{2,0}$	0.60		[77]
A_{d2}	6.2×10^{16}	s^{-1}	[77]
E_{d2}	132	kJ mol^{-1}	[77]
$\Delta V_{d2}^{\#}$	0	cm^3 mol^{-1}	[77]
$p_{0,d2}$	27.6	MPa	[77]
A_{p2}	6.4×10^5	L mol^{-1} s^{-1}	[48, 78]
E_{p2}	15	kJ mol^{-1}	[59, 79], typical value
$\Delta V_{p2}^{\#}$	−25	cm^3 mol^{-1}	[59, 79, 80], typical value
$p_{0,p2}$	2.2	MPa	[78]
$A_{p/\sqrt{t},2}$	6.5×10^9	L$^{1/2}$ mol$^{-1/2}$ s$^{-1/2}$	[30, 48]
$E_{p/\sqrt{t},2}$	70	kJ mol^{-1}	[30, 48]
$\Delta V_{t2}^{\#}$	15	cm^3 mol^{-1}	[59, 79–81], typical value
$p_{0,p/\sqrt{t},2}$	23.4	MPa	[30]
k_{fp2}/k_{p2}	2.0×10^{-6}		fitted
Kinetic parameters – continuous phase			
$f_{1,0}$	0.60		[77]
A_{d1}	6.3×10^{16}	s^{-1}	[77]
E_{d1}	132	kJ mol^{-1}	[77]
$\Delta V_{d1}^{\#}$	0	cm^3 mol^{-1}	[77]
$p_{0,d1}$	27.6	MPa	[77]
A_{p1}	6.4×10^5	L mol^{-1} s^{-1}	[48, 78]
E_{p1}	15	kJ mol^{-1}	[59, 79], typical value
$\Delta V_{p1}^{\#}$	−25	cm^3 mol^{-1}	[59, 79, 80], typical value
$p_{0,p1}$	2.2	MPa	[78]
$A_{p/\sqrt{t},1}$	6.5×10^9	L$^{1/2}$ mol$^{-1/2}$ s$^{-1/2}$	[30, 48]
$E_{p/\sqrt{t},1}$	70	kJ mol^{-1}	[30, 48]
$\Delta V_{t1}^{\#}$	15	cm^3 mol^{-1}	[59, 79–81], typical value
$p_{0,p/\sqrt{t},1}$	23.4	MPa	[30]
k_{fp1}/k_{p1}	2.0×10^{-6}		fitted
Free-volume parameters			
D_0	7.66×10^{-4}	cm^2 s^{-1}	by fitting data from [82]
E	8.21×10^2	J mol^{-1}	by fitting data from [82]
$K_{1,VDF}/\gamma$	1.04×10^{-3}	cm^3 g^{-1} K^{-1}	by fitting data from [82]
$K_{1,PVDF}/\gamma$	6.22×10^{-5}	cm^3 g^{-1} K^{-1}	by fitting data from Solvay
$K_{2,VDF} - T_{g,VDF}$	−0.6	K	by fitting data from [82]
$K_{2,PVDF} - T_{g,PVDF}$	330	K	by fitting data from Solvay
V_{VDF}^{*}	0.690	cm^3 g^{-1}	[83, 84]
$V_{CO_2}^{*}$	0.589	cm^3 g^{-1}	[68, 69]
V_{PVDF}^{*}	0.565	cm^3 g^{-1}	[83, 84]
V_{FH,CO_2}^{ref}	0.231	cm^3 g^{-1}	[68, 69]
a_{CO_2}	8.76×10^{-4}	K^{-1}	[68, 89]

Table 6.3 (continued)

Parameter	Value	Unit	Source
$\zeta_{VDF/PVDF}$	0.484		[83, 85]
ζ_{VDF/CO_2}	0.284		[83, 85]
Thermodynamic parameters			
ε^*_{VDF}	2632	$J\ mol^{-1}$	by fitting data from Solvay
ε^*_{PVDF}	6652	$J\ mol^{-1}$	[86]
$\varepsilon^*_{CO_2}$	2536	$J\ mol^{-1}$	[74]
υ^*_{VDF}	7.76	$cm^3\ mol^{-1}$	by fitting data from Solvay
υ^*_{PVDF}	20.16	$cm^3\ mol^{-1}$	[86]
$\upsilon^*_{CO_2}$	4.41	$cm^3\ mol^{-1}$	[74]
r_{VDF}	6.03		by fitting data from Solvay
r_{CO_2}	6.60		[74]
ρ^*_{PVDF}	1.920	$g\ cm^{-3}$	[86]
δ_{VDF/CO_2}	0.925 (0.964)		by fitting data from [87]
$\delta_{VDF/PVDF}$	0.900 (0.980)		fitted
$\delta_{CO_2/PVDF}$	0.925 (0.940)		by fitting data from [87]
η_{VDF/CO_2}	0.838 (0.896)		by fitting data from [87]
$\eta_{VDF/PVDF}$	1.000 (1.000)		assumed
$\eta_{CO_2/PVDF}$	0.840 (0.870)		by fitting data from [87]
K_{DEPDC}	1.0		[58]
a	−0.3		by fitting data from [45, 76]
a^f_p	0.25	$m^2\ g^{-1}$	fitted

Table 6.4 Recipes and operation conditons of the experimental batch runs used in case study II.

	Value	Unit
Temperature	50	°C
Initial pressure	13.3/20.4/33.2	MPa
Reactor volume	2.0	L
c^0_{VDF}	1.0/3.1/6.2	$mol\ L^{-1}$
c^0_{DEPDC}	5.0	$mmol\ L^{-1}$

The polymerization proceeds in two reaction loci, the continuous phase and the polymer particles. At low conversion, the first locus is dominant and low-molecular-weight chains are predominantly produced. Since no gel effect is operative in supercritical phase, no acceleration is verified in the first part of the conversion versus time curve and the first mode of the MWD is quite narrow, corresponding to a polydispersity of about 1.5. On the other hand, the relevance of the polymerization in the particles is increases rapidly and becomes dominant by the first hour. This behavior explains the increase in polymerization rate and the high molecular weight of the second MWD mode (much lower ter-

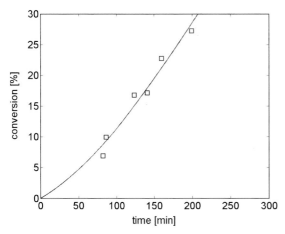

Fig. 6.8 Conversion as a function of reaction time calculated by the model (line) and measured experimentally under the conditions of the base case (□) (from [48]).

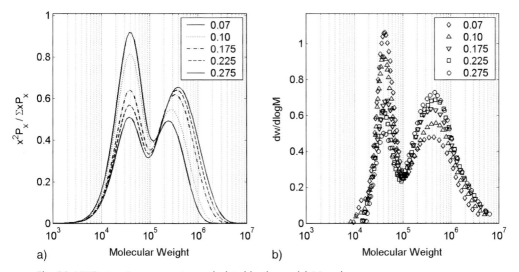

a)   b)

Fig. 6.9 MWD at various conversions calculated by the model (a) and measured experimentally (b) under the conditions of the base case (from [48]).

mination rates are operative). Moreover, the broadness of the same MWD mode is due to chain transfer to polymer. In fact, such a reaction should take place to a significant extent only inside the particles, where there is a high enough concentration of the polymer.

The effect of changing the initial pressure on reaction rate and MWD is shown in Figs. 6.10 and 6.11, respectively. Even though the experimental data exhibit some dependence upon pressure (increasing reaction rate and chain lengths at increasing pressure), the model predictions are much less affected and are practically independent. The latter observation is in agreement with the findings by Hsiao et al. [2] for the dispersion polymerization of MMA in $scCO_2$. On the other hand, by changing the system pressure, the kinetic rate constants as well as the interphase partitioning of both monomer and CO_2 are affected. Higher pressures usually increase the propagation rate constant and decrease the termination rate constant (see activation volumes in Table 6.4), which in turn leads to increased polymerization rates and higher molecular weights. With regard to partitioning, higher pressures increase the sorption of CO_2 in the dispersed phase, which results in dilution and, therefore, in the reduction of all diffusion limitations upon the different reactions. As a consequence, higher termination and initiation rates are to be expected, thus resulting in shorter lengths of the chains produced. Moreover, Saraf et al. [32] reported a decrease in the monomer partition coefficient between polymer and continuous phase at increasing pressure, from which reduced polymerization rate and, again, lower molecular weights would be expected. Summarizing therefore, changes of the initial pressure of the polymerization system affects several factors leading to changes in the polymerization rate and the molecular weights of the chains produced, these changes being in opposite directions. Looking at the model predictions, it is seen that the experimentally observed increase in the polymerization rate is reproduced, even though it is less distinct (see Fig. 6.10). On the other hand, higher molecular weights with increasing pressure are only predicted for the chains produced in the continuous phase (see Fig. 6.11a) whereas the experimental MWD exhibits a shift of both modes as shown in Fig. 6.11b. Despite this small discrepancy, the agreement between model prediction and experimental observation is considered to be satisfactory.

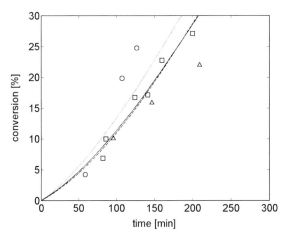

Fig. 6.10 Conversion as a function of reaction time predicted by the model (lines) and measured experimentally with $c_{VDF}^0 = 3.1$ mol L^{-1} at 50 °C and different densities: $\rho = 0.65$ kg L^{-1} ($\triangle$ dashed); 0.79 kg L^{-1} ($\square$ solid); 0.89 kg L^{-1} ($\bigcirc$ dotted) (from [48]).

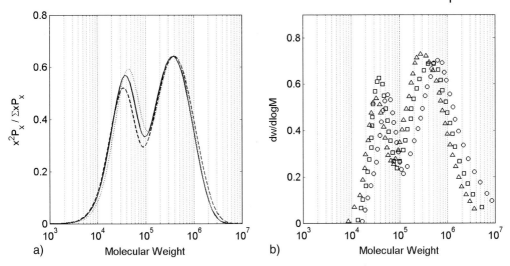

Fig. 6.11 MWD predicted by the model (a) and measured experimentally (b) with $c_{VDF}^0 = 3.1$ mol L^{-1} at 50 °C and different densities: $\rho = 0.65$ kg L^{-1} ($\triangle$ dashed); 0.79 kg L^{-1} ($\square$ solid); 0.89 kg L^{-1} ($\bigcirc$ dotted) (from [48]).

The effect of changing the initial monomer concentration is shown in Figs. 6.12 and 6.13. Remarkable agreement between the model predictions and the experimental data is found. In the case of the lowest monomer concentration, the reaction rate does not exhibit any acceleration, thus indicating that the polymerization is almost exclusively occurring in the continuous phase. However, the main reaction locus is shifted toward the dispersed phase at increasing monomer concentration, which in turn results in a significant rate increase. The same conclusion can be drawn in an even more convincing way when we look at the MWD results shown in Fig. 6.13. At increasing monomer concentration, a second mode grows in the high-molecular-weight region as a result of the reaction in the dispersed phase. Note that this second mode becomes dominant when some acceleration in the reaction rate is found, and this joint behavior strongly supports the two-loci mechanism underlying the developed model. It is worth noting that changing the specific surface area for the different reaction conditions could improve the fitting only to a very limited extent, indicating that the system stability is quite constant under all conditions examined.

As mentioned above, the model was adapted to continuous reactors (CSTR) in order to further validate its prediction ability. In particular, the effect of changing the inlet monomer concentration on the MWD was investigated. Simulations were carried out using the identical set of model parameter values already considered for the batch simulations above. However, because of the higher operating temperature in the CSTR reactions (see Table 6.5) with respect to the batch experiments, the binary interaction parameters used in the frame of the Sanchez-Lacombe equation of state have to be updated. This has been done

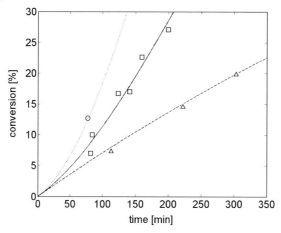

Fig. 6.12 Conversion as a function of reaction time predicted by the model (lines) and measured experimentally at 50 °C and 20.4 MPa using different monomer concentrations: c_{VDF}^0 =1.0 mol L^{-1} ($\triangle$ dashed); 3.1 mol L^{-1} ($\square$ solid); 6.2 mol L^{-1} ($\bigcirc$ dotted) (from [48]).

Table 6.5 Recipes and operation conditions of the experimental CSTR runs used in case study II [31].

	Value	Unit
Temperature	75	°C
Pressure	27.7	MPa
Reactor volume	800	mL
Residence time	21	min
c_{VDF}^{IN}	0.77/1.68/2.79/3.53	mol L^{-1}
c_{DEPDC}^{IN}	2.94/2.95/2.84/3.32	mmol L^{-1}

using experimental sorption/swelling data of CO_2 in PVDF as well as density data as a function of pressure for the binary mixture of VDF/CO_2 at the appropriate temperature. The obtained values are reported in parentheses in Table 6.3. Note that the higher operating temperature was not a problem when evaluating the kinetic rate constants, since Arrhenius laws were available.

The comparison between model predictions and experimental data is given in Fig. 6.14. Keeping in mind that the latter are obtained in a completely different polymerization system, at largely different conditions, and without any parameter tuning, the agreement is surprisingly good. Moreover, the CSTR data confirm the observations made above when analyzing the batch data, namely that the main reaction locus is shifted toward the dispersed phase at increasing monomer concentration, leading to a growing mode at higher molecular weights.

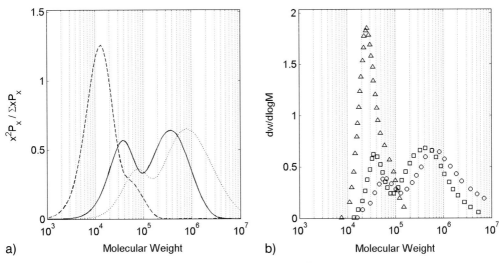

Fig. 6.13 MWD predicted by the model (a) and measured experimentally (b) at 50 °C and 20.4 MPa using different monomer concentrations: $c_{VDF}^0 = 1.0$ mol L^{-1} ($\triangle$ dashed); 3.1 mol L^{-1} ($\square$ solid); 6.2 mol L^{-1} ($\bigcirc$ dotted) (from [48]).

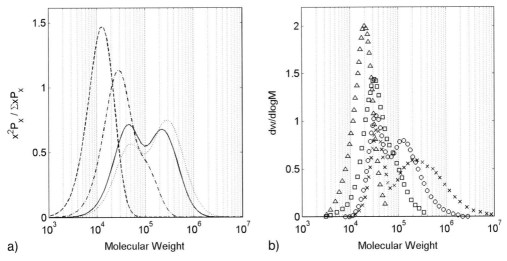

Fig. 6.14 MWD predicted by the model (a) and measured experimentally (b) under the operating conditions reported in Table 6.5: $c_{VDF}^{IN} = 0.77$ mol L^{-1} ($\triangle$ dashed); 1.68 mol L^{-1} ($\square$ dash-dotted); 2.79 mol L^{-1} ($\bigcirc$ solid); 3.53 mol L^{-1}.

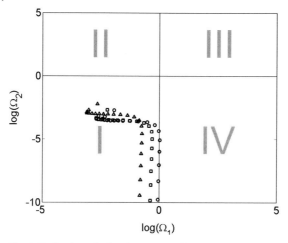

Fig. 6.15 Ω_p plot calculated at 50 °C and 20.4 MPa using different monomer concentrations: c^0_{VDF} = 1.0 mol L^{-1} ($\triangle$); 3.1 mol L^{-1} ($\square$); 6.2 mol L^{-1} ($\bigcirc$).

To summarize, the model was able to describe the experimental data obtained in batch as well as in CSTR with reasonable accuracy, thus validating the initial schematization of the process. Before concluding, it is now worth repeating the Ω analysis developed for MMA polymerizationin order to check the role of the interphase transport of active chains and, therefore, to validate the previous analysis concerning the reaction loci. Since bimodal MWDs indicate that two re-action loci are operative, Ω_1 values near unity may be expected for most of the active chains, whatever their length. The Ω traces for the reactions carried out in the batch reactor at three different monomer concentrations and calculated at 25% of conversion are shown in Fig. 6.15. As expected, with the exception of the shortest chains (degree of polymerization less than 20), most growing chains exhibit Ω_1 values very close to 1, thus confirming their ability to diffuse from the continuous to the dispersed phase, where the same chains will experi-ence a higher growth rate and some chain transfer to polymer. Note that the calculated Ω_1 values approach 1 as the system becomes more concentrated: therefore, the increasing bimodal nature of the final MWD at increasing mono-mer concentration is fully explained, once more indicating the benefit of the simplified analysis based on characteristic times.

6.6
Concluding Remarks and Outlook

In this chapter, a comprehensive model of heterogeneous polymerization in supercritical carbon dioxide has been applied to two different systems, with the aim of elucidating the key mechanisms underlying the process. The main model equations are the population balances accounting for the evolution of the concentration of the different polymer species as a function of the chain length. Moreover, the calculation of interphase equilibria and volumetric behavior has been based on the Sanchez-Lacombe equation of state, a popular thermodynamic model suitable for polymer systems and high pressure. The particle size distribution is not accounted for, and a constant number of particles with variable but uniform size is assumed. Moreover, all kinetic parameters (reaction rate constants, diffusion coefficients) have been evaluated using state-of-the-art relationships, thus accounting for chain length and system viscosity effects. Finally, most parameter values have been found in the literature or estimated from independent sources, in order to ensure a meaningful model validation.

Despite the large differences between the two cases examined (corresponding to dispersion and precipitation polymerization conditions), a unifying modeling approach has been found when accounting for the mass transport of all species between continuous and dispersed phase. The agreement between model predictions and experimental results is a strong validation of the assumed reaction scheme, accounting for the polymerization reaction in both phases. The fundamental interplay between the particle surface area and the polymer microstructure (i.e., the molecular weight distribution) has been clearly highlighted. A pair of meaningful parameters, defined as the ratios of the characteristic times of termination and diffusion out of each phase, appeared to be quite useful in predicting the system behavior at least in a semi-quantitative way. Therefore, a rough evaluation of these quantities could represent an effective tool to identify the main features of the polymerization system under examination.

Acknowledgment

The financial support of BBW (Swiss Federal Office for Education and Science, Contract No. 02-0131) and ETH (ETHZ – internal research project "Dispersion Polymerization in Supercritical Carbon Dioxide") is gratefully acknowledged. Moreover, we thank Solvay-Solexis for providing the experimental VDF batch data.

Notation

Symbols

a	root mean square end-to-end distance per square root of monomer units	cm
a_p^f	final specific particle surface area	$m^2 \, g^{-1}$
$a_{p,j}^f$	total particle surface area per volume of phase j	m^{-1}
A_p	total particle surface area	m^2
A_{dj}	pre-exponential factor of initiator decomposition in phase j	s^{-1}
A_{pj}	pre-exponential factor of propagation in phase j	$L^{1/2} \, mol^{-1/2} \, s^{-1/2}$

$A_{p/\sqrt{t},j}$	pre-exponential factor of the ratio $k_{pj}/\sqrt{k_{tj}}$ in phase j	L mol^{-1} s^{-1}
A_{tj}	pre-exponential factor of termination in phase j	L mol^{-1} s^{-1}
c_i^0	initial concentration of component i	mol L^{-1}
c_i^{IN}	inlet concentration of component i	mol L^{-1}
D_0	pre-exponential factor, see Eq. (21)	cm^2 s^{-1}
D_i	self-diffusion coefficient of component i	cm^2 s^{-1}
$D_{i,0}$	self-diffusion coefficient of component i at zero conversion	cm^2 s^{-1}
$D_{x,com}$	center-of-mass diffusion coefficient of a chain of length x	cm^2 s^{-1}
$D_{x,j}$	self-diffusion coefficient of a chain of length x in phase j	cm^2 s^{-1}
E	critical energy to overcome attractive forces, see Eq. (21)	J mol^{-1}
E_{dj}	activation energy of initiator decomposition in phase j	kJ mol^{-1}
E_{pj}	activation energy of propagation in phase j	kJ mol^{-1}
$E_{p/\sqrt{t},j}$	activation energy of the ratio $k_{pj}/\sqrt{k_{tj}}$ in phase j	kJ mol^{-1}
E_{tj}	activation energy of termination in phase j	kJ mol^{-1}
$f_{i,0}$	initiator efficiency in phase j at zero conversion	
f_i	initiator efficiency in phase j	
I	total amount of initiator	mol
I_j	amount of initiator in phase j	mol
$[I_j]$	initiator concentration in phase j	mol L^{-1}
$I_j^{\bullet}$	amount of activated initiator in phase j	mol
j_c	entanglement spacing	
k_{dj}	initiator decomposition rate constant in phase j	s^{-1}
k_{fmj}	chain transfer to monomer rate constant in phase j	L mol^{-1} s^{-1}
k_{fpj}	chain transfer to polymer rate constant in phase j	L mol^{-1} s^{-1}
k_{ij}	binary interaction parameter of the 1-parameter SL model	
k_{Ij}	initiation rate constant in phase j	L mol^{-1} s^{-1}
k_{pj}	propagation rate constant in phase j	L mol^{-1} s^{-1}
$k_{pj,0}$	propagation rate constant in phase j at zero conversion	L mol^{-1} s^{-1}
k_{tcj}	termination by combination rate constant in phase j	L mol^{-1} s^{-1}
k_{tdj}	termination by disproportionation rate constant in phase j	L mol^{-1} s^{-1}
$k_{tj,0}$	termination rate constant in phase j at zero conversion	L mol^{-1} s^{-1}
$k_{x,j}$	local mass transfer coefficient of a chain of length x in phase j	cm s^{-1}
$K_{1,i}$	free-volume parameter of component i	cm^3 g^{-1} K^{-1}
$K_{2,i}$	free-volume parameter of component i	K
K_i	partition coefficient of component i	
$K_{x,j}$	overall mass transfer coefficient of a chain of length x referred to phase j	cm s^{-1}
m_j^0	total initial mass of component i	g
m_x	partition coefficient of a chain of length x	

M	total amount of monomer	mol
M_j	amount of monomer in phase j	mol
$[M_j]$	monomer concentration in phase j	mol L^{-1}
M_n	number average molecular weight	g mol^{-1}
M_w	weight average molecular weight	g mol^{-1}
N_A	Avogadro number	mol^{-1}
N_p	particle number	
p	pressure	MPa
$p_{0,dj}$	reference pressure at which k_{dj} has been obtained	MPa
$p_{0,pj}$	reference pressure at which k_{pj} has been obtained	MPa
$p_{0,p/\sqrt{t,j}}$	reference pressure at which $k_{pj}/\sqrt{k_{tj}}$ has been obtained	MPa
$p_{0,pj}$	reference pressure at which k_{tj} has been obtained	MPa
$P_{x,j}$	amount of terminated polymer chains of length x in phase j	mol
$[P_{x,j}]$	concentration of terminated polymer chains of length x in phase j	mol L^{-1}
$[P_{x,j}^*]$	hypothetical concentration of terminated polymer chains of length x in phase j in equilibrium with the corresponding bulk concentration in the other phase	mol L^{-1}
r_i	number of lattice sites occupied by component i	
r_p	particle radius	cm
R	ideal gas constant	J mol^{-1} K^{-1}
$R_{x,j}$	amount of radical chains of length x in phase j	mol
$[R_j]$	total radical concentration in phase j	mol L^{-1}
$[R_{x,j}]$	concentration of radical chains of length x in phase j	mol L^{-1}
$[R_{x,j}^*]$	hypothetical concentration of radical chains of length x in phase j in equilibrium with the corresponding bulk concentration in the other phase	mol L^{-1}
S	total amount of solvent	mol
S_j	amount of solvent in phase j	mol
$[S_j]$	solvent concentration in phase j	mol L^{-1}
t	time	s
T	temperature	K
$T_{g,j}$	glass transition temperature of component i	K
$V_{FH,i}$	specific hole-free volume of component i	cm^{-3} g^{-1}
V_i^*	specific critical hole-free volume of component i	cm^{-3} g^{-1}
V_j	volume of phase j	L
$\Delta V_{dj}^{\#}$	activation volume of initiator decomposition in phase j	cm^3 mol^{-1}
$\Delta V_{pj}^{\#}$	activation volume of propagation in phase j	cm^3 mol^{-1}
$\Delta V_{tj}^{\#}$	activation volume of termination in phase j	cm^3 mol^{-1}
x, y	chain length	
$x_{max,j}$	maximum average length of chains produced in phase j	
$x_{act,j}$	actual average length of chains produced in phase j	
X	conversion	

X^f final conversion

Greek Symbols

a	chain partitioning parameter, see Eq. (6.13)	
a_i	coefficient of thermal expansion of component i	K^{-1}
$\tilde{a}_i$	ratio between coefficients of thermal expansion of component i above and below T_g	
γ	overlap factor, see Eqs. (6.21 to 6.23)	
$\delta(x - x_0)$	Kronecker delta function	
δ_{ij}	binary interaction parameter of the 2-parameter SL model	
ε_i^*	characteristic interaction energy of component i	$J\ mol^{-1}$
η_{ij}	binary interaction parameter of the 2-parameter SL model	
μ_i^j	chemical potential of component i in phase j	$J\ mol^{-1}$
ξ_{ij}	ratio between molar volumes of jumping units of i and j, Eq. (6.21)	
ρ	density	$kg\ L^{-1}$
ρ_i	density of component i	$g\ cm^{-3}$
ρ_i^*	characteristic density of component i,	$g\ cm^{-3}$
σ_i	Lennard-Jones diameter of component i	cm
$\tau_{diff,j}$	characteristic time of diffusion out of phase j	s
$\tau_{p,j}$	characteristic time of propagation in phase j	s
υ_i^*	characteristic volume of component i	$cm^3\ mol^{-1}$
$\Omega_j(x)$	ratio between characteristic times of termination and interphase mass transport of a chain of length x in phase j	
ω_i	weight fraction of component i	

References

1 J. M. DeSimone, E. E. Maury, Y. Z. Menceloglu, J. B. McClain, T. J. Romack, J. R. Combes, Science, **1994**, *265*, 356.

2 Y.-L. Hsiao, E. E. Maury, J. M. DeSimone, S. Mawson, K. P. Johnston, Macromolecules, **1995**, *28*, 8159.

3 Y.-L. Hsiao, J. M. DeSimone, J. Polym. Sci. Part A Polym. Chem., **1997**, *35*, 2009.

4 K. A. Schaffer, T. A. Jones, D. A. Canelas, J. M. DeSimone, S. P. Wilkinson, Macromolecules, **1996**, *29*, 2704.

5 C. Lepilleur, E. J. Beckman, Macromolecules, **1997**, *30*, 745.

6 M. L. O'Neill, M. Z. Yates, K. P. Johnston, C. D. Smith, S. P. Wilkinson, Macromolecules, **1998**, *31*, 2838.

7 M. L. O'Neill, M. Z. Yates, K. P. Johnston, C. D. Smith, S. P. Wilkinson, Macromolecules, **1998**, *31*, 2848.

8 M. Z. Yates, G. Li, J. J. Shim, S. Maniar, K. P. Johnston, K. T. Lim, S. Webber, Macromolecules, **1999**, *32*, 1018.

9 M. R. Giles, J. N. Hay, S. M. Howdle, R. J. Winder, Polymer, **2000**, *41*, 6715.

10 P. Christian, S. M. Howdle, Macromolecules, **2000**, *33*, 237.

11 M. R. Giles, S. J. O'Connor, J. N. Hay, R. J. Winder, S. M. Howdle, Macromolecules, **2000**, *33*, 1996.

12 G. Li, M. Z. Yates, K. P. Johnston, S. M. Howdle, Macromolecules, **2000**, *33*, 4008.

13 U. Fehrenbacher, O. Muth, T. Hirth, M. Ballauf, Macromol. Chem. Phys., **2000**, *201*, 1532.

14 U. Fehrenbacher, M. Ballauf, Macromolecules, **2002**, *35*, 3653.

15 M. Okubo, S. Fujii, H. Maenaka, H. Minami, Colloid Polym. Sci., **2002**, *280*, 183.

16 Sumitomo **1968**, French Pat 1,524,533.

17 K. Fukui, T. Kagiya, H. Yokota, Y. Toriuchi, F. Kuniyoshi **1970**, U.S. Pat 3,522,228.

18 H. Hartmann, W. Denzinger **1986**, Canadian Pat 1,262,995.

19 T.J. Romack, E.E. Maury, J.M. DeSimone, Macromolecules, **1995**, *28*, 912.

20 H.Q. Hu, M.C. Chen, J. Li, G.M. Cong, Acta Polymerica Sinica, **1998**, *6*, 740.

21 Q. Xu, B. Han, H. Yan, Polymer, **2001**, *42*, 1369.

22 A.I. Cooper, W.P. Hems, A.B. Holmes, Macromol. Rapid Commun., **1998**, *19*, 353.

23 X.R. Teng, X.C. Hu, H.L. Shao, Polymer Journal, **2002**, *34*, 534.

24 X.R. Teng, H.L. Shao, X.C. Hu, J. Appl. Polym. Sci., **2002**, *86*, 2338.

25 M. Okubo, S. Fujii, H. Maenaka, H. Minami, Colloid Polym.Sci., **2002**, *280*, 1084.

26 M. Okubo, S. Fujii, H. Maenaka, H. Minami, Colloid Polym.Sci., **2003**, *281*, 964.

27 T.J. Romack, B.E. Kipp, J.M. DeSimone, Macromolecules, **1995**, *28*, 8432.

28 T.J. Romack, J.M. DeSimone, T.A. Treat, Macromolecules, **1995**, *28*, 8429.

29 P.A. Charpentier, K.A. Kennedy, J.M. DeSimone, G.W. Roberts, Macromolecules, **1999**, *32*, 5973.

30 P.A. Charpentier, J.M. DeSimone, G.W. Roberts, Ind. Eng. Chem. Res., **2000**, *39*, 4588.

31 M.K. Saraf, S. Gerard, L.M. Wojcinski II, P.A. Charpentier, J.M. DeSimone, G.W. Roberts, Macromolecules, **2002**, *35*, 7976.

32 M.K. Saraf, L.M. Wojcinski II, K.A. Kennedy, S. Gerard, P.A. Charpentier, J.M. DeSimone, G.W. Roberts, Macromol. Symp., **2002**, *182*, 119.

33 A.J.Paine, Macromolecules, **1990**, *23*, 3109.

34 O. Prochazka, J. Stejskal, Polymer, **1992**, *33*, 3658.

35 S.F. Ahmed, G.W. Poehlein, Ind. Eng. Chem. Res., **1997**, **36**, *2597*.

36 S.F. Ahmed, G.W. Poehlein, Ind. Eng. Chem. Res., **1997**, **36**, *2605*.

37 J.M. Saenz, J.M. Asua, Colloids and Surfaces A: Physiochemical and Engineering Aspects, **1999**, **153**, *61*.

38 P.H.H. Araujo, J.C. Pinto, Braz. J. Chem. Eng., **2000**, *17*, *383*.

39 C. Chatzidoukas, P. Pladis, C. Kiparissides, Ind. Eng. Chem. Res., **2003**, *42*, 743.

40 C. Kiparissides, G. Daskalakis, D.S. Achilias, E. Sidiropoulou, Ind. Eng. Chem. Res., **1997**, *36*, 1253.

41 P.A. Mueller, G. Storti, M. Morbidelli, Chem. Eng. Sci., **2005**, *60*, 377.

42 P.A. Mueller, G. Storti, M. Morbidelli, Chem. Eng. Sci., **2005**, *60*, 1911.

43 I. Sanchez, R. Lacombe, J. Phys. Chem., **1976**, *80*, 2352.

44 I. Sanchez, R. Lacombe, Macromolecules, **1978**, *11*, 1145.

45 S.K. Kumar, S.P. Chhabria, R.C. Reid, U.W. Suter, Macromolecules, **1987**, *20*, 2550.

46 I. Sanchez, R. Lacombe, J. Phys. Chem., **1976**, *80*, 2352.

47 I. Sanchez, R. Lacombe, Macromolecules, **1978**, *11*, 1145.

48 P.A. Mueller, G. Storti, M. Morbidelli, M. Apostolo, R. Martin, Macromolecules, **2005**, in press.

49 S. Kumar, D. Ramkrishna, Chem. Eng. Sci., **1996**, *51*, 1311.

50 S. Beuermann, M. Buback, Prog. Polym. Sci., **2002**, *27*, 191.

51 B.R. Morrison, M.C. Piton, M.A. Winnik, R.G. Gilbert, D.H. Napper, Macromolecules, **1993**, *26*, 4368.

52 J.S. Vrentas, J.L. Duda, J. Polym. Sci.: Polym. Phys. Ed., **1977**, *15*, 403.

53 J.S. Vrentas, J.L. Duda, J. Polym. Sci.: Polym. Phys. Ed., **1977**, *15*, 417.

54 M.C. Griffiths, J. Strauch, M.J. Monteiro, R.G. Gilbert, Macromolecules, **1998**, *31*, 7835.

55 S. Fortini, Th. Meyer, in 8th International Workshop on Polymer Reaction Engineering, DECHEMA Monographs, Vol. 138, Frankfurt am Main, **2004**.

56 W.K. Lewis, W.G. Whitman, Ind. Eng. Chem. Res., **1924**, *16*, 1215.

57 G.T. Russell, D.H. Napper, R.G. Gilbert, Macromolecules, **1988**, *21*, 2141.

58 J. Shen, Y. Tian, G. Wang, M. Yang, Macromol. Chem., **1991**, *192*, 2669.

59 G. Odian, Principles of Polymerization, McGraw-Hill, New York, **1970**.

60 H.K. Mahabadi, K.F. O'Driscoll, Journal of Macromolecular Science-Chemistry, **1977**, *A11*, 967.

61 S. Beuermann, M. Buback, G.T. Russell, Macromol. Rapid Commun., **1994**, *15*, 351.

62 M. Buback, C. Kowollik, Macromolecules, **1998**, *31*, 3211.

63 D. Kukulj, T.P. Davis, R.G. Gilbert, Macromolecules, **1994**, *31*, 994.

64 M.D. Zammit, T.P. Davis, D.M. Haddleton, K.G. Suddaby, Macromolecules, **1997**, *30*, 1915.

65 Z.B. Guan, J.R. Combes, Y.Z. Menceloglu, J.M. DeSimone, Macromolecules, **1993**, *26*, 2663.

66 M.A. Quadir, J.M. DeSimone, A.M. van Herk, A.L. German, Macromolecules, **1998**, *31*, 6481.

67 A. Faldi, M. Tirell, Macromolecules, **1994**, *27*, 4184.

68 S. Alsoy, J.L. Duda, AIChE J., **1998**, *44*, 582.

69 R.R. Gupta, K.A. Lavery, T.J. Francis, J.R.P. Webster, G.S. Smith, T.P. Russell, J.J. Watkins, Macromolecules, **2003**, *36*, 346.

70 J.M. Zielinski, J.L. Duda, AIChE J., **1992**, *38*, 405.

71 M. Matheson, E. Auer, E. Bevilacqua, E. Hart, J. Am. Chem. Soc., **1949**, *71*, 497.

72 B.B. Kine, R.W. Novak, in: H.F. Mark, N.M. Bikales, C.G. Overberger, G. Meges, J.I. Kroschwitz (Eds.), Encyclopedia of Polymer Science and Engineering, Vol. 1, John Wiley & Sons, **1985**, pp. 234–299.

73 T. Boublik, V. Fried, E. Hala, The Vapour Pressure of Pure Substances, Vol. 17, Elsevier, Amsterdam, **1984**.

74 M.B. Kiszka, M.A. Meilchen, M.A. McHugh, Journal of Applied Polymer Science, **1988**, *36*, 583.

75 A. Rajendran, B. Bonavoglia, N. Forrer, G. Storti, M. Mazzotti, M. Morbidelli, in press on Ind. Eng. Chem. Res., **2004**.

76 B. Bonavoglia, U. Verga, G. Storti, M. Morbidelli, Paper presented at AIChE Annual Meeting, San Francisco, CA, Nov 16–21, **2003**.

77 P.A. Charpentier, J.M. DeSimone, G.W. Roberts, Chem. Eng. Sci., **2000**, *55*, 5341.

78 M. Apostolo, V. Arcella, G. Storti, M. Morbidelli, Macromolecules, **2002**, *35*, 6154.

79 J. Brandrup, E.H. Immergut, Polymer Handbook, 4th edn, John Wiley & Sons, New York, **1998**.

80 Y. Ogo, M. Yokawa, Macromol. Chem., **1977**, *178*, 453.

81 Y. Ogo, T. Kyotani, Macromol. Chem., **1978**, *179*, 2407.

82 T.E. Daupert, R.P. Danner, H.M. Sibul, C.C. Stebbin, Physical and Thermodynamic Properties of Pure Chemicals. Data Compilation, Taylor & Francis, Washington, **1989**.

83 S.U. Hong, Ind. Eng. Chem. Res., **1995**, *34*, 2536.

84 R.N. Howard, J. Macromol. Sci. Rev. Macromol. Chem., **1970**, *C4*, 191.

85 S.T. Ju, J.L. Duda, J.S. Vrentas, Ind. Eng. Chem. Prod. Res. Dev., **1981**, *20*, 330.

86 B.J. Briscoe, O. Lorge, A. Wajs, P. Dang, Journal of Polymer Science: Part B: Polymer Physics, **1998**, *36*, 2435.

87 K.A. Kennedy, PhD Thesis, North Carolina State University, **2003**.

7
Inverse Emulsion Polymerization in Carbon Dioxide

Eric J. Beckman

7.1
Introduction

Emulsion polymerization refers to one type of heterogeneous polymerization where two phases exist (continuous and discontinuous). The monomer in question is relatively poorly soluble in the continuous phase and therefore partitions strongly to the discontinuous phase. The discontinuous phase is composed of a mixture of micelles (aggregates of surfactant, or soap molecules, of order 20 nm in size) and monomer droplets (partially stabilized by surfactant, of order hundreds of nanometers to microns in size). The micelles are thermodynamically stable constructs, while the droplets would eventually settle from solution if stirring of the system ceased. Emulsion polymerization occurs when an initiator that is soluble in the continuous phase is added to the system and forms radicals, either through thermal, radiative, or redox initiation processes. These radicals diffuse to the micelle interface and encounter monomer, starting the polymerization (Fig. 7.1). Monomer diffuses through the continuous phase from the droplets to the micelles, feeding the reaction. When a second radical enters the micelle, the growing chain is rapidly terminated.

Emulsion polymerization is of interest for several reasons [1]:

1. The kinetics of the reaction are such that one can generate polymer at a high rate with simultaneous high molecular weight. In a typical bulk or solution polymerization, one elevates the rate of reaction via increases to initiator concentration, yet this also reduces the average kinetic chain length. In an emulsion polymerization, one can increase the rate of reaction either by elevating the initiator or the surfactant concentration. Given two variables to adjust, it becomes possible to increase the rate while maintaining molecular weight.
2. Operating the polymerization in a heterogeneous mode allows one to more easily absorb the heat of the polymerization produced by vinyl monomers.
3. If the final product (polymer chains in micelles, known as a latex) is stable, then this product can be used in interesting ways, such as in coating processes.

Supercritical Carbon Dioxide: in Polymer Reaction Engineering
Edited by Maartje F. Kemmere and Thierry Meyer
Copyright © 2005 WILEY-VCH Verlag GmbH & Co. KGaA, Weinheim
ISBN: 3-527-31092-4

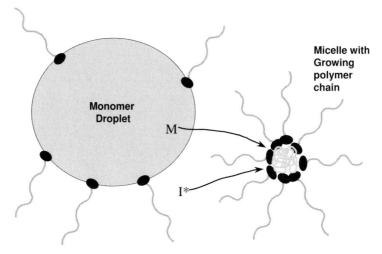

Fig. 7.1 Schematic of emulsion polymerization.

Emulsions are generally classified as normal (oil-in-water) or inverse (water-in-oil); in this paper, we will only consider inverse emulsions, as our continuous phase is a non-aqueous fluid (primarily CO_2). Typical monomers used in inverse emulsion polymerization are water soluble, and include (meth)acrylic acid, (meth)acrylamide, vinyl pyrrolidone, N-vinyl formamide, vinyl sulfonic acid salts, etc. Typical continuous phases used commercially are hydrocarbons, oftentimes inexpensive fluids such as kerosene or naphtha. In this chapter we consider emulsion polymerizations in near-critical or supercritical continuous phases.

Regarding choice of continuous phase, in theory we might operate an inverse emulsion polymerization in any fluid with a relatively low critical temperature (for example, approximately 373 K and below), including alkanes and alkenes, CFCs, HFCs, CO_2, N_2O, Xe, and dimethyl ether. However, practical considerations greatly reduce this list; the flammability of alkanes, the danger inherent in the use of N_2O (it is a strong oxidant), the cost of HFCs and Xe, and the ozone-depleting capacity of CFCs have meant that only CO_2 is seriously considered as a replacement for conventional oil-based continuous phases. We consequently focus our attention on carbon dioxide.

Restricting our continuous phase to carbon dioxide means that the monomers in question must be relatively CO_2-phobic (in other words hydrophilic or at least very polar). Note that this greatly reduces the number of monomers that would be viable candidates for inverse emulsion polymerization in CO_2 (or in another supercritical continuous phase); indeed even some water-soluble monomers such as acrylic acid are miscible with CO_2 under relatively mild conditions. We will examine the issues surrounding monomer selection and surfactant design as they relate to the phase behavior of the system in the next section.

The potential benefits of operating an inverse emulsion polymerization in CO_2 are two-fold. In a conventional inverse emulsion polymerization of a water-soluble monomer, a hydrocarbon (kerosene, for example) is used as the continuous phase (CP). Following polymerization, the polymer must be recovered and separated from the CP, because the polymers are typically used by the customer as aqueous solutions (in many cases where the presence of organic solvent is not permitted). Separation of polymer from CP occurs by breaking the emulsion (through addition of salt, for example), then stripping the solvent from the material using a wiped film evaporator. Thus, recovery of the polymer following polymerization is an energy-intensive and multi-step process. Conducting the same polymerization in CO_2 allows for easy separation; dropping the pressure will lead to rapid precipitation of dry polymer powder. In addition to easy product recovery, CO_2 affords the added benefit that it does not participate in chain transfer reactions. Most solvents exhibit some degree of chain transfer, where a proton is abstracted from a solvent molecule, terminating a polymer chain (and possibly starting another). Chain transfer lowers molecular weight. However, it has been shown by DeSimone and colleagues [2] that CO_2 does not participate in chain transfer.

7.2
Inverse Emulsion Polymerization in CO₂: Design Constraints

As noted above, not all water-soluble monomers can be polymerized in an inverse emulsion in CO_2, simply because they are completely miscible with CO_2 (which would lead to a dispersion polymerization, a kinetically and thermodynamically distinct type of heterogeneous polymerization). In order to conduct an inverse emulsion polymerization in CO_2, the phase behavior of the system must follow the somewhat simplified schematic shown in Fig. 7.2. Here, we see that the surfactant to be used is soluble in CO_2 at moderate pressures, allowing it to dissolve and stabilize the micelles in which the polymer is formed. The monomer exhibits a liquid-liquid phase boundary at substantially higher pressures than that for the surfactant, such that the monomer will tend to partition into the micelles at the operating pressure rather than simply dissolve in the continuous phase. The monomer should not be *completely* immiscible with the CP, because it *does* have to diffuse from the large droplets into the micelles during polymerization. Although easy to sketch, the schematic shown in Fig. 7.2 is exceptionally difficult to achieve in practice, which means that inverse emulsion polymerization in CO_2 has not been widely practiced. Vinyl monomers are usually miscible with CO_2 at modest pressures, while the design of highly CO_2-soluble surfactants has proven to be a complex task. Thus, it is usual for the phase boundary for the surfactant to appear at much higher pressures than that of the monomer in CO_2, meaning that only dispersion polymerizations are viable heterogeneous options. The dispersion polymerization regime, which is described in Chapter 6 of this book, exhibits significantly different kinetic behavior. In

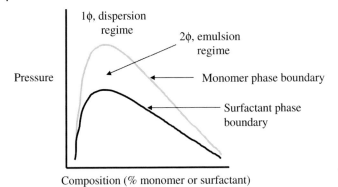

Fig. 7.2 Phase behavior requirements for viable emulsion polymerization in CO_2.

summary, it has proven relatively difficult to create a system where the phase boundary for the surfactant in CO_2 exists at lower pressures than the phase boundary for the monomer in CO_2.

7.3
Surfactant Design for Inverse Emulsion Polymerization

As noted above, surfactants used in inverse emulsion polymerizations in CO_2 must dissolve in CO_2 and create micelles that solubilize the monomer. Therefore, one end of these bifunctional molecules must interact favorably (in a thermodynamic sense) with CO_2, while the other interacts favorably with the monomer. Creation of suitable *hydrophiles* for these molecules has been straightforward, while design of the CO_2-philic portion has been difficult, and even at present cannot be considered to be an entirely solved problem. Thus, we consider the design of CO_2-philes in more depth.

Unfortunately, CO_2 is a feeble solvent [3]; although it can solubilize low-molecular-weight, volatile compounds at pressures below 100 bar, polar and high-molecular-weight materials are usually poorly soluble at tractable pressures. An ability to design *a priori* highly CO_2-soluble ("CO_2-philic") compounds would render emulsion polymerization (and other applications!) in CO_2 much easier to perform. CO_2's solvent power has in the past been likened to that of toluene (based on FT-IR spectroscopy [4]), to hexane [5] and to pyridine [6] (based on thermodynamic solubility parameter calculations), and to acetone (based on hydrogen bond accepting tendency [7]). These descriptions have all been discarded over the years when experimental data revealed them to be oversimplifications.

7.3.1

Designing CO$_2$-philic Compounds: What Can We Learn from Fluoropolymer Behavior?

In 1992, DeSimone and colleagues [8] published the first report of a truly "CO$_2$-philic" material, when they showed that a poly(perfluoroalkyl acrylate) with over 2500 repeat units was miscible with CO$_2$ at pressures below 150 bar. By contrast, Heller and colleagues [9] had earlier noted that typical non-fluorinated polymers with fewer than 25 repeat units were *insoluble in CO$_2$ at 200 bar*. Subsequent work showed that attachment of fluorinated "ponytails" to chelating agents, surfactants, and catalyst ligands generally enhanced the solubility of such compounds in CO$_2$ [10]. The problem of CO$_2$'s weak solvent strength seemed to have been solved: *fluorination = CO$_2$-philicity*. Indeed, as noted below, fluorinated surfactants are the only amphiphiles that have proven to be technically successful in supporting emulsion polymerizations.

Unfortunately, the use of fluorinated ponytails to achieve CO$_2$ solubility is relatively expensive, and towards the end of the 1990s it also became somewhat environmentally problematic. Indeed, the sales value of most of the polymers generated using emulsion polymerization is of order 2 Euro/kg, while the value of the fluorinated materials typically employed as CO$_2$-philes can easily be 100 times greater. Further, *some*, but interestingly *not all* fluorinated alkane, acrylate, and ether polymers are miscible with CO$_2$ at much lower pressures than their non-fluorous counterparts [8, 11, 12]. Attempts to explain the CO$_2$-philic character of fluorinated CO$_2$-philes have focused on determining whether there exist any specific interactions between CO$_2$ and these molecules. Yee et al. used FTIR to investigate mixtures of CO$_2$ and hexafluoroethane [13], but found no evidence of specific attractive interactions between the F atoms and CO$_2$. The authors consequently attributed the observed enhanced solubility of fluorocarbons to weak solute-solute interactions. However, when Dardin et al. compared ^{1}H and ^{19}F NMR chemical shifts of *n*-hexane and perfluoro-*n*-hexane in CO$_2$ [14], they observed a chemical shift in the C$_6$F$_{14}$ spectra, which they ascribed to C$_6$F$_{14}$-CO$_2$ van der Waals interactions. By contrast, Yonker et al. showed (using ^{1}H and ^{19}F NMR) that neither fluoromethane (CH$_3$F) nor trifluoromethane (CHF$_3$) exhibit significant specific attractive interactions with CO$_2$ [15].

Theoretical studies have also resulted in contradictory findings. Using restricted Hartree-Fock level *ab initio* calculations, Cece et al. suggested that there exist specific interactions between CO$_2$ and the fluorines of C$_2$F$_6$ [16]. Han and Jeong [17], however, disagreed with these results, noting that Cece et al. did not take into account basis set superposition error (BSSE) corrections during their calculations. Using similar *ab initio* calculations, but accounting for BSSE corrections, Diep et al. [18] reported no evidence of CO$_2$-F interactions in perfluorinated compounds. Raveendran and Wallen computationally investigated the effect of stepwise fluorination on the CO$_2$-philicity of methane in an effort to address the existence of F-CO$_2$ interactions. In partially fluorinated systems, the fluorine atom acts as a Lewis base toward an electron-deficient carbon atom of CO$_2$, and the hydrogen atoms, having increased their positive charge because of the neighboring fluorine, act as Lewis acids toward the electron-rich oxygen atoms of CO_2 [19].

Fried and Hu used MP2 calculations (6-31++G** basis set) in an effort to identify the nature of specific interactions between CO_2 and fluorinated substituent groups on polymers [20]. They reported that quadrupole-dipole interactions are important contributors to the total energy of interaction. In experimental studies by McHugh et al., the favorable miscibility of fluorocarbons has also been attributed to polar-quadrupole interactions [21]. The authors noted that fluorination imparts solubility to the polymer *provided* that polarity is also introduced to the polymer via such fluorination. Too high a level of fluorination produces an adverse effect on miscibility due to dominance of dipole-dipole interactions between the polymer chains [22].

Clearly, there is considerable controversy in the literature surrounding the origin of the miscibility of some fluorinated polymers in CO_2, yet there do seem to be some interesting lessons to be learned from this work regarding CO_2-phile design, namely:

- The presence of fluorine creates molecules with weak self-interaction, rendering miscibility with CO_2 possible at lower pressures.
- Electronegative fluorine may exhibit specific interactions with CO_2's electron-poor carbon, lowering miscibility pressures.
- The presence of fluorine will affect the acidity of neighboring protons, allowing for the possibility of specific interactions between these protons and CO_2's oxygen atoms.

7.3.2
Non-Fluorous CO_2-Philes: the Role of Oxygen

We have explored the possibility that one could design a non-fluorous CO_2-phile, basing our early designs on intriguing literature on interactions between oxygen-containing functional groups and CO_2. For example, Kazarian and coworkers reported the existence of Lewis acid-Lewis base interactions (via FT-IR) between CO_2 and the oxygen of a carbonyl [23]. In this and other studies, it was shown that the carbonyl oxygen interacts with the carbon atom of CO_2, where the geometry and strength of the interaction may vary depending on adjacent groups [24–26]. The use of oxygen-containing functional groups appeared particularly advantageous in CO_2-phile design because it allows for the creation of specific interactions with CO_2 while minimizing the strength of self-interactions of the solute.

We subsequently showed that addition of carbonyl-containing functional groups lowers the miscibility pressures of silicones [27] in CO_2; the combination of weak self-interaction (silicones) and CO_2-carbonyl groups interaction is clearly favorable. Addition of ether groups to a silicone backbone also lowers miscibility pressures; our subsequent calculations suggested equal strength of interaction between ether oxygen:CO_2 and carbonyl oxygen:CO_2 pairs, allowing for strategic combinations. The particular utility of acetates in lowering miscibility pres-

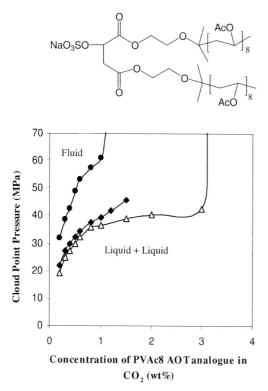

Fig. 7.3 Phase behavior of PVAc8 AOT analog/CO_2 mixtures at 298 K. $W=0$ (+); $W=10$ (◆); $W=50$ (●). The structure of the polymer is shown.

sures in CO_2 was demonstrated by Wallen [28] and by our group [29] using acetate-functional saccharides and polysaccharides. As shown in Fig. 7.3, a surfactant formed from vinyl acetate will dissolve in CO_2 at accessible pressures, although they are still relatively high.

To our CO_2-phile molecular "wish list" of weak self-interaction and oxygen: CO_2 interactions we added high flexibility/low softening point, assuming that this characteristic would enhance the entropy of mixing of the compound with CO_2. We then designed a series of ether-carbonate copolymers that exhibited lower miscibility pressures than the fluorinated polyethers we had employed as CO_2-philes for over a decade [30]. Nevertheless, our set of guidelines remained simply guidelines; true *a priori* design was not possible. In particular, we noted that very small changes to structure led to dramatic and unpredictable changes in phase behavior, a frustrating situation. For example, it has been known for almost a decade [21] that polymethyl acrylate (PMA) and polyvinyl acetate (PVAc) exhibit miscibility pressures in CO_2 that differ by hundreds of bar – this result would not be predicted by any group contribution thermodynamic model currently in use without purely empirical adjustments. Indeed, the PMA/PVAc effect is preserved even when we attach the ester group in either an "acrylate" or "acetate" fashion to other polymer backbones (Fig. 7.4). Finally, while polyvinyl acetate exhibits relatively accessible miscibility pressures, we have found that the addition of a single methylene unit (polyallyl acetate) creates a material that is, for all intents and purposes, insoluble. Again, traditional thermodynamic models provide no guidance here.

We believe that at least part of the answer to these puzzles lies in CO_2's ability to act as both Lewis acid and Lewis base, coupled with subtle effects of neighboring substituents on the acidity of certain protons. For example, Wallen and colleagues [31], in an analysis of interactions between acetate groups and CO_2, found that the acidity of the methyl acetate protons allows for binding of CO_2 through both its carbon and oxygen atoms. We [32] have found experimentally that copolymers of vinyl acetate (VAc) and tetrafluoroethylene (TFE) exhibit lower miscibility pressures than either of the homopolymers. Not surprisingly, calculations made (using MP2/aug-cc-pVDZ level of theory) on the geometry and strength of the interactions between CO_2 and various dyads (TFE-VAc, VAc-VAc, etc.) in the copolymer showed that the presence of the difluoromethylene groups in the backbone render neighboring protons more acidic. This neighbor effect allows quadradentate binding between CO_2 and the TFE-VAc dyad (Fig. 7.5). Given our results with TFE-VAc copolymers, can one design a struc-

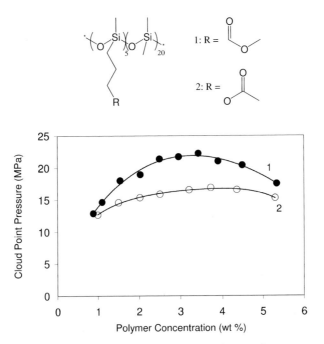

Fig. 7.4 Phase behavior of two functionalized silicone polymers in carbon dioxide at 295 K. Structures are shown.

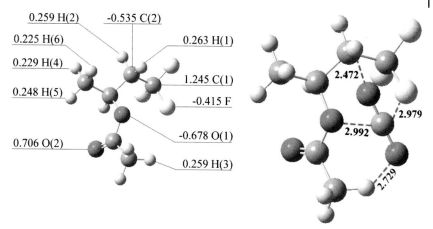

0.259 H(2) -0.535 C(2)

0.225 H(6) 0.263 H(1)

0.229 H(4)

0.248 H(5) 1.245 C(1)

-0.415 F

-0.678 O(1)

0.706 O(2)

0.259 H(3)

2.472

2.979

2.992

2.729

Fig. 7.5 Above: Charge distribution on model for TFE-VAc dyad calculated using Gaussian 98 package, revision A11; note charges on protons next to carbonyl and CF$_3$ group. Below: Optimized binding geometries for the CO$_2$ – TFE-VAc dyad using MP2/6-31+g(d) level of theory; optimized configurations were then used to calculate the more accurate single-point binding energies using the aug-cc-pVDZ basis set with counterpoise corrections.

ture that might incorporate the benefits of this copolymer, yet without the need for fluorine?

In summary, combinations of theory and experiment are rapidly advancing our ability to design cost-effective CO$_2$-philic materials and hence surfactants that are both economical and effective. We would propose the following molecular characteristics for a "CO$_2$-phile":

- Flexible, high free volume materials. Eastoe and colleagues [33] have previously demonstrated that increasing free volume through functional group changes will enhance solubility of compounds in CO$_2$. Elevating free volume and flexibility can, for example, be accomplished through branching and use of ether linkages in the main chain.

- Weak self-interactions. O'Neill and colleagues [34] have previously noted that most of the CO$_2$-philes known exhibit relatively weak self-interaction, as evidenced by low cohesive energy density. We have found, for example, that while tertiary amines interact more strongly with CO$_2$ than do carbonyls, the stronger self-interaction of the amine-containing compounds actually elevates their miscibility pressures in CO$_2$ versus oxygen-containing analogs.

- Multidentate interactions between CO$_2$ and solute functional groups, where interactions involve both the carbon and oxygens in CO$_2$. Oxygen-containing functional groups are advantageous in that they interact with CO$_2$, adjust the acidity of neighboring protons, and add comparatively little to the strength of self-interaction of the solute.

Inexpensive, effective surfactant design is probably the key issue that must be dealt with in order to render emulsion polymerization in carbon dioxide a viable process, and hence advances in the understanding of the solvent behavior of CO_2 must continue to drive this application forward.

7.4
Inverse Emulsion Polymerization in CO_2: Results

Adamsky and Beckman [35] first examined the emulsion polymerization of acrylamide in CO_2 using a fluoroether-functional surfactant to stabilize the system. Because acrylamide is a solid, it was added to the system as an aqueous solution, rendering post-polymerization examination of particle size problematic. However, given that neither acrylamide nor water are appreciably soluble in CO_2, while the perfluoropolyether (PFPE) surfactant is, this system could indeed be considered an inverse emulsion polymerization. Adamsky followed up his initial work with an evaluation of the effect of surfactant structure on the rate of polymerization and the molecular weight of the polyacrylamide. A series of nonionic surfactants were generated by esterifying a poly(hexafluoropropylene oxide) carboxylic acid (2500 MW, Krytox functional fluid, Miller-Stephenson) with polyethylene glycols of varying molecular weights (200–1500). Polymerizations were conducted at 333 K, typically at a starting pressure of 34.5 MPa. Not surprisingly, as the chain length of the polyethylene glycol block increased, the phase boundary of the surfactant shifted to higher pressures; the 1500 MW analog was in fact poorly soluble under the polymerization conditions. Although the analog with PEG with a molecular weight of 200 was the easiest to solubilize, it exhibited very little effect on the outcome of the polymerization. The rate using PEG200 did not vary as surfactant concentration increased, while the molecular weight dropped, possibly because of chain transfer to surfactant. For the cases of the PEG600, PEG900, and PEG1500 amphiphiles, the rate of polymerization varied as $[S]^{1.25}$, close to the expected range of 0.4 to 1.2 from Smith-Ewart kinetics, and close to the value of 1 found by Vanderhoff for acrylamide polymerization in an alkane continuous phase [36] using alkyl-PEG surfactants. This behavior may be due to better anchoring of the surfactants with longer PEG chains to the growing polymer particles – the fact that the PEG200 surfactant is more CO_2-philic may prompt it to exist as unimers in the continuous phase, rather than forming micelles and stabilizing monomer droplets. In all cases, molecular weight dropped as surfactant concentration increased, $MW \approx [S]^{-0.35}$, again very similar to Vanderhoff's observed exponent of –0.3 found using conventional alkyl-PEG diblock surfactants in a hydrocarbon continuous phase.

We subsequently examined the heterogeneous polymerization of N-vinyl formamide (NVF) in CO_2, using a series of PFPE-functional sorbitol surfactants [37] to stabilize the system and AIBN (azobisisobutyrnitrile) as initiator. NVF is a liquid monomer with a sizable phase envelope at 338 K (Fig. 7.6), allowing

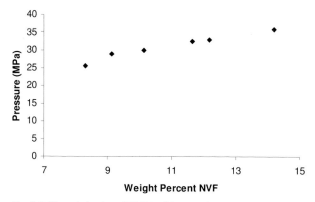

Fig. 7.6 Phase behavior of NVF in CO_2 at 338 K.

polymerization to commence in either the dispersion (monomer completely miscible) or emulsion (monomer partially miscible) regimes. The dispersion regime was accessed by operating the polymerizations at 34.5 MPa and 338 K, where the system was initially homogeneous and clear, while emulsion polymerizations were operated at 20 MPa and 338 K. In the dispersion regime, the rate of polymerization varied as $R \approx [S]^a$, where the exponent was 0.6 for a sorbitol surfactant with a 2500 MW PFPE chain and 1.1 for a sorbitol surfactant with a 5000 MW PFPE chain [38]. One might argue that the latter surfactant demonstrated a better balance between anchoring to the polymer particles and CO_2-philicity (the analog to a hydrophobic-lipophilic balance), but we also observed that the polymerization was faster in the absence of surfactant, where the system operated simply as a precipitation polymerization. It was difficult to draw any conclusions as to trends in molecular weight and surfactant structure. Polymerizations of NVF were also run in the emulsion regime, below the phase boundary of the monomer in CO_2 where the system was initially heterogeneous (cloudy). In contrast to polymerizations in the dispersion regime, here there was a clear advantage to working with the surfactants, in that the rate was typically 2 to 4 times higher with surfactant than without. Here again, the dependence of rate on surfactant concentration varied as the surfactant structure was changed, as shown in Table 7.1.

Here again, we might argue that the 5000 MW PFPE-functional material exhibits the best balance between CO_2-philicty and anchoring to the polymer particle (hydrophilicity). Interestingly, the behavior of the 2500 PFPE surfactant varied as the amount of monomer in the system varies. At an initial concentration of 7% NVF, the behavior of the 2500 MW perfluoropolyether surfactant is as shown in Table 7.1; however when the monomer concentration is increased to 17 wt%, the behavior changes, such that the rate varies as $[S]^{0.2}$ – it is not clear why this behavior occurs.

It should be noted that while the surfactants used by Adamsky and Singley were able to support emulsion polymerization in CO_2, the latex formed during

Table 7.1 Correlation between polymerization rate and surfactant
concentration during polymerization of NVF in CO_2 in the emulsion
regime ([AIBN]=0.15%, [NVF]=7.3%, T=338 K, P=20 MPa)

Surfactant	Rate: [S] correlation
1 2500 MW PFPE chain on sorbitol	$R \approx [S]^{-0.35}$
1 5000 MW PFPE chain on sorbitol	$R \approx [S]^{0.87}$
1 7500 MW PFPE chain on sorbitol	$R \approx [S]^{0.18}$

these reactions was not stable – the polymer particles settled to the bottom of
the reactor if stirring ceased. This demonstrates another challenge for surfactants used in heterogeneous polymerization in CO_2; CO_2's relatively low density
and viscosity (compared to water or organic solvents) prompt more rapid set-
tling of particles than conventional organic solvents or water.

Fink synthesized a series of silicone-based surfactants [39] and used these to
examine the emulsion polymerization of vinyl pyrrolidone (VP) in CO_2. These
monomers are liquids under ambient conditions, and hence phase behavior in
CO_2 was measured to determine under what conditions one could operate in an
emulsion polymerization mode (Fig. 7.7). As is the case with NVF, the phase
behavior of 1-vinyl-2-pyrrolidone offers the possibility for distinct polymerization
regimes. Above pressures of ca. 28 MPa at 338 K, VP is miscible with carbon dioxide in all proportions. Below 28 MPa, VP and CO_2 will either phase split into
monomer-rich and CO_2-rich phases, or exist as a single phase, depending upon
the initial VP concentration and system pressure. Consequently, a polymerization can be conducted initially in the single phase regime above 28 MPa, leading to a purely dispersion mechanism, or initially within the two-phase dome in
P–x space, leading to an inverse emulsion polymerization. Finally, a polymerization can be conducted where pressure and composition are chosen to initially

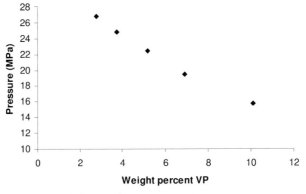

Fig. 7.7 Phase behavior of VP in CO_2 at 338 K.

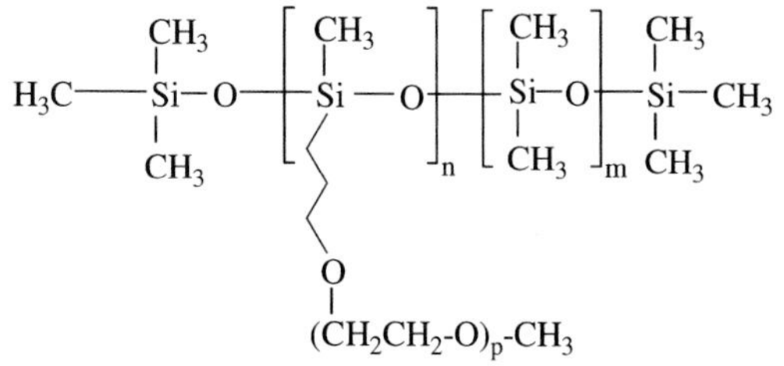

Fig. 7.8 Silicone surfactants evaluated during polymerization of VP in carbon dioxide.

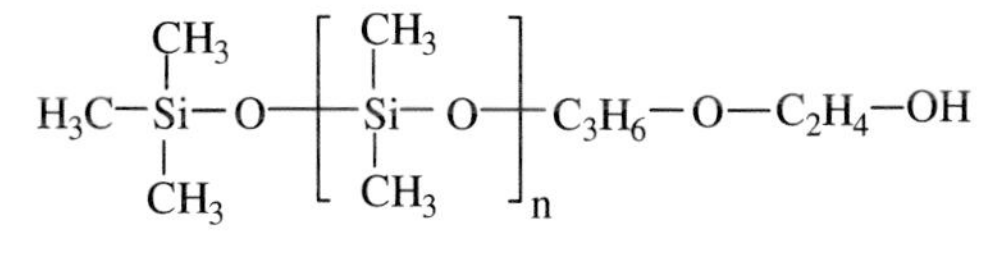

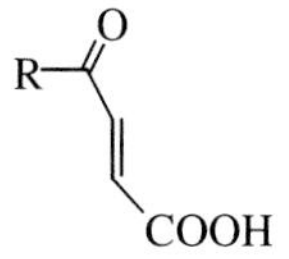

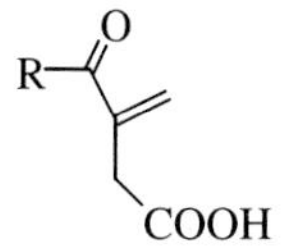

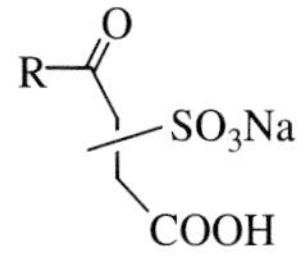

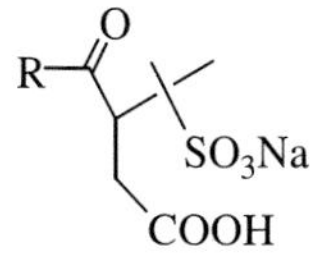

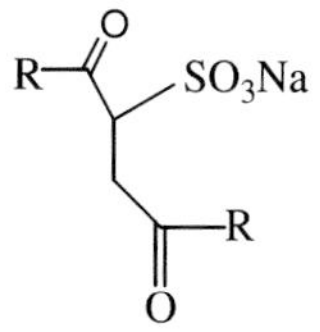

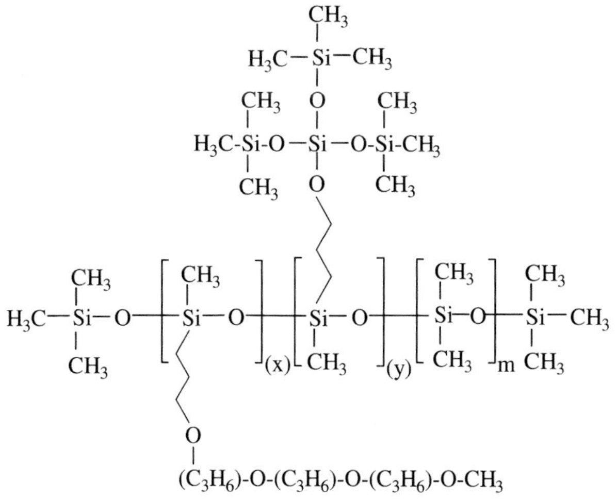

Table 7.2 Summary of structural variables for surfactants 3 a–d and 4 a–c shown in Fig. 7.8.

Code number	Topology	Silicone chain length	CO_2-phobe type	CO_2-phobe chain length	Number of CO_2-phobic chains
3a	Comb	28 repeats	Polypropylene oxide	3	1 [a]
3b	Comb	28 repeats	Polypropylene oxide	3	2 [a]
3c	Comb	28 repeats	Polypropylene oxide	3	3 [a]
3d	Comb	28 repeats	Polypropylene oxide	3	6 [a]
4a	Comb	175 repeats	Polyethylene oxide	3	8
4b	Comb	175 repeats	Polyethylene oxide	7	8
4c	Comb	175 repeats	Polyethylene oxide	12	8

a) For surfactants 3 a–3 d, when "*n*" CO_2-phobic chains (where *n* varies between 1 and 6) are attached to the backbone, (6-*n*) short silicone chains (4 repeats, branched) are also attached to block all reactive hydromethyl siloxane groups.

produce a single phase (for example, at 10 wt% VP at 20 Mpa and 338 K), such that the emulsion regime is entered as monomer is depleted from the continuous phase during the polymerization.

The silicone surfactants (Fig. 7.8) were generated and evaluated because they are known to be significantly less expensive than analogous fluorinated materials; however, silicones are also significantly less CO_2-philic than fluoroalkyls. Interestingly, unlike most materials in mixtures with CO_2, silicones do exhibit UCST type behavior (decreasing phase boundary pressures as temperature increases), and hence it was felt that they might be useful for a heterogeneous polymerization at 338 K. However, silicones are not nearly as CO_2-philic as their fluorinated counterparts, and hence the phase boundaries for these compounds (in CO_2) occur at higher pressures than does the phase boundary of VP. Here, however, the monomer acts as a co-solvent for the surfactants, allowing amphiphile miscibility at lower pressures and permitting the support of heterogeneous polymerization.

Most of the surfactants in Fig. 7.8 were able to support dispersion polymerization (polymerization that was performed at pressures above the phase boundary of the monomer in CO_2) such that the rate of polymerization was higher with surfactant than without (Table 7.3). Differences in the behavior of the various surfactants was ascribed to differences in the CO_2-phile:hydrophile balance; surfactants must have the correct balance between the ability to anchor to the polymer particles and the ability to mix readily with CO_2 and hence stabilize growing particles. Attempts to operate the polymerization purely in the emulsion regime (within the phase envelope of the monomer) failed to produce useful results (very poor yields resulted). This was likely due to the fact that (a) lower monomer concentration and lower pressure were employed to access this regime, given the phase behavior of VP in CO_2, and (b) at these conditions, neither parameter was high enough to solubilize the silicone surfactant. Several "hybrid" polymerizations were therefore attempted. As can be seen from the VP phase diagram in Fig. 7.7, if we commence polymerization at an initial pressure of 22 MPa, 338 K, and 16.5% VP, we will begin the reaction in the dispersion regime. However, as monomer is depleted from the system through polymerization, we could ultimately enter the emulsion regime. Such polymerizations using surfactant 4b showed that rate is ca. $[S]^{0.34}$; polymer molecular weight also increased as surfactant concentration increased ($M \approx [S]^{0.23}$). These values are lower than what one would expect from a typical inverse emulsion polymerization, but this may be entirely due to the fact that the silicone surfactants will begin to precipitate from solution as the monomer is consumed. Thus, in summary, although silicones are significantly less expensive than fluorinated amphiphiles, they were not sufficiently CO_2-philic to be employed as stabilizers in an inverse emulsion polymerization in CO_2. Here again we see that a surfactant design combining low cost and high solubility is crucial.

Recently, Ye and DeSimone [40] showed that a diblock copolymer of a fluorinated acrylate and a glucose-containing hydrophilic block will support the emulsion polymerization of N-ethyl acrylamide in CO_2. Here, the use of a fluorinated azo-initiator that partitions strongly to the CO_2 phase (avoiding the monomer droplets) led to the generation of very fine (submicron) particles of polymer. Fluoroacrylates are known to be the most CO_2-philic materials found to date, and hence their use allows for achievement of the crucial phase behavior conditions shown in Fig. 7.2.

Table 7.3 The polymerization of polyvinyl pyrrolidone (16.3 wt.-%, 1% AIBN, ca. 22 MPa, 338 K) with 2.5% surfactant: yield and relative rate of polymerization in carbon dioxide.

Surfactant	Yield (%)	Relative rate
no surfactant	69	1
2 a	77	3.7
2 b	81	3.9
2 c	80	2.4
3 a	84	5.3
3 b	86	4.8
3 c	84	3.0
3 d	71	0.5
4 a	72	2.1
4 b	76	3.5
4 c	79	2.3

a) Versus the polymerization without surfactant as a reference.

7.5
Future Challenges

Any future application of inverse emulsion polymerization in CO_2 will depend entirely upon the design of suitable surfactants, combined with choosing suitable monomer systems. The surfactant and monomer must exhibit phase behavior in CO_2 according to Fig. 7.2 to operate in the emulsion polymerization regime. At present, fluorinated surfactants allow for technical success, yet are not sufficiently economical to allow for commercial success as well.

References

1 Odian, G., principles of Polymerization, 3rd edition (1991), John Wiley and Sons.
2 Kendall J. L.; Canelas D. A.; Young J. L.; DeSimone J. M. *Chem. Rev.* **1999**, 99, 543.
3 Kirby, C. F.; McHugh, M. A. Phase Behavior of Polymers in Supercritical Fluid Solvents. *Chem. Rev.* **1999**, 99, 565; Kauffman, J. F. Quadrupolar Solvent Effects on Solvation and Reactivity of Solutes Dissolved in Supercritical CO_2. *J. Phys. Chem. A* **2001**, 105, 3433.
4 Hyatt, J. A. Liquid and supercritical carbon dioxide as organic solvents. *J. Org. Chem.* **1984**, 49, 5097.
5 Allada, S. R. Solubility parameters of supercritical fluids. *Ind. Eng. Chem. Proc. Des. Dev.* **1984**, 23, 344.
6 Giddings, J. C.; Myers, M. N.; McLaren, L.; Keller, R. A. High-pressure gas chromatography of nonvolatile species. Compressed gas is used to cause migration of intractable solutes. *Science* **1968**, 162, 67.
7 Walsh, J. M.; Ikonomou, G. D.; Donohue, M. D. Supercritical phase behavior: the entrainer effect. *Fluid Phase Equilibria* **1987**, 33, 295.
8 DeSimone, J. M.; Guan, Z.; Elsbernd, C. S. Synthesis of fluoropolymers in supercritical carbon dioxide. *Science* **1992**, 257, 945.
9 Heller, J. P.; Dandge, D. K.; Card, R. J.; Donaruma, L. G. Direct thickeners for mobility control of carbon dioxide floods. *Soc. Petr. Eng. J.* **1985**, 25, 679.

10 Beckman, E. J. Supercritical and Near-critical CO_2 in Green Chemical Synthesis and Processing *J. Supercrit. Fluids* **2004**, 28, 121.
11 Hoefling, T.; Stofesky, D.; Reid, M.; Beckman, E. J.; Enick, R. M. The incorporation of a fluorinated ether functionality into a polymer or surfactant to enhance carbon dioxide solubility. *J. Supercrit. Fluids* **1992**. 5, 237.
12 Hoefling, T. A.; Newman, D. A.; Enick, R. M.; Beckman, E. J. Effect of structure on the cloud-point curves of silicone-based amphiphiles in supercritical carbon dioxide. *J. Supercrit. Fluids* **1993**, 6, 165.
13 Yee, G. G.; Fulton, J. L.; Smith, R. D. Fourier transform infrared spectroscopy of molecular interactions of heptafluoro-1-butanol or 1-butanol in supercritical carbon dioxide and supercritical ethane. *J. Phys. Chem.* **1992**, 96, 6172.
14 Dardin, A.; DeSimone, J. M.; Samulski, E. T. Fluorocarbons Dissolved in Supercritical Carbon Dioxide. NMR Evidence for Specific Solute-Solvent Interactions. *J. Phys. Chem. B* **1998**, 102, 1775.
15 Yonker, C. R.; Palmer, B. J. Investigation of CO_2/Fluorine Interactions through the Intermolecular Effects on the 1H and 19F Shielding of CH_3F and CHF_3 at Various Temperatures and Pressures. *J. Phys. Chem. A* **2001**, 105, 308.
16 Cece, A; Jureller, S. H.; Kerscher, J. L.; Moschner, K. F. Molecular Modeling Approach for Contrasting the Interaction of Ethane and Hexafluoroethane with Carbon Dioxide. *J. Phys. Chem.* **1996**, 100, 7435.

17 Han, Y.-K.; Jeong, H. Y. Comment on "Molecular Modeling Approach for Contrasting the Interaction of Ethane and Hexafluoroethane with Carbon Dioxide". *J. Phys. Chem. A* **1997**, 101, 5604.

18 Diep, P; Jordan, K. D.; Johnson, J. K.; Beckman, E. J. CO_2-Fluorocarbon and CO_2-Hydrocarbon Interactions from First-Principles Calculations. *J. Phys. Chem. A* **1998**,102, 2231.

19 Raveendran, P; Wallen, S. L. Exploring CO_2-Philicity: Effects of Stepwise Fluorination. *J. Phys. Chem. B.* **2003**, 107, 1473.

20 Fried, J. R.; Hu, N. The molecular basis of CO_2 interaction with polymers containing fluorinated groups: computational chemistry of model compounds and molecular simulation of poly[bis-(2,2,2-trifluoroethoxy)phosphazene]. *Polymer* **2003**, 44, 4363.

21 Rindfleisch, F.; DiNoia, T. P.; McHugh, M. A. Solubility of Polymers and Copolymers in Supercritical CO_2. *J. Phys. Chem.* **1996**, 100, 15581.

22 McHugh, M. A.; Park, I.-H.; Reisinger, J. J.; Ren, Y.; Lodge, T. P.; Hillmyer, M. A. Solubility of CF2-Modified Polybutadiene and Polyisoprene in Supercritical Carbon Dioxide. *Macromolecules* **2002**. 35, 4653.

23 Kazarian, S. G.; Vincent, M. F.; Bright, F. V.; Lioota, C. L.; Eckert, C. A. Specific Intermolecular Interaction of Carbon Dioxide with Polymers. *J. Am. Chem. Soc.* **1996**, 118, 1729.

24 Nelson, M. R.; Borkman, R. F. Ab initio calculations on CO_2 binding to carbonyl groups. *J. Phys. Chem.* **1998**, 102, 7860.

25 Raveendran, P.; Wallen, S. L. Cooperative C-HO Hydrogen Bonding in CO_2-Lewis Base Complexes: Implications for Solvation in Supercritical CO_2. *J. Am. Chem. Soc.* **2002**, 124, 12590.

26 Kilic, S.; Michalik, S.; Wang, Y.; Johnson, K. J.; Enick, R. M., Beckman, E. J. Effect of Grafted Lewis Base Groups on the Phase Behavior of Model Poly(dimethyl siloxanes) in CO_2. *Ind. Eng. Chem. Res.* **2003**, 42, 6415.

27 Fink, R.; Hancu, D.; Valentine, R.; Beckman, E. J.; Toward the development of "CO_2-philic" hydrocarbons: I. The use of side-chain functionalization to lower the miscibility pressure of poly(dimethyl siloxane) in CO_2. *J. Phys. Chem.* **1999**, 103, 6441.

28 Raveendran, P.; Wallen, S. L. *J. Am. Chem. Soc.* **2002**. 124, 7274.

29 Potluri, V. J.; Xu, J.; Enick, R. M.; Beckman, E. J.; Hamilton, A. D. Peracetylated sugar derivatives show high solubility in liquid and supercritical carbon dioxide. *Org. Lett.* **2002**. 4, 2333.

30 Sarbu, T.; Styranec, T.; Beckman, E. J. Non-fluorous polymers with very high solubility in supercritical CO_2 down to low pressures. *Nature* **2000**, 405, 165.

31 Raveendran, P.; Wallen, S. L. Cooperative C-H···O Hydrogen Bonding in CO_2-Lewis Base Complexes: Implications for Solvation in Supercritical CO_2. *J. Am. Chem. Soc.* **2002**, 124, 12590.

32 Baradie, B.; Shoichet, M. S.; Shen, Z.; McHugh, M.; Hong, L.; Wang, Y.; Johnson, J. K.; Beckman, E. J.; Enick, R. M.; The synthesis and solubility of linear poly(tetrafluoroethylene-co-vinyl acetate) in dense CO_2: experimental and molecular modeling results. *Polymer*, in press.

33 Eastoe, J.; Paul, A.; Nave, S.; Steytler, D. C.; Robinson, B. H.; Rumsey, E.; Thorpe, M.; Heenan, R. K. Micellization of Hydrocarbon Surfactants in Supercritical Carbon Dioxide. *J. Am. Chem. Soc.* **2001**, 123, 988.

34 O'Neill, M. L.; Cao, Q.; Fang, M.; Johnston, K. P.; Wilkinson, S. P.; Smith, C. D.; Kerschner, J. L.; Jureller, S. H. Solubility of Homopolymers and Copolymers in Carbon Dioxide. *Ind. Eng. Chem., Res.* **1998**, 37, 3067.

35 Adamsky, F. A.; Beckman, E. J. Inverse Emulsion Polymerization of Acrylamide in Supercritical Carbon Dioxide *Macromolecules* **1994**, 27, 312

36 Vanderhoff, J. W.; DiStefano, F. V.; El-Aasser, M. S.; O'Leary, R.; Shaffer, O. M.; Visioli, D. L. *J. Disp. Sci. Tech.* **1984**, 5, 323-363.

37 Singley, E. J.; Liu, W.; Beckman, E. J. Phase Behavior and Emulsion Formation of Fluoroether Amphiphiles in Carbon Dioxide, *Fluid Phase Equilibria* **1997**, 128, 199.

38 Singley, E. J. Development of fluoroether amphiphiles and their applications in heterogeneous polymerizations in super-critical carbon dioxide, PhD thesis, Univ. of Pittsburgh 1997.

39 Fink, R.; Beckman, E. J. Phase Behavior of Silicone-Based Amphiphiles in Super-critical Carbon Dioxide, *J. Supercrit. Fluids* **2000**, 18, 101.

40 Ye, W.; DeSimone, J. M. Emulsion Poly-merization of N-ethyl Acrylamide in Supercritical Carbon Dioxide, *Macromole-cules* **2005**, 38, 2180-2190.

8
Catalytic Polymerization of Olefins in Supercritical Carbon Dioxide

Maartje Kemmere, Tjerk J. de Vries, and Jos Keurentjes

8.1
Introduction

Polyolefins are the largest group of synthetic polymers and comprise a number of homopolymers including polyethylene (PE), polypropylene (PP), and poly(iso-butene) as well as many copolymers. Most of the copolymers, in particular elastomers, are currently synthesized in organic solvents. An illustrative example is the production of EPDM in hexane. Typically, the process involves a solution polymerization in a CSTR, in which 20 wt% of polymer is dissolved in an excess of hexane [1]. From an environmental point of view, these processes are undesirable, because of inevitable losses of the solvent to the environment. Another major drawback of polymerizations in organic solvents is the inefficient removal and recovery of the solvents and monomers. Often the solvent recovery requires more effort and energy than the actual polymerization. Therefore, a strong incentive exists in the polymer industry to make processes more sustainable, and this is also driven by legislation. As a result of this, process concepts are needed in which the use of organic solvents is avoided or where the solvents are replaced by environmentally benign alternatives.

Alternative slurry and gas phase processes for the production of EPDM are an improvement with respect to the solvent recovery step [2]. However, in gas phase processes the possibility of incorporating large amounts of heavier monomers is limited because of the low vapor pressure of these monomers. Furthermore, conventional slurry processes use aliphatic diluents such as *iso*-butane or, in some processes, supercritical propane, which are highly flammable.

In this chapter, we explore the potential of supercritical carbon dioxide (scCO$_2$) as an alternative reaction medium for the production of EPDM and other elastomers. A new process for the catalytic polymerization of olefins in scCO$_2$ has been developed, for which the phase behavior, the catalyst system, various polymerization reactions, and a preliminary process design have been considered.

Supercritical Carbon Dioxide: in Polymer Reaction Engineering
Edited by Maartje F. Kemmere and Thierry Meyer
Copyright © 2005 WILEY-VCH Verlag GmbH & Co. KGaA, Weinheim
ISBN: 3-527-31092-4

8.2
Phase Behavior of Polyolefin-Monomer-CO$_2$ Systems

For the application of supercritical carbon dioxide as a medium for the production of polyolefins, it is important to have reliable thermodynamic data for the systems involved. Knowledge of the phase behavior of the reaction mixture is crucial to properly choose process variables such as temperature and pressure in order to achieve maximum process efficiency. For this reason, the ethylene-poly(ethylene-co-propylene) (PEP)-CO$_2$ system has been taken as a representative model system [3]. The effect of molecular weight as well as the influence of CO$_2$ on the phase behavior has been studied experimentally by cloud-point measurements. In addition, the Statistical Associating Fluid Theory (SAFT) has been applied to predict the experimental results.

8.2.1
Cloud-Point Measurements on the PEP-Ethylene-CO$_2$ System

The phase behavior of the binary system ethylene-PEP and the ternary system ethylene-PEP-CO$_2$ have been determined in an optical high-pressure cell designed for pressures up to 400 MPa and temperatures up to 450 K. Fig. 8.1 gives a schematic view of the method used for cloud-point measurements.

The cell, which is provided with sapphire windows and magnetic stirring, is a modification of the one described by Van Hest and Diepen [4]. A detailed description of this apparatus and the experimental techniques used is given by De Loos et al. [5]. The cloud-point pressures of mixtures of known composition have been measured as a function of temperature by visual observation of the onset of phase separation of the homogeneous phase by lowering the pressure (cloud-point isopleths). The cloud-points have been determined with an absolute error of ±0.03 K in temperature and ±0.1 MPa in pressure.

The sample preparation has been carried out at ambient pressure and temperature. The properties of the PEP-polymers used for the cloud-point measurements are listed in Table 8.1, for which the molecular weights have been measured with high temperature gel permeation chromatography (GPC, dynamic viscosity). The estimated relative error in the amounts of the components in the mixtures is <2% for PEP 8.7, <0.1% for PEP 51 and ethylene, and <0.7% for CO$_2$.

In Fig. 8.2, the experimental cloud-point isopleths of the PEP51-ethylene and PEP8.7-ethylene systems are presented. The results show that the cloud-point

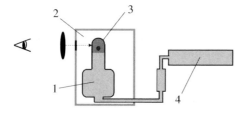

Fig. 8.1 Schematic view of the Cailletet apparatus used for the cloud-point measurements, with: 1. mercury, 2. water, 3. sample, 4. pressure system.

Table 8.1 Characterization of the poly(ethylene-co-propylene) used for the cloud-point measurements.

Sample	M_n (kg mol⁻¹)	M_w (kg mol⁻¹)	M_z (kg mol⁻¹)	Me/100C [b]
PEP8.7 [a]	8.7	24	47	22
PEP51 [a]	51	120	210	21

a) Polymer obtained from the Polymer Technology Laboratory of Eind-hoven University of Technology.
b) Me/100C is the number of methyl branches per 100 carbon atoms determined from the overall propylene content of the polymer.

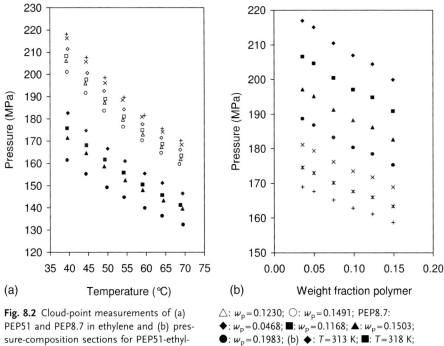

(a) Temperature (°C)

(b) Weight fraction polymer

Fig. 8.2 Cloud-point measurements of (a) PEP51 and PEP8.7 in ethylene and (b) pressure-composition sections for PEP51-ethylene system. (a) PEP51: +: w_p=0.0357; ×: w_p=0.0498; ◇: w_p=0.0756; □: w_p=0.0989; △: w_p=0.1230; ○: w_p=0.1491; PEP8.7: ◆: w_p=0.0468; ■: w_p=0.1168; ▲: w_p=0.1503; ●: w_p=0.1983; (b) ◆: T=313 K; ■: T=318 K; ▲: T=323 K; ●: T=328 K; ×: T=333 K; ✳: T=338 K; +: T=343 K.

pressure decreases on increase in temperature or polymer weight fraction (w_p) or decrease in molecular weight of the polymer at the given experimental conditions. A similar trend has been observed in systems of linear polyethylene and ethylene [5]. The experimental cloud points for the ternary system ethylene-PEP51-CO₂ are presented in Fig. 8.3. From these data it can be concluded that an increase in the weight fraction of CO₂ leads to higher cloud-point pressures for the two different polymer weight fractions. This effect can be attributed to

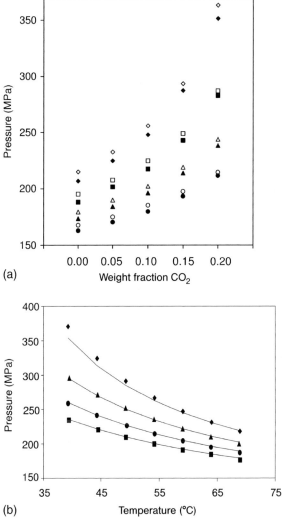

Fig. 8.3 Cloud-points for the ethylene-PEP51-CO$_2$ system (a) with 5 wt% PEP51 (open symbols) and 10 wt% PEP51 (closed symbols). ◆, ◇: T=313 K; ■,□: T=323 K; ▲,△: T=333 K; ●,○: T=343 K.

(b) Symbols show measured cloud-points of 5 wt% PEP51 in x wt% CO$_2$ and (95–x) wt% ethylene. (◆) x=20; (▲) x=15; (●) x=10; (■) x=5. Lines show SAFT modeling.

the lower permittivity and the higher quadrupole moment of CO$_2$ when compared to ethylene. Both properties lower the dispersion interaction with the polymer, because the polymer has a much higher permittivity. The increase in the slope (dP/dT) of the isopleths with increasing weight fraction of CO$_2$ shown in Fig. 8.3b can be attributed to the higher quadrupole moment of CO$_2$, as quadrupole moments are more temperature sensitive. More detailed experimental results of the investigated systems are given by De Vries et al. [3].

In general, the results show that when CO_2 is added to the system ethylene-PEP, it acts as a strong antisolvent. Therefore, these results imply that the olefin process in supercritical CO_2 in principle involves a precipitation polymerization.

8.2.2
SAFT Modeling of the PEP-Ethylene-CO₂ System

To describe the measured cloud-points, the SAFT equation of state (eos) has been used. The SAFT eos [6] is based on the perturbation theory (see Chapter 3), and, in spite of the rather complex derivation of the model equations, the basic idea and the application of the model is less complex. The SAFT eos can be written as a sum of Helmholtz energies. The first contribution is the Helmholtz energy of an ideal gas, followed by a correction for a mixture of hard spheres, a correction for chain formation, and a correction for the dispersion and association forces:

$$a = a_{\text{ideal gas}} + a_{\text{hard sphere}} + a_{\text{chain}} + a_{\text{dispersion}}(+a_{\text{association}}) \tag{1}$$

Because of the absence of strong specific interactions in our systems, such as hydrogen bonding, the association contribution is omitted in the calculations. Without the association term, the SAFT eos requires three parameters for pure components: a segment volume, $v^{\circ\circ}$, a segment-segment interaction energy, u°/k, and the number of segments, m. The dispersion energy is incorporated in the model as a square-well potential. To describe multicomponent systems, a binary interaction parameter, k_{ij}, is needed for each pair of components. This k_{ij} is used as a correction of the segment-segment interaction energy between the two different segments. In this work, mixing rules based on the Van der Waals one-fluid theory [7] have been applied to describe the mixture parameters:

$$v_i^0 = \frac{1}{8} \left(v_i^{0\frac{1}{3}} + v_j^{0\frac{1}{3}} \right)^3 \tag{2}$$

$$u_{ij} = (1 - k_{ij}) \sqrt{u_i u_j} \tag{3}$$

$$m = \Sigma_i x_i m_i \tag{4}$$

The procedure to obtain the pure component parameters and binary interaction parameters for the ethylene-PEP-CO₂ system has been described in detail previously [3]. The pure-component parameters for the small molecules (carbon dioxide and ethylene) have been obtained by fitting to experimental vapor pressure data and saturated liquid densities. The procedure to obtain parameters for large molecules such as polymers is less evident. For PEP, the set of pure component parameters has been obtained by fitting the parameters to PEP *PVT* data [8] by minimization of the residual squares of calculated and measured densi-

Table 8.2 SAFT pure-component and binary interaction parameters for the PEP-CO_2-ethylene system.

System	u^0/k (K)	v^{00} (kg mol^{-1})	m	e/k [a]	M (g mol^{-1})	Source
Carbon dioxide	216.08	13.578	1.417	40	44.01	[6]
Ethylene	212.06	18.157	1.464	10	28.054	[6]
PEP51	469.86	21.618	$0.034365 \cdot M$	10	51000	[3]
PEP8.7	469.86	21.618	$0.034365 \cdot M$	10	8700	[3]

Binary interaction parameter		
Ethylene-PEP51	$k_{12} = -0.0011847 \cdot T(K) + 0.27616$	[3]
Ethylene-CO_2	$k_{13} = 0.000491 \cdot T(K) - 0.0669$	[3]
PEP51-CO_2	$k_{23} = -0.001043 \cdot T(K) + 0.49125$	[3]

a) Parameter for temperature dependency of the interaction energy, u.

ties [3]. Binary interaction parameters have been determined using vapor-liquid or liquid-liquid composition equilibrium data. The polymer molecular mass is set at 51000, as the influence of the molecular mass on the calculations is negligible. The determination of the binary interaction parameter for PEP-CO_2, k_{23}, is complicated, because PEP is insoluble in pure CO_2. Therefore, k_{23} was fitted to all cloud points of the ternary PEP51-CO_2-ethylene system [9]. Both the pure-component and temperature-dependent binary interaction parameters SAFT-parameters are summarized in Table 8.2.

The calculated ternary cloud points containing 5 wt% PEP 51 are given in Fig. 8.3 b. The results show that the predictions agree very well with the experimental data. The SAFT calculations for systems containing 10 wt% PEP 51 are not given, but showed an even better agreement with the cloud-point measurements [9]. The results indicate that using binary data and one single ternary measurement, the behavior of ternary systems containing other amounts of CO_2 can adequately be predicted using the SAFT equation of state.

8.3
Catalyst System

Traditionally, catalysts for polyolefin production are based on early transition metal complexes, which are highly oxophilic. A point of concern for polymerization in scCO$_2$ is the compatibility of the catalyst with the mildly acidic CO_2. The acidity of CO_2 poisons the early transition metal catalysts used for conventional olefin polymerizations, and therefore these catalysts cannot be used in scCO$_2$. Since late transition metal-based catalysts are less sensitive to hetero-atom functional groups [10–12], they are more likely to be effective polymerization catalysts in scCO$_2$. Additionally, these catalysts are interesting for their high tolerance toward impurities and their ability to copolymerize polar monomers [12–14].

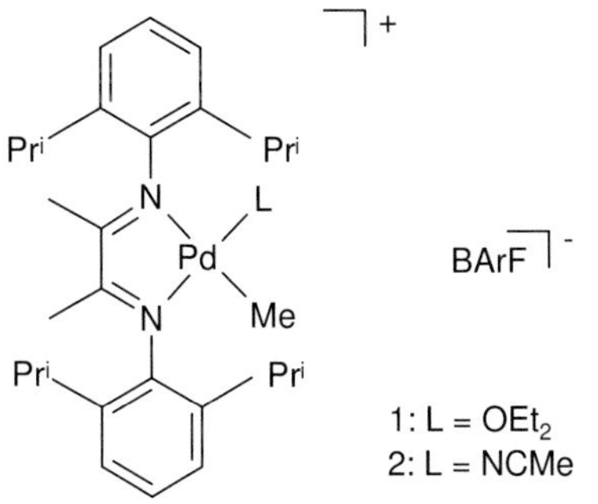

Fig. 8.4 Homogeneous diimine palladium complex used for the polymerizations in scCO$_2$, BArF ≡ tetrakis-(3,5-bis(trifluoromethyl)phenyl)-borate. Catalyst precursor 1 is air and temperature sensitive, whereas precursor 2 is not.

1: L = OEt$_2$
2: L = NCMe

However, compared to the early transition metal-based olefin catalysts, the research on late transition metal catalysts has just begun. Moreover, catalysts based on transition metals are rarely used for polymerizations in scCO$_2$ and certainly not for olefin polymerization. In this study, we have used a homogeneous diimine palladium complex, also known as the Brookhart system (see Fig. 8.4) [14, 15]. The catalyst has been synthesized according to literature procedures [14, 15]; for more details see De Vries [9].

8.3.1
Solubility of the Brookhart Catalyst in scCO$_2$

An important issue for polymerization in scCO$_2$ is the solubility of the catalyst in this medium. In order to know the active catalyst concentration in the scCO$_2$ system, solubility measurements have been carried out using UV spectrometry [16]. The solubility of the Brookhart catalyst in scCO$_2$ was determined in a 1.4 mL high-pressure view cell equipped with sapphire windows and a heating jacket (New Way Analytics, custom-made; optical distance 0.75 cm). The cell was filled with 3 mg (~2 μmol) solid catalyst. The air was carefully flushed with CO$_2$ at low pressure. The cell was then heated, filled with CO$_2$ at high pressure, and placed in a UV spectrophotometer. Based on a calibration in acetonitrile, the relative error of the catalyst solubility measurements has been estimated to be in the order of 5%.

Fig. 8.5 shows the maximum solubility of the catalyst at 308 and 313 K, respectively. Although scCO$_2$ is a poor solvent for ionic compounds, the solubility was found to be in the order of 1×10^{-4} mol/L. The relatively high solubility can be ascribed to the bulky anion BArF. Generally, the solubility of an anion in hydrocarbon solvents is known to increase with increasing distribution of charge; for example, the solubility increases in the order BF$_4$, PF$_6$, SbF$_6$, BarF, and this is likely to apply for the apolar scCO$_2$ as well. However, the rate of solubilization was low, as it typically took 1 h to reach equilibrium.

From an application point of view, the solubility of the palladium-diimine catalyst is rather limited. In general, the intrinsic activity of a catalyst is higher when it is dissolved or highly dispersed. Moreover, to allow for efficient catalyst recycle, it is

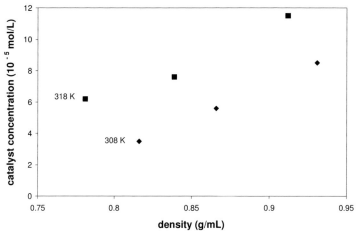

Fig. 8.5 Solubility of catalyst precursor 2 as a function of CO$_2$ density. The corresponding pressures for both temperatures are 15.0, 19.9, and 30.2 MPa, respectively.

important to develop more CO$_2$-philic catalysts [17]. To increase the solubility, the catalyst can be made more CO$_2$-philic by lowering the self-cohesion of the catalyst. For this purpose, several CO$_2$-philic moieties can be attached to the catalyst, including poly(propylene oxide) chains [18], perfluorinated alkane tails [19–22], and siloxanes [23]. Although perfluor chains are mostly used to enhance the catalyst solubility, poly(propylene-oxide) tails are expected to work just as well and are much cheaper. A route to synthesize a modified diimine ligand is proposed in Fig. 8.6.

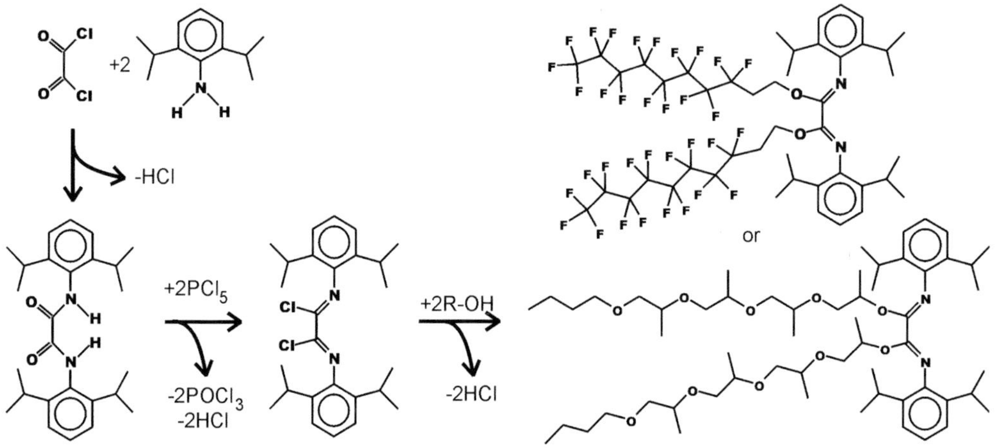

Fig. 8.6 Synthetic route to a modified diimine ligand with poly(propylene-oxide) or perfluoroalkane tails in order to enhance solubility in scCO$_2$.

Other approaches are to use different catalyst systems with a higher solubility or to work with a highly dispersed or supported catalyst. The typical low solubility of the Brookhart catalyst in scCO$_2$ due to the electric charge on the catalyst can be avoided by using a neutral catalyst. Section 8.3.2 describes preliminary experiments using a neutral palladium-based catalyst for the polymerization of ethylene and norbornene in dichloromethane and scCO$_2$. In Section 8.3.3, preliminary experiments on the ring-opening metathesis polymerization of norbornene in scCO$_2$ are discussed using a supported catalyst.

8.3.2
Copolymerization of Ethylene and Norbornene Using a Neutral Pd-Catalyst

Preliminary experiments of a neutral palladium-based catalyst for the polymerization of ethylene and norbornene in dichloromethane and scCO$_2$ have been conducted. The neutral palladium-based catalyst CODPdMeCl was synthesized according to literature procedures [24]. The ligand was prepared using a similar synthetic route to that used by Wang et al. [25]. For further details on the catalyst synthesis and the polymerization procedure, see De Vries [9].

Both homo- and co-polymerizations of norbornene and ethylene with this neutral palladium catalyst have been succesfully performed in dichloromethane and compressed CO$_2$. The results of the polymerizations are summarized in Table 8.3. Homopolymerization of ethylene in dichloromethane results in a viscous liquid. Because of the low molecular weight of the oligomer, a sample of the contents of the reactor was analyzed by ^{1}H NMR to determine the yield. The ^{1}H NMR of the vacuum-dried oligomer indicated a branch content of approximately 250 methyls per 1000 CH$_2$ groups. The ^{13}C NMR spectrum of the vacuum-dried oligomer showed that the oligomer contained methyl, ethyl, propyl, butyl, and longer branches [9]. The homo-polymerization of norbornene (experiment N2) resulted in a yellowish powder that did not dissolve in any common solvent, not even in trichlorobenzene at 398 K. A tough glassy polymer was obtained when ethylene and norbornene were in the reaction mixture (experiment N3 and N4). The polymer cannot be a homopolymer of ethylene, because this would suggest a high-molecular-weight polyethylene with very low branching. This is inconsistent with the

Table 8.3 Results of the copolymerization of ethylene and norbornene with the neutral palladium catalyst.

Exp.	Catalyst (10^{-5} mol)	Solvent	Ethylene (mol L^{-1})	Norbornene (mol L^{-1})	Yield (g)	M_n [a] (g mol^{-1})	M_w/M_n (–)
N1	2.0	CH$_2$Cl$_2$	1.12	–	0.71	320	1.3
N2	1.4	CH$_2$Cl$_2$	–	2.3	0.13		
N3	1.2	CH$_2$Cl$_2$	0.17	1.1	0.59		
N4	1.0	CO$_2$	5.54	0.085	0.16		

a) GPC against polystyrene standards.

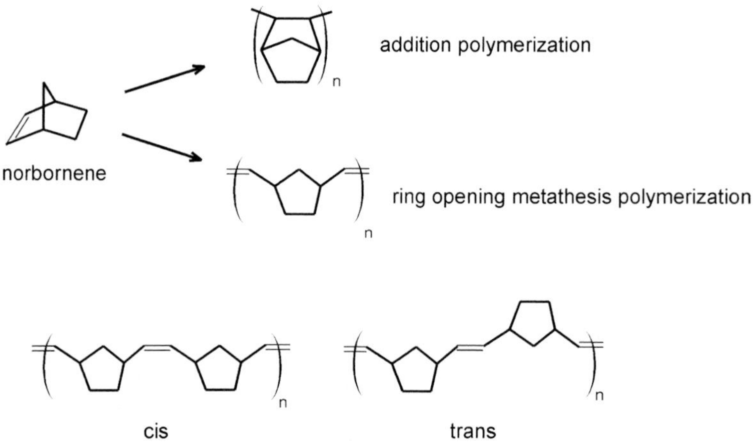

Fig. 8.7 Structure of homopolymers of norbornene by addition polymerization and ring-opening metathesis polymerization.

viscous liquid-like polyethylene as obtained in experiment N1. Furthermore, the polymers could be softened by dichloromethane, which is not possible with the homopolymer of norbornene (experiment N2). Therefore, copolymerization of ethylene and norbornene is assumed. The type of homopolymerization of norbornene is probably addition polymerization (see Fig. 8.7).

8.3.3
Ring-Opening Metathesis Polymerization of Norbornene Using an MTO Catalyst

To investigate the effect of a supported catalyst, preliminary experiments on the ring-opening metathesis polymerization of norbornene have been performed using a supported methyltrioxorhenium (MTO) catalyst in supercritical carbon dioxide [26]. For details on the catalyst synthesis and the presumed structure of the active metathesis catalyst, we refer to Morris et al. [27]. The effect of different solvents on the ring-opening metathesis polymerization (ROMP) of norbornene has been investigated. Perfluorodecalin is used as an analog for polymerization in $scCO_2$, since norbornene is soluble in both solvents, while the polymer is not soluble in perfluorodecaline or in $scCO_2$. Polynorbornene is probably not very soluble in chlorobenzene either. For further details on the polymerization procedure, see De Vries [9].

The results of the polymerizations are summarized in Table 8.4. Polymerizations conducted in perfluorodecaline result in small floating agglomerated polymer particles, whereas one chunk of rubber-like polymer has been obtained in the polymerization in CO_2. During the polymerization of norbornene in chlorobenzene, a viscous gel is formed.

Broad molecular-weight distributions are observed in all media. The number average molecular weight and turnover numbers are much lower in $scCO_2$ and

Table 8.4 Results of ring-opening metathesis polymerization of norbornene in different solvents.

Media [a]	Norbornene (mol L^{-1})	TON [b]	M_n [c] (kg mol^{-1})	M_w/M_n	T_g (K)	cis:trans [d] (%:%)
Perfluoro-decaline	1.1	133	62	4.8	291	77:23
CO$_2$	1.8	189	40	3.3	283	84:16
Chloro-benzene	1.8	1184	100	3.7	309	78:22 [e]

a) Reaction conditions: 323 K; initial pressure in CO$_2$ was 15 MPa.
b) Turnover number: mol norbornene/mol MTO.
c) SEC-DV in trichlorobenzene at 423 K, universal calibration.
d) Determined by ^{1}H NMR.
e) Polynorbornene is only partially dissolved.

Fig. 8.8 Scanning electron microscopy photograph of SiO$_2$·Al$_2$O$_3$ support and supported catalyst after polymerization. (a) Support, (b) polymerization in chlorobenzene, (c) polymerization in perfluorodecaline, (d) polymerization in CO$_2$.

in perfluorodecaline than in chlorobenzene. This can be caused by the low plasticization by scCO$_2$ and perfluorodecaline of the tough rubber-like polynorbornene as compared to the significant plasticization by chlorobenzene. The lack of plasticization in scCO$_2$ and perfluorodecaline could sterically hinder polymer formation. Mass transfer limitation of norbornene in the polymer phase toward the catalyst could also play a role. Similar cis/trans ratios are observed for all polymers, which indicate that there is no solvent effect on the polymerization mechanism.

Scanning electron microscopy (SEM) has been used to observe the structure of the polymers deposited on the supported catalyst after polymerization (see Fig. 8.8). The supported catalyst used in the polymerization in chlorobenzene is covered with a thick coating of polynorbornene (Fig. 8.8b), and the original structure of the support is hardly visible (Fig. 8.8a). A thin layer of polynorbornene is deposited on the catalyst support in perfluorodecaline and scCO$_2$ (Fig. 8.8c and d, respectively). However, the particles produced in scCO$_2$ seem to be more agglomerated than those produced in perfluorodecaline. This can be attributed to a small plasticizing effect in scCO$_2$ or the lower T_g of the polymer produced in scCO$_2$, which also explains the higher turnover number in scCO$_2$ compared to perfluorodecaline.

8.4
Polymerization of Olefins in Supercritical CO$_2$ Using Brookhart Catalyst

To explore the possibilities of olefin polymerization in scCO$_2$, several monomer and catalyst systems have been investigated. Initially, the proof of principle was established by the polymerization of 1-hexene using the Brookhart catalyst in scCO$_2$. Subsequently, the polymerization of ethylene was studied in detail. Finally, the copolymerization of ethylene with methyl acrylate was assessed.

8.4.1
Catalytic Polymerization of 1-Hexene in Supercritical CO$_2$

In the first polymerization experiments in scCO$_2$, our interest has been mainly focused on whether CO$_2$ would hinder the active catalyst or not and whether high-molecular-weight polymers could be produced [16]. For this purpose, the homopolymerization of 1-hexene was taken as a representative example.

The polymerization reactions with the Brookhart catalyst were conducted in a 75 mL stainless steel high-pressure reactor equipped with sapphire windows, a heating jacket, and a magnetic stirring bar. The solid catalyst precursor 1, as shown in Fig. 8.4, is air and temperature sensitive. Therefore, this precursor was put in the reactor in a glass ampulla, which broke under pressure; catalyst precursor 2 (stable at room temperature in air) was placed directly in the reactor. The air was carefully flushed with CO$_2$ at low pressure and the reactor was heated. The reactor was first filled with CO$_2$, and subsequently 1-hexene was added. All the experiments were performed at a reaction temperature of 308 K. After the desired reaction time, the polymerization was short-stopped with a small amount of concentrated HCl, and the reactor contents were vented through a polymer trap. To provide a reference experiment to the olefin polymerization in scCO$_2$, polymerizations of 1-hexene in CH$_2$Cl$_2$ were performed at 308 K under an argon atmosphere. A Schlenk flask was filled with catalyst pre-

Table 8.5 Polymerization of 1-hexene in CO_2 and in CH_2Cl_2 for 2 h.

Experiment	Catalyst precursor	Catalyst $(10^{-5}mol\ L^{-1})$	Solvent	Pressure (MPa)	1-Hexene (mol L^{-1})	TOF[a] $(1\ h^{-1})$	M_n[b] (kg mol^{-1})	M_w/M_n (–)
H1	1	42	CO_2	23.9	1.7	450	135	2.1
H2	2	13	CH_2Cl_2	–	2.3	810	101	1.7
H3	2	13	CO_2	19.3	2.1	990	103	2.0
H4	2	30	CH_2Cl_2	–	2.7	1010	93	1.5
H5	2	30	CO_2	19.4	2.9	560	102	1.5

a) Turnover frequency: mol monomer converted/mol catalyst/h.
b) GPC against polystyrene standards.

cursor. CH_2Cl_2 was added, and the solution was heated to the required reaction temperature. Subsequently, 1-hexene was added. After the desired reaction time, the polymerization was terminated by injecting a small amount of concentrated HCl. The polymer was isolated by evaporating the solvent and monomer residue and drying under vacuum.

The results of the polymerizations in CO_2 and in dichloromethane are listed in Table 8.5. In all cases, a rubbery, high-molecular-weight polymer was obtained. When experiments H2 and H3 are compared, the values of the turnover frequency (TOF) show that the activity of the catalyst is similar in both solvents, despite the large difference in phase behavior of the reaction mixtures. In dichloromethane, both the catalyst and the polymers are soluble. As expected, the polymer does not dissolve in $scCO_2$. Consequently, the polymerizations performed in CO_2 were all precipitation polymerizations, as was observed in the experiments. Although the differences in phase behavior are rather large, the polymers produced are very similar, in both molecular weight and molecular-weight distribution. This behavior indicates that there is no diffusion limitation, despite the precipitation in CO_2, suggesting strong swelling of the polymer, either by CO_2 or by the monomer used. More importantly, a similar molecular weight indicates that the active catalyst does not exhibit complexation behavior with CO_2, since this would have resulted in a lower molecular weight.

To study the effect of catalyst concentration, two sets of experiments were performed, one set with a relatively high catalyst concentration in CH_2Cl_2 and CO_2 (experiments H4 and H5, respectively) and one set with a low catalyst concentration (experiments H2 and H3, respectively). From the solubility measurements shown Fig. 8.5, it appears that all polymerizations in CO_2 were performed above the maximum solubility of the catalyst. Even in experiment H3, with the lowest catalyst concentration, only half of the added amount of catalyst was dissolved in the initial stage of the polymerization. Obviously, the catalyst solubilization was not limiting in experiment H3 in $scCO_2$, because the activity was similar to experiment H2 in CH_2Cl_2. Comparing experiments H4 and H5 with the higher catalyst concentration, a difference in TOF is observed, which can be explained by the slow solubilization and low solubility of the catalyst in CO_2, since the TOF was calculated based on the added amount of catalyst and not on the dissolved amount.

8.4.2
Catalytic Polymerization of Ethylene in Supercritical CO_2

Polymerization reactions of ethylene were carried out in $scCO_2$ at different pressures, temperatures, and monomer concentrations using the Brookhart catalyst. In addition, the pressure decrease upon polymerization was modeled to determine the reaction rate. Moreover, the polyethylenes have been analyzed in detail by differential scanning calorimetry (DSC) as well as hydrogen and carbon nuclear magnetic resonance (^{1}H and ^{13}C NMR) in order to determine the branching of the polymer produced.

8.4.2.1 Experimental Procedure for Polymerization Experiments
The precipitation polymerizations were performed in a 536 mL high-pressure reactor, equipped with two sapphire windows, a magnetically coupled stirrer, a Pt-100, a pressure sensor, and a heating jacket (see Fig. 8.9). Before a polymerization experiment, the air in the reactor was removed by repeatedly applying vacuum and filling with 0.2 MPa argon (3 times). Secondly, vacuum was applied and approximately 5 MPa of ethylene was added directly from the cylinder into the reactor. Subsequently, CO_2 was added to the reactor up to 1 MPa below the final polymerization pressure. The catalyst was put in the catalyst injection port, which was placed under a vacuum to remove air. Finally, the catalyst was

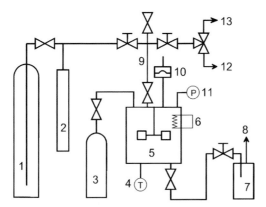

Fig. 8.9 Experimental set-up for ethylene polymerizations in $scCO_2$:
(1) CO_2 cylinder, (2) syringe pump, (3) ethylene cylinder, (4) Pt-100
resistance thermometer, (5) reactor with sapphire windows and
magnetically coupled stirrer, (6) heating jacket, (7) polymer trap,
(8) vent, (9) catalyst injection port, (10) rupture disc, (11) pressure
transducer connected to a computer, (12) argon, (13) vacuum pump.

Table 8.6 Experimental conditions of the ethylene polymerizations.

Experiment	E1	E2	E3	E4	E5	E6
Solvent	CO_2	CO_2	CO_2	CO_2	CO_2	CH_2Cl_2
Initial pressure (MPa)	20.72	10.66	20.57	10.84	22.13	0.69
Temperature (K)	313	313	323	323	313	313
Catalyst (10^{-6} mol L^{-1})	40	39	40	38	41	120
Ethylene (g L^{-1})	84	84	75	75	38	35

flushed into the reactor with 30 MPa CO_2 until the desired polymerization pressure was obtained. During the polymerization the pressure and temperature were recorded. After the reaction time, the pressure was relieved through a polymer trap and the polymer was collected by rinsing the reactor several times with heptane. The polymer solution was evaporated in a rotary evaporator.

The reference polymerization experiment in dichloromethane was carried out in a glass reactor equipped with a magnetically coupled stirrer. The following procedure was used: the reactor was filled with 50 mL dichloromethane and gas-washed with ethylene by applying vacuum and filling with ethylene (3 times). The catalyst was flushed into the reactor by a head pressure of ethylene and subsequently the polymerization started. After the reaction, the polymer solution was evaporated in a rotary evaporator. The various polymerization conditions used are summarized in Table 8.6.

8.4.2.2 Determination of Reaction Rate

In conventional polymerization processes in organic solvents, it is possible to follow the reaction rate by correlating the decrease in pressure of a supply of gaseous monomer to the conversion. Heller describes a method in which the decrease in pressure is correlated to the reaction rate with a virial equation of state [28]. A similar method can be used for reactions in supercritical media, which are often subject to a pressure change upon reaction. In this study, a model was developed to determine the reaction rate indirectly based on the measured pressure during polymerization and based on a description of the phase behavior of the polymer and supercritical fluid phase [9, 29].

The model assumes that the reaction mixture consists of a polymer phase swollen with ethylene and CO_2, and an ethylene-CO_2 phase. The assumption that no polymer dissolves in the ethylene-CO_2 phase within the experimental conditions has previously been confirmed experimentally for a similar polymer [3]. The swollen polymer phase, i.e. the polymer-ethylene-CO_2 system, is modeled using the Statistical Associating Fluid Theory (SAFT) eos [6, 7], see Section 8.2.2. The supercritical phase, i.e. the ethylene-CO_2 system, is either modeled with the Lee-Kessler-Plöcker (LKP) eos [30] or with the Peng-Robinson (PR) eos [31], because the use of the SAFT eos for the simulation of both phases results in physically inconsistent behavior [9]. The temperatures used in this work are just above the critical temperature of CO_2, and SAFT is known to overpredict the critical properties [7]. Both PR and LK(P) are able to describe the critical properties of pure components, although LK(P) describes the density much better [9].

The PR eos is one of the most commonly used cubic eos. The eos uses three parameters for the description of pure components: the critical temperature, T_c, the critical pressure, P_c, and an acentric factor, ω. The PR eos can be written as follows in which variable a is a function of T, T_c, P_c and ω and variable b is a function of T_c and P_c:

$$P = \frac{RT}{v - b_m} - \frac{a_m}{v^2 + 2b_m v - b_m^2} \tag{5}$$

An interaction parameter, k_{ij}, is used to describe mixture properties. The following mixing rules are applied for the PR eos:

$$a_m = \Sigma_i \Sigma_j x_i x_j \sqrt{a_i a_j} (1 - k_{ij}) \tag{6}$$

$$b_m = \Sigma_i x_i b_i \tag{7}$$

The LKP eos is equivalent to the original Lee-Kesler (LK) eos, except for the mixing rules. It uses the same pure-component parameters as the PR eos, i.e. T_c, P_c and ω. The LKP eos can be described in terms of compressibility factors in which $ref1$ is a reference fluid with an acentric factor of zero and $ref2$ is a fluid with a high acentric factor, i.e. n-octane. The compressibility factors of the reference fluids are a function of the reduced temperature and pressure of the mixture of interest. The fluid mixture of interest is calculated by a linear combination of the mixture acentric factor:

$$Z = Z_{ref1} + \frac{\omega_m - \omega_{ref1}}{\omega_{ref2} - \omega_{ref1}} (Z_{ref2} - Z_{ref1}) \tag{8}$$

To obtain the mixture parameters, T_{cm}, P_{cm} and ω_m, respectively, the following mixing rules are applied using a binary interaction parameter for the critical temperature:

$$T_{cij} = k_{ij} \sqrt{T_{ci} T_{cj}} \tag{9}$$

$$V_{cij} = \frac{1}{8} \left(V_{ci}^{\frac{1}{3}} + V_{cj}^{\frac{1}{3}} \right)^3 \tag{10}$$

$$V_{cm} = \Sigma_i \Sigma_j x_i x_j V_{cij} \tag{11}$$

$$T_{cm} = \frac{1}{V_{cm}^{\frac{1}{4}}} \Sigma_i \Sigma_j x_i x_j V_{cij}^{\frac{1}{4}} T_{cij} \tag{12}$$

$$\omega_m = \Sigma_i x_i \omega_i \tag{13}$$

Table 8.7 Pure-component parameters and binary interaction
parameters for the LKP and PR eos.

Model	System	T_c (K)	P_c (MPa)	ω	Source
LKP, PR	CO_2	304.1	7.38	0.225	[32]
LKP, PR	C_2H_4	282.4	5.04	0.089	[32]
Binary interaction parameter					
LKP	C_2H_4-CO_2		0.957		[33]
PR	C_2H_4-CO_2		0.0541		[33]

$$P_{cm} = (0.2905 - 0.085\ \omega_m)\ \frac{RT_{cm}}{V_{cm}} \tag{14}$$

In this work, the eos models were used to calculate the pressure in the reactor
as a function of the conversion. A third-order polynome was used to correlate
pressure with conversion, which in turn was applied to calculate the reaction
rate. The iteration scheme and the programs, written in Mathematica, are described by De Vries [9]. The pure-component parameters and the interaction parameters necessary for the SAFT and PR/LKP eos models are given in Tables
8.2 and 8.7, respectively.

8.4.2.3 Results of the Ethylene Polymerizations

The polymerizations of ethylene were performed at two different temperatures and
two overall pressures (approximately 10 and 20 MPa). The polymerizations in
scCO$_2$ start homogeneous and rapidly become heterogeneous when the polymer
precipitates. In contrast to the experiments performed in scCO$_2$, the polymerization
in dichloromethane remains homogeneous throughout the reaction. The polymers
resulting from all polymerizations are clear and transparent rubbery materials.

Based on the results shown in Table 8.8, it can be concluded that high-molecular-weight polymers were produced in all polymerization runs. The molecular
weights determined by light scattering are substantially higher than the molecular
weights obtained using polystyrene standards. Agreement of light-scattering molecular weights with molecular weights relative to polystyrene is only expected for
polymers of very similar structure. At equivalent hydrodynamic volume, polystyrene would be expected to have a larger molecular weight than a linear polyethylene, so that the larger molecular weight determined by light scattering indicates
branching. Moreover, the molecular weights obtained from light scattering are
considered to be absolute molecular weights. From the results shown in Table
8.6, it can be concluded that a high pressure and high temperature lead to the lowest molecular weight (polymer E3), whereas a low pressure and a low temperature
lead to the highest molecular weight in scCO$_2$ (polymer E2), i.e. 210 and 332 kg
mol^{-1}, respectively. This effect is most likely caused by a difference in chain trans-

Table 8.8 Experimental results of ethylene polymerizations.

Exp. No.	Temperature (K)	Reaction time (h)	Initial mass of ethylene (g)	Polymer yield (g)	M_n [a] (kg mol^{-1})	M_w [b] (kg mol^{-1})	M_w/M_n [b] (–)
E1	313	48.7	45.2	23.5	182	294	2.3
E2	313	70.4	45.1	24.9	191	332	1.9
E3	323	28.2	40.4	13.0	123	210	2.0
E4	323	38.2	40.1	19.6	129	250	1.8
E5	313	96.1	20.2	12.7	129	300	2.2
E6	313	24.0	18.8	5.78	159	379 [a]	2.4 [a]

a) GPC against polystyrene standards.
b) Absolute molecular weight based on light scattering.

Table 8.9 Simulated results of ethylene polymerization.

Exp. No.	Final pressure (MPa)			Polymer yield (g)		
	Experiment	SAFT-LKP	SAFT-PR	Experiment	SAFT-LKP	SAFT-PR
E1	15.44	15.25	16.54	23.5	22.5	30.3
E2	9.75	9.52	9.47	24.9	19.5	18.5
E3	17.97	17.98	18.51	13.0	13.1	16.5
E4	10.00	9.94	9.92	19.6	18.2	17.9
E5	17.67	18.26	19.39	12.7	14.8	21.1

fer rate, which involves a dissociative displacement mechanism according to molecular modeling studies of Musaev et al. [34].

Apparently, the overall pressure does not influence the resulting polymer, as the molecular weight distribution of the resulting polymer is similar at different pressures. The pressure keeps decreasing even after long reaction times during all the polymerization runs (28–96 h), which indicates that some of the catalyst remains active (Fig. 8.10a). It is unlikely that the pressure decrease was caused by small leaks, as leak tests at 313 K with pure CO_2 at 10.8 and 20.4 MPa have shown that the pressure decrease due to small leaks is less than 0.01 and 0.04 MPa, respectively, over a period of >65 hours. The higher yields of the polymerization runs E1 and E2 as compared to runs E3 and E4 can be attributed to the lower reaction temperature and the longer reaction times. The lower reaction temperature causes the catalyst to remain active for a longer period of time because of a slower deactivation of the catalyst. At higher temperatures, the binding of ethylene is weaker. In this situation, the catalyst is unstable and will deactivate more easily. Lowering the initial ethylene concentration (run E5) results in a somewhat broader molecular weight distribution (M_w/M_n) and a lower yield, which indicates that the resistance for mass transfer of ethylene through the polymer towards the catalyst increases.

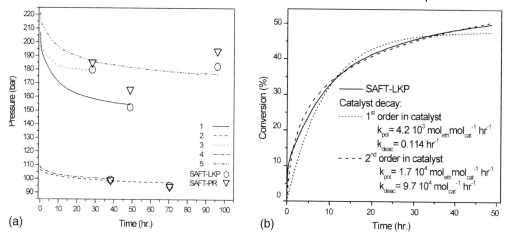

Fig. 8.10 (a) Experimentally observed pressure change as a function of time for ethylene polymerization runs E1 through E5 and prediction with the SAFT-PR and SAFT-LKP model.
(b) Conversion versus time curve of run E1 as calculated by the SAFT-LKP model. The deactivation rate of the catalyst was modeled as first and second order in catalyst concentration.

8.4.2.4 Monitoring Reaction Rate Using SAFT-LKP and SAFT-PR

The measured pressure-time curves have been combined with the conversion-pressure curves calculated by either the SAFT-LKP or the SAFT-PR model as described in Section 8.4.2.2 to yield the conversion as a function of time (see, e.g., Fig. 8.10b) [9]. The shape of the conversion-versus-time curves of all polymerizations is similar. At the start of the polymerization, high conversion rates were observed which gradually decrease. The reaction rate does not become zero, which indicates that the catalyst is still active at the end of the experiment. As the volume of the polymer phase is approximately 4–8 vol% according to the SAFT model, the choice for the LKP or PR eos model to describe the supercritical phase largely determines the outcome of the calculations.

Since the yield is a function of pressure decrease, the calculated yield based on the observed pressure decrease should in principle correspond with the observed yield. Similarly, the calculated pressure decrease using the observed yield should correspond with the observed pressure decrease. The discrepancies then give a good indication of the accuracy of the models. The calculations are summarized in Table 8.9 and Fig. 8.10. The deviations in the model calculations of run E1 and E3 (relatively high pressure) are very small for the SAFT-LKP model. The SAFT-PR model, however, shows a significant deviation. This is consistent with the deviation in density as a function of pressure for pure components for the PR eos [9].

Both models show some deviation from the experiments for run E2 and E4 (relatively low pressure). This discrepancy can partly be explained by the smaller overall pressure change during the course of the polymerization, which increases the error in the calculated conversion. The discrepancy can further be explained by the limited calculated swelling. Swelling experiments of the polymer produced in run E1 with pure CO_2 show that the polymer swelling is approximately 20 and 25 vol% at 100 and 20 MPa, respectively. However, the calculated swelling under these conditions using the SAFT-LKP and SAFT-PR models is less than 4 vol%. Because the difference in density between the swollen polymer phase and the supercritical phase is much higher at lower pressures, a higher difference between experiment and modeling can be expected at lower pressures as sorption and swelling will have a significant influence.

As the model calculations using the SAFT-LKP approach are considered to be the most accurate, they have been used to fit a rate equation. The rate of polymerization using the palladium-based catalyst is known to be zero-order in monomer concentration:

$$r_{pol} = k_{pol} \, n_{cat} \tag{15}$$

From a mechanistic point of view, the catalyst decay is likely to be first order in catalyst:

$$r_{deact} = k_{deact} \, n_{cat} \tag{16}$$

However, a first order fit in catalyst decay results in a poor description of the polymerization rate. The fitted polymerization rate is either too low at the start of the polymerization or too low at the end (see Fig. 8.10b). Fitting the observed polymerization rate to Eqs. (15) and (17), which is second order in catalyst decay, gives a much better representation:

$$r_{deact} = k_{deact} \, n_{cat}^2 \tag{17}$$

It is not likely that two catalyst molecules react with each other and become inactive. A more reasonable explanation for the second order in catalyst decay is that the catalyst is starved because of mass transfer limitations in the precipitated polymer. This can be explained as follows: the resting state of the catalyst is an olefin complex. Without complexation with an electron-donating molecule, the catalyst is highly unstable and is susceptible to deactivation. The initially high concentration of catalyst causes rapid depletion of monomer near the catalyst, which is encapsulated by polymer molecules, thus inducing rapid catalyst decay. When more catalyst becomes inactive, mass transfer limitations to the catalyst decrease and subsequently catalyst decay decreases. This explains why the apparent order in catalyst is higher than one.

8.4.2.5 Topology of Synthesized Polyethylenes

The Brookhart catalyst is known to produce highly branched polyethylenes, because of a process called chain walking. Branches ranging from methyl to hexyl and longer as well as so-called *sec*-butyl-ended branches (branch structures on branches) are reported in the literature [35]. In general, short chain branching (SCB) is used to modify the chemical and physical properties of the polymer, such as glass transition temperature and crystallinity. As a result of the chain-walking effect, the Brookhart catalyst in principle eliminates the necessity to use *α*-olefins in order to introduce SCB in polyethylenes.

In this work, DSC measurements as well as ^{1}H and ^{13}C NMR characterization were performed in order to determine the topology of the synthesized polyethylenes [36]. Table 8.10 shows the glass transition temperatures of the polyethylenes formed. The transitions appear to be equally broad, but the polyethylenes synthesized in scCO₂ all have a lower T_g than the polyethylene produced in dichloromethane. Although the polyethylenes are highly amorphous, a small enthalpy change of melting can still be observed in the range 223–263 K [36]. This shows that the polyethylenes produced in scCO₂ are even less crystalline than the polyethylene produced in dichloromethane. Compared to 100% crystalline polyethylene, which has a melting enthalpy of 294 J g^{-1}, the polyethylenes produced in scCO₂ and dichloromethane have a crystallinity of about 1% and 2%, respectively. Although crystallinities of branched polyethylenes are higher than calculated from the value for 100% crystalline polyethylene because of poor quality of the polymer crystals in branched polyethylene, the relative difference between the polyethylenes produced in scCO₂ and dichloromethane remains the same. Furthermore, the melting enthalpy of 100% crystalline polyethylene can be used to determine the crystallinity sufficiently accurately, which is supported by X-ray diffraction and DSC measurements on a series of branched polyethylenes by Mirabella and Bafna [37].

Fig. 8.11 shows a typical ^{1}H NMR spectrum of the polymers produced both in scCO₂ and dichloromethane. The peaks in the range from 0.76 to 0.94 ppm

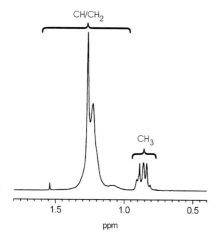

Fig. 8.11 Typical ^{1}H NMR spectrum of the polyethylenes produced by the palladium-based catalyst. Spectrum is a 300 MHz ^{1}H NMR spectrum of polymer E3 measured in CDCl₃ at 298 K.

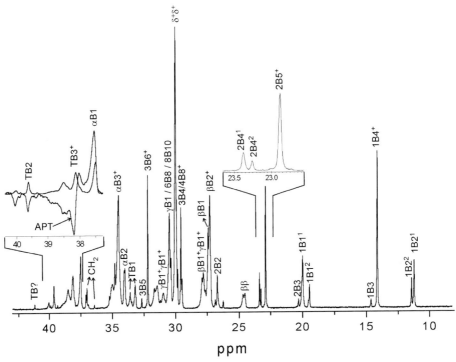

Fig. 8.12 Typical ^{13}C NMR spectrum of poly-ethylenes produced by the palladium-based catalyst. The spectrum is a 500 MHz ^{13}C NMR spectrum of sample E4 measured in 1,2,4-trichlorobenzene at 393 K. The labels "xBy" have the following meaning: where x is a number, it gives the position of the carbon in a branch of length y; where x is a Greek symbol, carbon in the main chain with 1, 2, 3 or 4 carbons from a branch, respectively is denoted as α, β, γ, δ. Positive peaks in the APT insert (attached proton test) mean CH$_2$ groups, negative peaks mean CH or CH$_3$ groups.

reflect the large amount of CH$_3$ groups, which indicate that the polymer is highly branched. Although the polyethylenes seem to dissolve completely in CDCl$_3$, phase segregation occurs after a few hours, rendering CDCl$_3$ unsuitable for reliable ^{13}C NMR measurements.

Fig. 8.12 shows the ^{13}C NMR spectrum in 1,2,4-trichlorobenzene of the polymer sample E4 produced in scCO$_2$ at 323 K, 10.84 MPa total pressure, and 75 g L^{-1} ethylene. The spectrum shows that the polymers produced by the palladium-based catalyst are indeed highly branched. The peaks resulting from resonances of carbons belonging to methyl, ethyl, butyl, pentyl, branches longer than pentyl (hexyl$^+$) and branches belonging to 3-methyl-butyl$^+$-ended branches have been assigned according to literature [35, 37–39]. Table 8.10 shows the number of branches per 1000 carbon atoms of the polyethylenes produced in scCO$_2$ and dichloromethane. The number of branches determined by ^{1}H NMR in CDCl$_3$ and ^{13}C NMR in 1,2,4-trichlorobenzene is very similar, which indicates that CDCl$_3$ can be used to accurately determine the number of branches.

Table 8.10 Characterization of the polyethylenes by DSC, ^{1}H and ^{13}C NMR.

Polymer sample	E1	E2	E3	E4	E5	E6
T_g (K)	204	204	204	205	204	206
ΔH_m (J g^{-1})	2.5	2.2	2.4	3.4	2.3	6.6
% Crystallinity	0.9	0.8	0.8	1.2	0.8	2.3
Branches total ^{1}H NMR [a]	106.5	107.8	107.4	105.1	106.5	99.2
Branches total ^{13}C NMR [a]	107.1	106.8	109.3	105.9	107.0	99.1
Methyl [a]	36.9	36.7	37.2	36.6	35.1	32.1
-3-methyl-ended branch [a]	6.8	8.2	7.7	7.4	8.1	6.0
Ethyl [a]	24.3	24.2	24.7	23.9	24.8	21.7
Propyl [a]	2.3	3.2	2.8	2.8	2.5	2.2
Butyl [a]	11.4	10.6	11.9	11.4	13.0	10.5
-5-methyl-ended branch [a]	4.5	3.1	3.0	3.0	3.4	2.2
Pentyl [a]	4.0	8.6	3.9	3.8	2.8	3.0
Hexyl+ [a]	28.1	23.5	28.8	27.5	28.8	29.6

a)   Branches per 1000 C.

The type and amount of branches present in the polyethylenes produced in scCO$_2$ is not influenced by the total pressure or the temperature. The solvent, however, appears to have a small influence on the total branch content of the polyethylenes. The total amount of branches in the polyethylene produced in scCO$_2$ is slightly higher. This increase in total branch content cannot be attributed to a certain type or length of branch, but is rather a general increase in the amount of all branches.

The DSC experiments illustrate that the polyethylenes produced in scCO$_2$ are more amorphous than the polyethylenes synthesized in dichloromethane, which is likely to originate from the slightly higher branch content as measured by ^{13}C NMR spectroscopy. A study by Held and Mecking [40] using a similar palladium-diimine catalyst has shown that branching is strongly dependent on the manner in which the polymerization is carried out. A branch content of 65, 66 and 105 branches per 1000 C atoms was measured in water, gas phase, and dichloromethane, respectively. It was concluded that the mobility of the growing polymer chain is reduced in the precipitation polymerization in the water and the gas phase, which leads to a decrease in the chain-walking process. Apparently, when the polymerizations are performed in scCO$_2$, carbon dioxide is able to plasticize the precipitated polymer to such an extent that chain walking is still possible.

The higher branch content measured for the polymers synthesized in CO$_2$ compared to the polymers synthesized in dichloromethane is not yet clear. The difference is probably not caused by the high hydrostatic pressure of CO$_2$, because the polymers produced at different CO$_2$ pressures were identical in branch content. In addition, within the investigated range, the concentration of ethylene does not change the branch content. Cotts et al. [35] have performed polymerizations in chlorobenzene at ethylene pressures ranging from 0.01 up to 3.5 MPa. It was

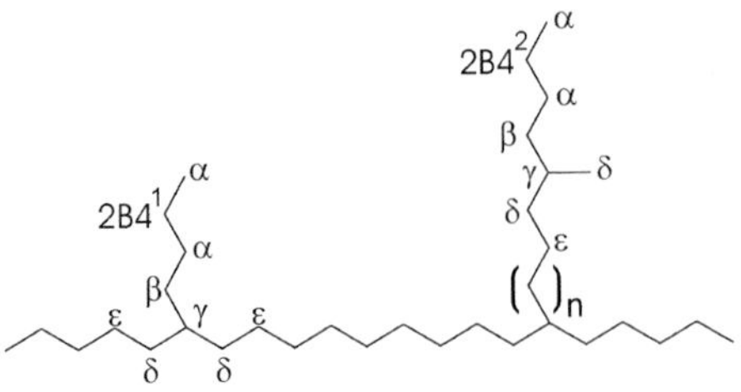

Fig. 8.13 Ordinary 2B4 branch and branch on branch structure.

shown that the short chain branching was only slightly decreased at higher ethylene pressures, from 99 down to 93 branches per 1000 carbon atoms. A higher branch content would be expected if the hydrostatic pressure or the ethylene concentration is the cause of the higher branch content in scCO$_2$. Finally, the temperature only affects the molecular weight and not the branching of the polymers. Therefore, the observation that the polyethylenes produced in scCO$_2$ have a higher branch content likely originates from the strong non-polar environment compared to the relatively polar dichloromethane.

There is a strong indication that a new branch-on-branch structure has been identified (see Fig. 8.13) [36]. According to the literature, the 2B4 peak occurs at 23.37 ppm. Near that shift an additional resonance can be observed at 23.26 ppm, which in all ^{13}C NMR spectra is equal in size to the 1B4$^+$ minus the 2B5$^+$ peak (total amount of "butyl" branches). The shift difference cannot be explained by configurational differences of a 2B4 peak. Therefore, the peak must originate from a different structure. The most probable structure is a 5-methyl-ended branch, because the 5-ethyl-ended branch is very likely to give a similar shift to that of an ordinary butyl branch, as it has the same type of carbons up to 4 carbons from the carbon in question. When the shifts are calculated using the modified Grant and Paul parameters [34] a similar trend is observed, although these parameters have been fitted to poly(ethylene-co-α-olefin)s, which do not contain structures such as the 5-methyl-ended branch. For an ordinary butyl branch and a 5-methyl-ended branch, a shift is calculated of 23.11 and 23.05 ppm, respectively.

8.4.3
Copolymerization of Ethylene and Methyl Acrylate in Supercritical CO$_2$

In this section, the polymerization of ethylene and a functional monomer is discussed. For this purpose copolymerization reactions of methyl acrylate (MA) and ethylene using a palladium-based catalyst have been carried out in compressed carbon dioxide at different monomer concentrations and monomer ratios. The incorporation of methyl acrylate and the molecular weight of the polymers have been compared to literature values of polymerizations conducted in dichloromethane.

Catalyst precursor 2 as shown in Fig. 8.4 was used for the copolymerization reactions of ethylene and methyl acrylate (0.05 mmol catalyst/L) [9, 41]. The inhibitor 2,6-di-*tert*-butyl-4-methylphenol, abbreviated as BHT, was added to prevent radical polymerization (1 mmol/L BHT) [42]. The batch polymerizations with MA were performed in a 536 mL high-pressure reactor, as described in Section 8.4.2.1 and Fig. 8.9 for 168 h. Because of the high critical temperature of MA, the mixture critical temperature is substantially higher than the critical temperature of pure CO_2, approximately 323 K. Therefore, the polymerization temperature was chosen above or below 323 K to avoid the unstable critical point. Because temperatures above 323 K induce rapid catalyst deactivation, the polymerization temperature was set to 298 K.

The molecular-weight distributions were determined using GPC against polystyrene standards and ^1H NMR was used to determine the comonomer incorporation.

The results of the copolymerizations of MA and ethylene are summarized in Table 8.11. The copolymerizations of MA and ethylene in compressed CO_2 are initially homogeneous and become heterogeneous when polymer is formed. The NMR spectra indicate that after insertion of MA, chain walking occurs, resulting in a functional group located at the end of a branch [9], which has also been observed in polymerizations in dichloromethane [14].

In dichloromethane it was shown that the incorporation of MA is directly proportional to the concentration of MA in solution at a head pressure of 0.2 MPa ethylene [14]. Fig. 8.14a shows that the incorporation of MA is also directly proportional to the ratio of monomer concentrations of MA and ethylene in solu-

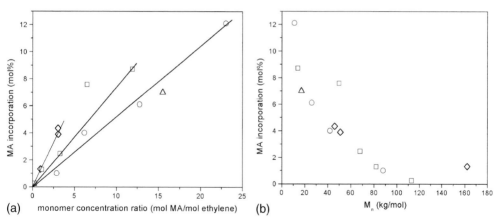

(a)   monomer concentration ratio (mol MA/mol ethylene)   (b) M_n (kg/mol)

Fig. 8.14 (a) Incorporation of methyl acrylate in copolymer as a function of concentration ratio of methyl acrylate and ethylene. ($\diamond$) CO_2 at 298 K using catalyst precursor 1; ($\bigcirc$ [14], $\square$ [42]) literature values in dichloromethane at 308 K using catalyst precursor 2; ($\triangle$ [14]) literature value in dichloromethane at 298 K using catalyst precursor 1. Lines are drawn for visual aid.

(b) Number average molecular weight of copolymer as a function of methyl acrylate incorporation. ($\diamond$) CO_2 at 298 K using catalyst precursor 1; at 298 K using catalyst precursor 1; ($\bigcirc$ [14], $\square$ [42]) literature values in dichloromethane at 308 K using catalyst precursor 2. ($\triangle$ [14]) literature value in dichloromethane at 298 K using catalyst precursor 1.

Table 8.11 Results of copolymerizations of ethylene and methyl acrylate in compressed carbon dioxide.

Exp. No.	MA (mol L^{-1})	Ethylene (mol L^{-1})	p [a] (MPa)	M_n (kg mol^{-1})	M_w/M_n (–)	Branches (CH$_3$/ 1000CH$_2$)	MA (mol%)	TON [b] MA	TON Ethylene
MA1	1.82	1.94	22.1	162	1.9	122	1.3	127	9448
MA2	2.67	0.88	22.4	51	2.0	126	3.9	117	2895
MA3	2.05	0.68	18.4	46	1.7	129	4.4	64	1404

a) Initial total pressure.
b) Turnover number, mol monomer/mol catalyst.

tion. The difference in MA incorporation between the two literature sources in dichloromethane can be attributed to the use of the BHT inhibitor.

The inhibitor effectively suppresses the formation of MA homopolymer, which decreases the catalyst activity and incorporation of MA into the polymer [42]. The incorporation of MA in the copolymer is approximately twice as large in CO$_2$ at the same monomer ratio. Therefore, less MA is needed for similar incorporation of MA in the copolymer. The rate-determining step in the polymerization is the insertion of the monomers. As a consequence, the incorporation of MA and ethylene is determined by the ratio of the catalyst-ethylene complex and catalyst-MA complex (see Fig. 8.15). Apparently, the polymerization temperature does not influence the incorporation of MA in the polymer, because the polymerization performed by Brookhart et al. [14] at 298 K follows the same trend as the polymerizations at 308 K.

The difference in solvent properties can affect the chemical potential of the catalyst complexes and monomers. The latter effect is expected to be small because of the absence of strong interactions between MA, CO$_2$, and ethylene. A higher incorporation of MA in the copolymer in CO$_2$ can be caused by an energetically more favorable catalyst-MA complex compared to the catalyst-ethylene complex. Another effect of the non-polar CO$_2$ is that the local concentration of the polar MA is probably higher around the catalyst than in the bulk to stabilize the electrical charge of the cationic palladium complex and the counterion in

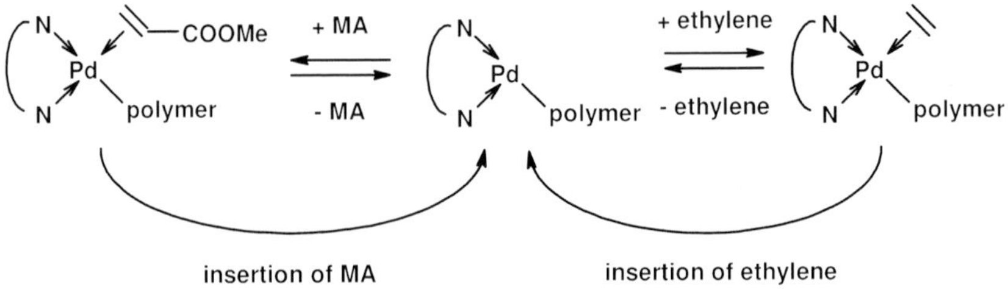

Fig. 8.15 Mechanistic description of growth of the copolymer by insertion of MA and ethylene.

CO_2. In dichloromethane, these charges are primarily stabilized by the polar dichloromethane. The higher local concentration of MA in the CO_2-system could explain the higher incorporation of MA into the polymer.

In Fig. 8.14b, the incorporation of MA in the polymer is given as a function of the number-averaged molecular weight, M_n. The experiments show an increase in M_n as the MA incorporation decreases. Both the polymerizations in CO_2 and dichloromethane follow the same trend, indicating that the chain-transfer due to inserted MA [14] is similar in both reaction media.

In conclusion, copolymerizations of MA and ethylene can be carried out in compressed CO_2. In comparison with dichloromethane, similar molecular weights are obtained. The incorporation of MA in the copolymer at the same monomer ratio is approximately twice as large in compressed CO_2, which can be attributed to a higher MA concentration near the catalyst or to an energetically more favorable catalyst-MA complex as compared to the catalyst-ethylene complex.

8.5
Concluding Remarks and Outlook

In this chapter, the possibility of using late transition metal catalysts to synthesize polyolefins in supercritical carbon dioxide was demonstrated [43]. The multicomponent phase behavior of polyolefin systems at supercritical conditions was studied experimentally by measuring cloud-point curves as well as by modeling polymer systems at supercritical conditions. The cloud-point measurements show that CO_2 acts as a strong antisolvent for the ethylene-PEP system, which implies that the polymerization concerned will involve a precipitation reaction. The model calculations prove that SAFT is able to describe the ethylene-PEP-CO_2 system accurately. Solubility measurements of the Brookhart catalyst reveal that the maximum catalyst solubility is rather low (in the order of 1×10^{-4} mol L^{-1}). However, a number of strategies are given to enhance this value.

Several catalyst systems and polymerization reactions were tested in $scCO_2$, including the homopolymerization of 1-hexene and ethylene. The results, discussed in Section 8.4.1, illustrate that it is possible to polymerize 1-hexene in $scCO_2$ using the Brookhart catalyst, yielding high-molecular-weight polymer. Based on a comparison with solution polymerization in CH_2Cl_2, the activity of this catalyst appears to be similar when $scCO_2$ is used as reaction medium instead. Moreover, no diffusion limitation of monomer occurs during the precipitation polymerization in $scCO_2$. In addition, a detailed study on the polymerization of ethylene in $scCO_2$ has emphasized that this reaction can effectively be carried out in CO_2 using the Brookhart catalyst. Observing the evolution of pressure during high-pressure reactions appears to be a reliable method to determine the reaction rate. Both the SAFT-LKP and the SAFT-PR model are able to derive a reaction rate from pressure change upon reaction in supercritical fluid systems. However, the SAFT-LKP model is more accurate than the SAFT-PR model, because of the better description of PVT behavior of the LKP eos in comparison with the PR eos. Fitting different rate equations to the observed re-action rate indicates that the polymerization rate is second order in catalyst de-cay. Moreover, highly branched polyethylene can be produced because of the plasticizing effect of $scCO_2$. The polymers obtained have an even higher branch content than polyethylenes produced in dichloromethane. Some preliminary re-sults indicate the possibility to copolymerize ethylene with functional mono-mers such as methyl acrylate and norbornene.

Based on the results obtained, it now becomes possible to address several im-portant aspects of the process design for the catalytic polymerization of olefins in $scCO_2$. In view of the scale of commercial polyolefin production, it seems plausible to expect that a transition from a batchwise operation (as on the labo-ratory scale) to a continuous process will be required. Because of the high pres-sures involved, a reactor of low diameter would be preferred because of the high wall thickness required. Large vessels would need extremely thick walls, which would be more expensive and would lead to inefficient removal of reaction heat. Another consideration is that although the polymer is highly plasticized by CO_2, it does not dissolve in $scCO_2$. Therefore, the polymerization is heteroge-neous. With respect to the catalyst system, currently the most promising cata-lysts capable of (co)polymerizing olefins in $scCO_2$ are based on nickel and palla-dium. Typically, these catalysts have lifetimes longer than 1 h. Therefore, the type of reactor should permit long residence times. In addition, the high stabili-ty of some of the late transition metal polymerization catalysts allows for effi-cient catalyst recycling [44].

In our opinion, the process concept for the catalytic polymerization of olefins in $scCO_2$ should include a loop reactor type of apparatus (Fig. 8.16), as de-scribed by Ahvenainen et al. for the polymerization of ethylene in supercritical propane [45]. In this type of reactor, the residence time can be adjusted to suffi-ciently long values to obtain a reasonable conversion. Extending the concept to

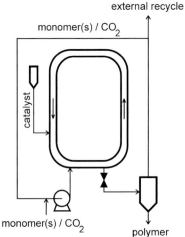

Fig. 8.16 Proposed process design for the catalytic polymerization of olefins [43].

several loop reactors in series enables the right conditions to be set in each reactor for fine tuning the properties of the polymer produced. As an example, different monomers or monomer ratios can be added at different locations in the system. In principle, this type of equipment also enables forced convection to be used, thus providing an efficient means to avoid heat and mass transfer limitations.

Two of the most interesting aspects of the catalysts used in this work are their ability to copolymerize ethylene and α-olefins with polar monomers and their inertness toward impurities, allowing for a relatively straightforward production of these types of polymers in scCO$_2$. Although the activity of these catalysts is still rather low for commercial use, it may be expected that this will improve significantly in the near future. This would enable the development of clean polyolefin production based on CO$_2$-technology, for which future applications may be expected in the production of EPDM and other elastomers.

Acknowledgments

This work was supported by a grant from the Dutch Polymer Institute. The phase behavior measurements were performed in the Applied Thermodynamics and Phase Equilibria Group of Delft University of Technology under the supervision of Dr. Theo de Loos. The authors would also like to thank Dr. Rob Duchateau and Peter Somers for their contribution to this work.

Notation

λ_{ij}	Interaction parameter	[–]
ω	Acentric factor	[–]
$v^{\circ\circ}$	Segment volume	[mL mol^{-1}]
u°/k	Segment-segment interaction energy	[K]
a	Temperature-dependent attractive parameter	[Nm^4 mol^{-2}]
b	Van der Waals hard sphere volume parameter	[m^3 mol]
k_{deact}	Catalyst deactivation rate constant	[1 h^{-1}] or [1 mol$_{cat}$ h^{-1}]
k_{ij}	Interaction parameter	[–]
k_{pol}	Propagation rate constant	[mol$_{eth}$ mol$_{cat}^{-1}$ h^{-1}]
m	Number of segments	[–]
n_{cat}	Catalyst concentration	[mol L^{-1}]
P	Pressure	[Pa]
P_A	Attractive pressure	[Pa]
P_c	Critical pressure	[Pa]
P_R	Repulsive pressure	[Pa]
T	Temperature	[K]
T_c	Critical temperature	[K]
Z	Compressibility factor	[–]

References

1 Borealis, *A/S Hydrocarbon Process* **1999**, *March*, 129.

2 Inventa-Fischer, *AG Hydrocarbon Process* **1977**, *56*(11), 189.

3 T. J. de Vries, P. J. A. Somers, Th. W. de Loos, M. A. G. Vorstman, J. T. F. Keurentjes, *Ind. Eng. Chem. Res.* **2000**, *39*, 4510.

4 J. A. M. van Hest, G. A. M. Diepen, *Solubility of naphthalene in supercritical methane*. in *The Physics and Chemistry of High Pressures, Symposium Papers*, Soc. Chem. Ind., London, **1963**, 10.

5 Th. W. de Loos, G. A. M. Diepen, *Macromolecules* **1983**, *16*, 111.

6 S. H. Huang, M. Radosz, *Ind. Eng. Chem. Res.* **1990**, *29*, 2284.

7 S. H. Huang, M. Radosz, *Ind. Eng. Chem. Res.* **1991**, *30*, 1994.

8 P. Zoller, D. J. Walsh, *Standard Pressure-Volume-Temperature Data for Polymers*, Technomic: Lancaster, Pennsylvania, **1995**, 69.

9 T. J. de Vries, *Late Transition State Metal Catalyzed Polymerizations of Olefins in Supercritical Carbon Dioxide*, PhD thesis, Technische Universiteit Eindhoven, ISBN 90-386-2605-3, Eindhoven, **2003**.

10 L. S. Boffa, B. M. Novak, *Chem. Rev.* **2000**, *100*, 1479.

11 S. D. Ittel, L. K. Johnson, M. Brookhart, *Chem. Rev.* **2000**, *100*, 1169.

12 T. R. Younkin, E. F. Conner, J. I. Henderson, S. K. Friedrich, R. H. Grubbs, D. A. Bansleben, *Science* **2001**, *287*, 460.

13 U. Klabunde, S. D. Ittel, *J. Mol. Catal.* **1987**, *41*, 123.

14 S. Mecking, L. K. Johnson, L. Wang, M. Brookhart, *J. Am. Chem. Soc.* **1998**, *120*, 888.

15 L. K. Johnson, C. M. Killian, M. Brookhart, *J. Am. Chem. Soc.* **1995**, *117*, 6414.

16 T. J. de Vries, R. DuChateau, M. A. G. Vorstman, J. T. F. Keurentjes, *Chem. Commun.* **2000**, 263.

17 M. A. Carroll, A. B. Holmes, *Chem. Commun.* **1998**, 1395.

18 S. Mawson, M. Z. Yates, M. L. O'Neill, K. P. Johnston, *Langmuir* **1997**, *13*(6), 1519.

19 S. Kainz, D. Koch, W. Baumann, W. Leitner, *Angew. Chem.* **1997**, *109*, 1699.

20 D. R. Palo, C. Erkey, *Ind. Eng. Chem. Res.* **1998**, *37*, 4203.

21 M. S. Super, E. Berluche, C. A. Costello, E. J. Beckman, *Macromolecules* **1997**, *30*, 368.

22 H. Hori, C. Six, W. Leitner, *Macromolecules* **1999**, *32*, 3178.

23 T. A. Hoefling, D. Stofesky, M. Reid, E. J. Beckman, R. M. Enick, *J. Supercrit. Fluids* **1992**, *5*, 237.

24 R. E. Rulke, J. M. Ernsting, A. L. Spek, C. J. Elsevier, P. W. N. M. van Leeuwen, K. Vrieze, *Inorg. Chem.* **1993**, *32*, 5769.

25 C. Wang, S. Friedrich, T. R. Younkin, R. T. Li, R. H. Grubbs, D. A. Bansleben, M. W. Day, *Organometallics* **1998**, *17*, 3149.

26 L. J. P van den Broeke, E. L. V. Goetheer, T. J. de Vries, J. T. F. Keurentjes, *Abstracts of Papers, 223rd ACS National Meeting* **2002**, April 7–11, Orlando, USA.

27 L. J. Morris, A. J. Downs, T. M. Greene, *Organometallics* **2001**, *20*, 2344.

28 D. Heller, *Chem. Ing. Tech.* **1992**, *64*, 725.

29 T. J. de Vries, M. F. Kemmere, J. T. F. Keurentjes, **2004**, submitted.

30 U. Plöcker, H. Knapp, J. Prausnitz, *Ind. Eng. Chem. Proc. Des. Dev.* **1978**, *17(3)*, 324.

31 D. Peng, D. B. Robinson, *Ind. Eng. Chem. Fund.* **1976**, *15*, 59.

32 R. C. Reid, J. M. Prausnitz, B. E. Poling, *The properties of gases and liquids 4th edn.* McGraw-Hill: London, **1987**.

33 J. Gmehling, U. Onken, W. Arlt, *Vapor-liquid equilibrium data collection;* DECHEMA: Frankfurt/Main, **1980**.

34 D. G. Musaev, R. D. J. Froese, K. Morokuma, *Organometallics* **1998**, *17*, 1850.

35 P. M. Cotts, Z. Guan, E. McCord, S. McLain, S. *Macromolecules* **2000**, *33*, 6945.

36 T. J. de Vries, M. F. Kemmere, J. T. F. Keurentjes, *Macromolecules* **2004**, *37*, 4241.

37 F. M. Mirabella, A. Bafna, *Pol. Sci. B Pol. Phys.* **2002**, *40*, 1637.

38 J.C. Randall, *J. Polym. Sci., Polym. Phys. Ed.* **1975**, *13*, 901.

39 W. Liu, D.G. Ray III, P.L. Rinaldi, *Macromolecules* **1999**, *32*, 3817.

40 A. Held, S. Mecking, *Chem. Eur. J.* **2000**, *6*, 4623.

41 M.F. Kemmere, T.J. de Vries, J.T.F. Keurentjes, *DECHEMA Mon.*, **2004**, *138*, 189.

42 J. Heinemann, R. Mülhaupt, P. Brinkmann, G. Luinstra, *Macromol. Chem. Phys.* **1999**, *200*, 384.

43 M.F. Kemmere, T.J. de Vries, M.A.G. Vorstman, J.T.F. Keurentjes, *Chem. Eng. Sci.* **2001**, *56*, 4197.

44 G. Verspui, F. Schanssema, R.A. Sheldon, *Angew. Chem. Int. Ed.* **2000**, *39*, 804.

45 A. Ahvenainen, K. Sarantila, H. Andtsjo, *WO Patent 92/12181*, **1992**.

9

Production of Fluoropolymers in Supercritical Carbon Dioxide

Colin D. Wood, Jason C. Yarbrough, George Roberts, and Joseph M. DeSimone

9.1
Introduction

Supercritical carbon dioxide has emerged as the most extensively studied super-critical fluid (SCF) medium for organic transformations and polymerization reactions. This stems from a list of advantages ranging from solvent properties to practical environmental and economic considerations. Increasing environmental regulation has prompted a tremendous effort to develop and apply environmentally benign alternative media to a variety of commercially significant, solvent-intensive processes. Carbon dioxide in the liquid and supercritical state is a potential alternative for many of the processes experiencing ever-increasing regulation, owing to its non-toxic, non-flammable nature. Additionally, its ubiquity makes it inexpensive and readily available. Several excellent reviews are available that discuss the utilization of carbon dioxide for many solvent- and waste-intensive industries [1–4]. Aside from the gains provided by CO_2 as a reaction medium in general, it finds particularly advantageous application in the synthesis and processing of fluoropolymers [5–7]. This being the case, the subject of this chapter is the synthesis of fluoropolymers in compressed carbon dioxide.

9.2
Fluoroolefin Polymerization in CO_2

9.2.1
Overview

Fluoropolymers are typically synthesized in aqueous polymerization systems (both emulsion and suspension), non-aqueous systems (Freon-113), or Freon-113/aqueous hybrid systems [8]. Such processes require the use of large quantities of water and CFCs (for non-aqueous polymerizations) or fluorinated surfactants (for emulsion polymerization). Aqueous suspension and dispersion poly-

Supercritical Carbon Dioxide: in Polymer Reaction Engineering
Edited by Maartje F. Kemmere and Thierry Meyer
Copyright © 2005 WILEY-VCH Verlag GmbH & Co. KGaA, Weinheim
ISBN: 3-527-31092-4

merizations of copolymers of TFE with a variety of comonomers including hexa-fluoropropylene (HFP) and various perfluoroalkyl vinyl ethers (PAVEs) typically exhibit high levels of carboxylic acid end groups [6, 9]. The presence of acid end-groups often proves deleterious to the intended properties and function of the polymer and introduces complications into certain post-polymerization, melt-processing steps. This problem is particularly prevalent in copolymers that include PAVE comonomers, which are an essential component in the control of crystallinity in TFE-based fluoroplastics. In order to prevent decomposition, dis-coloration, and emissions of hydrogen fluoride, polymers containing high levels of acid end-groups may require high-temperature hydrolysis and fluorination finishing steps.

A propagating polymeric radical with a PAVE active-radical center can have one of two possible reaction pathways. First, and most obvious, it can cross-pro-pagate to monomer, continuing the polymerization reaction, or it can undergo β-scission, resulting in an acid fluoride-terminated polymer and a perfluoroalkyl radical capable of initiating further polymerization. Essentially this is a chain-transfer-to-monomer step, the details of which are outlined in Scheme 9.1 [6, 9].

The use of CFCs circumvents these problems; however, CFCs have fallen un-der exceedingly strict regulation because of environmental concerns and, as a result, are no longer economically viable options as large-scale reaction media. Additionally, many of the fluorinated surfactants typically employed in aqueous

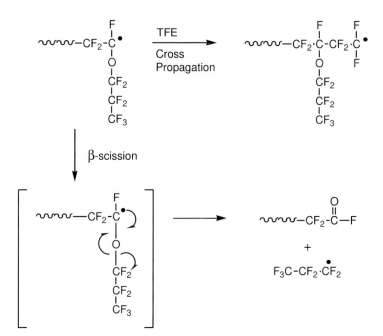

Scheme 9.1 Reaction sequence for β-scission in fluoroolefin polymerization [6, 9].

emulsion and suspension polymerizations are currently under scrutiny because of bioaccumulation and environmental persistence [10, 11]. These issues have collectively resulted in an increasingly urgent impetus for transition from the conventional fluoropolymer synthesis platforms to alternatives that meet the requirements of emerging public and regulatory demands.

Most commercially available fluoropolymers are prepared from a relatively small group of olefins including tetrafluoroethylene (TFE), chlorotrifluoroethylene (CTFE), vinylidene fluoride (VF2), hexafluoropropylene (HFP), ethylene, and perfluoroalkyl vinyl ethers (PAVEs) (Fig. 9.1). Many of these monomers are flammable, and some are explosive. For example, TFE is flammable when mixed with air and has a high propensity for explosion during expansion to a gas from its liquid phase under pressure. Further, TFE is highly explosive as a gas at elevated temperatures. In the presence of oxygen it will undergo auto-polymerization, a process sufficiently exothermic to ignite an explosion. It has been demonstrated that CO_2/TFE mixtures are far less susceptible to ignition, as TFE forms a "pseudo" azeotrope with CO_2 [12]. This makes handling and delivering monomer much safer.

Further, it is worth noting that fluoropolymers synthesized in sc-CO_2 exhibit significantly diminished levels of acid end-groups, leading to very high molecu-

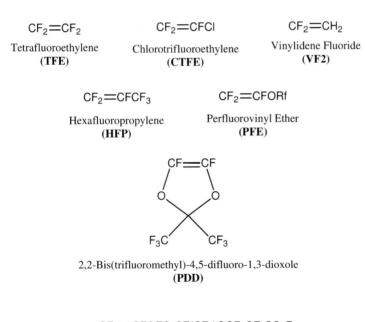

$$CF_2{=}CF_2$$

Tetrafluoroethylene
(**TFE**)

$$CF_2{=}CFCl$$

Chlorotrifluoroethylene
(**CTFE**)

$$CF_2{=}CH_2$$

Vinylidene Fluoride
(**VF2**)

$$CF_2{=}CFCF_3$$

Hexafluoropropylene
(**HFP**)

$$CF_2{=}CFORf$$

Perfluorovinyl Ether
(**PFE**)

$$CF{=}CF$$

$$F_3C \qquad CF_3$$

2,2-Bis(trifluoromethyl)-4,5-difluoro-1,3-dioxole
(**PDD**)

$$CF_2{=}CFOFC_2CF(CF_3)OCF_2CF_2SO_2F$$

Perfluorinated alkylvinyl ether sulfonyl fluoride
(**PSEVPE**)

Fig. 9.1 Common fluoroolefins used in fluoropolymer synthesis.

lar weight materials [9]. Indeed, in such instances chain transfer agents are necessary to control molecular weight and maintain melt-processability of the product [13]. IR analysis has confirmed that fluoropolymers synthesized under these conditions exhibit acid end-group levels an order of magnitude below (0 to 3 end-groups per 10^6 carbon atoms) those which are synthesized in conventional organic and aqueous reaction systems [9, 13]. The most likely explanation for this is that the tremendous plasticizing capability of CO_2 provides for efficient transport of monomer into the polymer phase, maintaining high effective concentrations of monomer in the vicinity of the active chain ends, thus favoring the bimolecular propagation step over the unimolecular β-scission process. Additionally, it has been demonstrated that dense CO_2 is exceptionally inert to free-radical chemistry in general [14].

9.2.2
TFE-based Materials

Tetrafluoroethylene (TFE)-based copolymers have become premium high-performance materials for a broad range of applications that demand chemical and thermal resistance and melt processing capability [15]. As mentioned earlier, caution is necessary when TFE is used as a monomer because of potential explosions. A range of TFE-based fluoropolymers have been successfully synthesized in CO_2. These include FEP, PFA, ETFE, TFE/vinyl acetate polymers, Nafion®-type materials, and Teflon® AF-type materials (Fig. 9.2). This demonstrates the broad applicability of liquid and supercritical CO_2 in the synthesis and processing of fluorinated polymers, which has led to the commercialization of a CO_2-based approach in the manufacturing of certain grades of polymers based on tetrafluoroethylene (Teflon™). DuPont has recently invested an initial $40 million in the construction of a Teflon FEP production facility in Fayetteville, North Carolina. The process employs CO_2 as the continuous phase instead of 1,1,2-trichloro-1,2,2-trifluoroethane or water and surfactant [16–19]. The use of carbon dioxide results in a surfactant-free system with no chain-transfer-to-solvent issues. This process not only avoids many environmental and safety concerns but also provides a product with superior properties. This is the first commercial example of fluoropolymer resins made using supercritical carbon dioxide as the solvent.

As mentioned earlier, carbon dioxide has been successfully employed as a solvent for the precipitation polymerization of FEP (TFE and HFP monomers) and TFE and PAVE to yield PFA (Fig. 9.2) utilizing a fully fluorinated initiator, bis(perfluoro-2-N-propoxypropionyl) peroxide (BPPP). A fluorinated initiator is used to prevent the formation of unstable end-groups, which will degrade upon processing at high temperatures [14]. A large excess of HFP was used for the preparation of FEP in order to achieve high incorporation in the resulting polymer because of the low rate of HFP incorporation during the polymerization. Additionally, it was found that polymers synthesized in CO_2 contained levels of acid end-groups as low as those observed in polymers synthesized in Freon-113

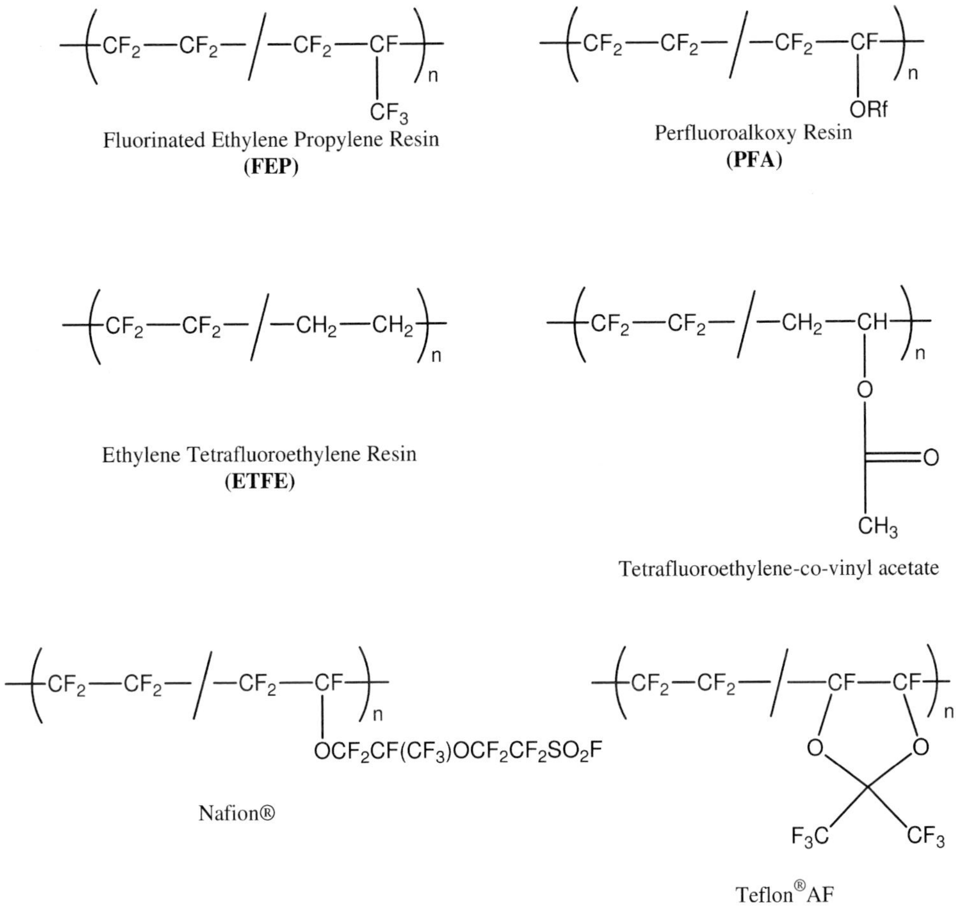

Fluorinated Ethylene Propylene Resin
(FEP)

Perfluoroalkoxy Resin
(PFA)

Ethylene Tetrafluoroethylene Resin
(ETFE)

Tetrafluoroethylene-co-vinyl acetate

Nafion®

Teflon®AF

Fig. 9.2 Tetrafluoroethylene-based fluoropolymers synthesized in carbon dioxide.

which were treated with a post-polymerization fluorination step [13]. This further underscores the advantages of utilizing CO_2 as a solvent for the synthesis of fluoropolymers, i.e. the several previously discussed environmental advantages, coupled with a product that can be easily isolated and does not require extensive post polymerization fluorination steps [6, 17]. In an analogous system, TFE and ethylene have also been copolymerized in CO_2 to yield ETFE [20].

Teflon® AF is an amorphous copolymer of tetrafluoroethylene (TFE) and 2,2-bis(trifluoromethyl)-4,5-difluoro-1,3-dioxole (PDD). It combines the properties of amorphous plastics, such as optical transparency and solubility in organic solvents, with those of perfluorinated polymers, including high thermal stability, excellent chemical stability, and low surface energy. Moreover Teflon® AF exhibits the lowest dielectric constant (1.90 for Teflon® AF 2400) and the lowest re-

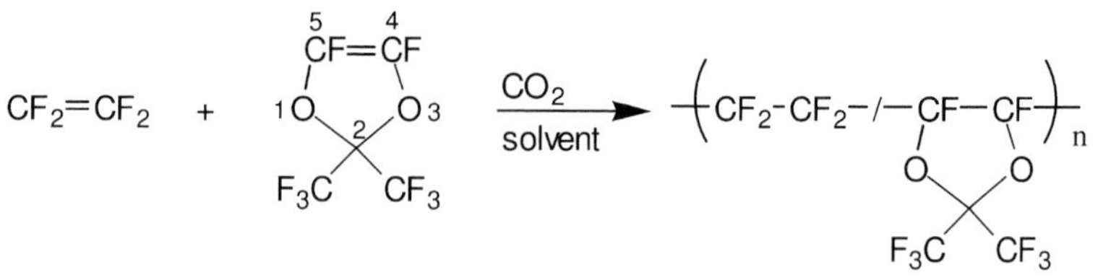

$$CF_2{=}CF_2 \quad + \quad \underset{\substack{F_3C \quad CF_3}}{\overset{\substack{5 \quad 4 \\ CF{=}CF \\ 1O \quad O3 \\ 2}}{}} \quad \xrightarrow[\text{solvent}]{CO_2} \quad \left(CF_2{-}CF_2{-}/{-}\underset{\substack{O \quad O \\ F_3C \quad CF_3}}{CF{-}CF} \right)_n$$

Scheme 9.2 Reaction scheme for the synthesis of Teflon®-AF in CO_2.

fractive index (1.29 for Teflon® AF 2400) known for any solid organic polymer [8]. As such, Teflon® AF is well-suited for use as an optical material. Teflon AF-based materials have also been synthesized in supercritical carbon dioxide (Scheme 9.2) [21]. A range of copolymers with various compositions and molecular weights were prepared in yields as high as 74%, and their properties were compared to commercially available Teflon® AF. The glass transition temperatures for the materials ranged from 67 to 334 °C. The synthesis of these copolymers in CO_2 has several key advantages compared to conventional polymerization techniques. The low reaction temperature and the use of a perfluorinated initiator results in copolymers with properties which are comparable to those of the analogous commercial product. However, an additional fluorination step is not necessary and the product is isolated directly from the reactor without contamination from solvents or surfactants. Moreover, the synthesis uses TFE/CO_2 mixtures instead of pure TFE, which has inherent safety advantages [12].

Copolymer syntheses of TFE and vinyl acetate have also been reported by Shoichet et al. in CO_2 at 23–26 MPa and 45 °C using diethyl peroxydicarbonate (EPDC) as the initiator. In this case TFE was not used as a mixture with CO_2 but from vacuum pyrolysis of PTFE [22]. In addition, a surfactant was used (Fluorad FC-171 – a mixture of fluorinated alkyl alkoxylates) in order to stabilize the dispersion polymerization process. Poly(TFE-*co*-vinyl acetate) copolymers are usually prepared by an aqueous emulsion polymerization process which typically results in a branched structure because of hydrogen abstraction from the backbone generating a branch point. However, when CO_2 was utilized as the continuous phase, linear polymer was reported. This was attributed to the fact that the abstraction of hydrogen by a propagating TFE radical is suppressed relative to propagation in carbon dioxide. Shoichet et al. have also reported several precipitation polymerizations of TFE/CTFE/VF2 with vinyl acetate in supercritical carbon dioxide in the absence of surfactant [23]. The yields of the TFE/vinyl acetate polymerizations were the highest, and the authors suggested that this was because of the high miscibility between the vinyl acetate, TFE, and propagating polymer chains in CO_2.

9.2.3
Ionomer Resins and Nafion®

There are extensive research efforts, both academically and industrially, to develop fuel cells as non-petroleum-based fuel sources. Proton exchange membranes are a fundamental component of a fuel cell. The membrane that is currently employed most commonly is a copolymer of tetrafluoroethylene and perfluoro-2-(2-fluorosulfonyl-ethoxy) propyl vinyl ether (PSEPVE) which is marketed by DuPont under the tradename of Nafion® (Fig. 9.2). Again, commercial syntheses of these and related materials are often carried out in non-aqueous media to avoid loss of expensive vinyl ether through partitioning into the aqueous phase, hydrolysis (to which these monomers are particularly susceptible), and the formation of acid end-groups (β-scission being the limiting synthetic parameter to synthesis of *high-molecular-weight* copolymers in any conventional medium). DeSimone and coworkers reported the use of sc-CO_2 for the synthesis of variants of Nafion®. As would be expected, the resulting polymers exhibited extremely low levels of acid end-groups relative to a commercially available control [6]. This observation was accompanied by increased molecular weights relative to commercial materials as evaluated by melt flow analysis.

9.2.4
VF2-based Materials

PVDF has been successfully synthesized in CO_2 in a continuous precipitation process using a continuous stirred tank reactor [24]. In such a process monomer, solvent (CO_2) and initiator are continuously added to the reactor. The advantage of a continuous process over a batch process is that the monomer and the CO_2 can be removed and recycled, leading to a high rate of polymerization. The authors later reported that, although this was a precipitation polymerization, a model based on the kinetics for a homogeneous polymerization fits the experimental data, confirming the reaction orders of 0.5 and 1 for the initiator and monomer respectively [25]. Further, the effect of VF2 flow rate, residence time, and temperature were investigated in this system [26]. It was found that monomer feed concentration had a pronounced effect upon the modality of the molecular weight distribution. At a high monomer concentration, the resulting polymer exhibited a bimodal molecular weight distribution. The authors suggested that at high monomer concentrations the kinetics may no longer adhere to the homogeneous kinetic model proposed earlier [25]. Several other parameters were investigated such as chain branching and agitation, but they were unable to explain the bimodality of the polymer.

9.2.5
VF2 and TFE Telomerization

DeSimone and coworkers were the first to report the homogeneous synthesis of fluorinated telomers in sc-CO_2 (Scheme 9.3) [27, 28]. Most conventional telomerization reactions are carried out in bulk because of the insolubility of fluoropolymers in all organic solvents other than CFCs.

Owing to the solubility of fluorinated oligomers in CO_2 and the utility of CO_2 as a medium for free-radical reactions, homogeneous telomerizations could be carried out through thermal decomposition of the telogen or using AIBN as a thermal initiator. These reactions thus offer further evidence of the inert behavior of CO_2 toward radical chemistry.

Table 9.1 lists reaction conditions for the AIBN-initiated telomerizations. Table 9.2 details the thermally initiated TFE telomerizations. It is interesting to note

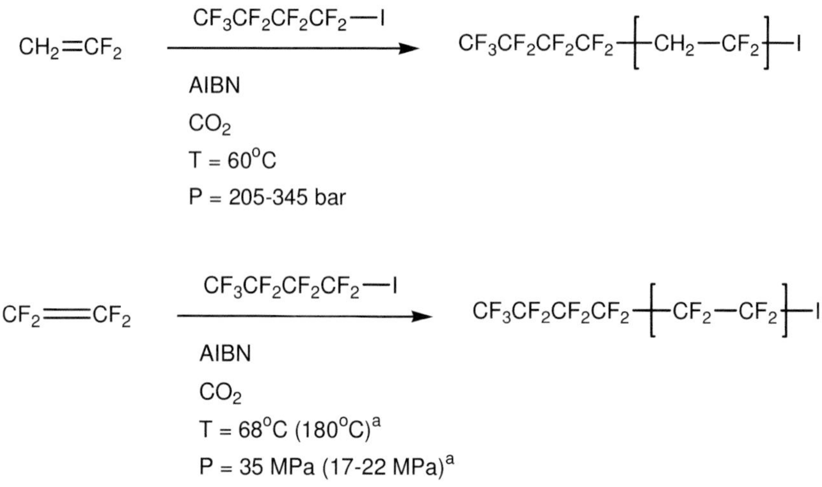

$$CH_2{=}CF_2 \quad \xrightarrow[\substack{\text{AIBN}\\ CO_2 \\ T = 60^\circ C \\ P = 205\text{-}345 \text{ bar}}]{CF_3CF_2CF_2CF_2{-}I} \quad CF_3CF_2CF_2CF_2{+}CH_2{-}CF_2{+}_{\!}I$$

$$CF_2{=}CF_2 \quad \xrightarrow[\substack{\text{AIBN}\\ CO_2 \\ T = 68^\circ C \,(180^\circ C)^a \\ P = 35 \text{ MPa} \,(17\text{-}22 \text{ MPa})^a}]{CF_3CF_2CF_2CF_2{-}I} \quad CF_3CF_2CF_2CF_2{+}CF_2{-}CF_2{+}_{\!}I$$

[a] Thermally initiated telomerizations of TFE were carried out at 180°C and 17-22 MPa.

Scheme 9.3 AIBN – initiated telomerization of vinylidene fluoride and tetrafluoroethylene in sc-CO_2.

Table 9.1 Reaction conditions and results for the AIBN-initiated telomerization of VF_2 in supercritical CO_2 at $T=60\,^\circ$C in a 10-mL stainless steel view cell for 24 h.

Initial pressure (MPa)	Mass of VF_2 (g)	Mass of telogen (C_4F_9I) (g)	M_w	M_w/M_n	Convn. (%)
34	3.8	1.6	608	1.05	35
28	4.0	2.2	588	1.05	32

Table 9.2 Reaction conditions and results for the thermally-initiated telo-merization of TFE at $T=180\,^{\circ}C$ in a 25-mL view cell (pressures varied based on telomer concentration, between 17 and 22 MPa).

[Monomer]/[Telomer]	Yield (%)	M_n	MWD(DPw/DPn)
1.6	88	570	1.35
1.5	87	590	1.38
1.8	86	630	1.38
2.2	78	650	1.44

that conversions and molecular weights were inconsistent and irreproducible for the AIBN-initiated reactions of TFE. However, thermal telomerizations of TFE gave decreasing yields with increasing monomer-to-telogen ratio, consistent with reactions where the telogen also acts as the initiator.

9.3
Fluoroalkyl Acrylate Polymerizations in CO_2

One of the more difficult issues associated with the use of CO_2 as a polymerization medium has been the limited solubility in CO_2 of most high-molecular-weight polymers. For this reason, much of the reported work on polymerizations typically concerns heterogeneous precipitation, dispersion, or emulsion processes. However, it has been demonstrated that oligomeric perfluoropolyethers and oligomeric poly(chlorotrifluoroethylene) oils are soluble in liquid CO_2 [1]. Additionally, it has been reported that highly fluorinated polyacrylates of high molecular weight ($>250\,000$) exhibit exceptional solubility in supercritical CO_2. Not surprisingly, since this discovery, it has been demonstrated that high-molecular-weight, amorphous fluoropolymers can be synthesized homogeneously in CO_2 utilizing free radical initiators [16, 29–34]. Most amorphous fluoropolymers exhibit resistance to common organic solvents, and therefore conventional approaches to synthesis and processing have depended on chlorofluorocarbon (CFC) platforms. Because of the previously mentioned environmental concerns with the use of CFC refrigerants and solvents, CFCs are no longer acceptable for large-scale commercial use. As an effective reaction medium for homogeneous polymerizations under mild conditions, CO is an increasingly attractive, inexpensive, and harmless platform option for the synthesis of highly fluorinated monomers (as domestic and international regulation of CFCs becomes ever more restrictive in the coming years). Various fluorinated acrylate monomers, such as 1,1-dihydroperfluorooctyl acrylate (FOA) were polymerized by this method (Scheme 9.4). In all cases, the polymerizations remained homogeneous throughout the reaction, illustrating the high solubility of the polymers in supercritical CO_2. It is also noted that reaction conditions for these polymerizations are mild in terms of temperature and pressure and not entirely unlike

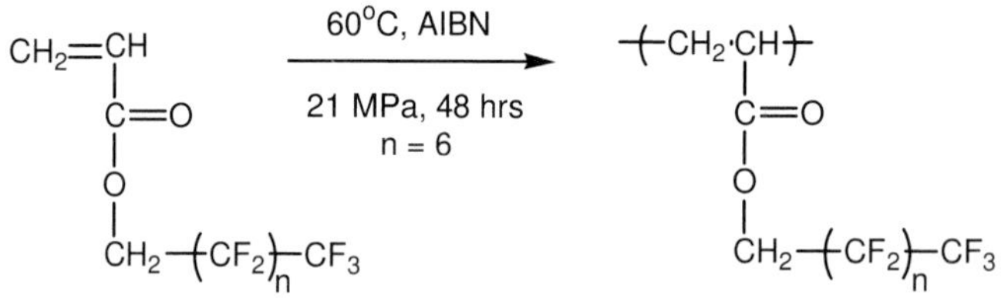

$$CH_2\!\!=\!\!CH \quad \xrightarrow[\substack{21\ MPa,\ 48\ hrs \\ n=6}]{60^\circ C,\ AIBN} \quad \text{---}(CH_2\!\cdot\!CH)\text{---}$$

Scheme 9.4 Free-radical polymerization of FOA in supercritical carbon dioxide.

those encountered in conventional solvent-based processes [16]. Further, studies have shown CO_2 to be inert to radical chemistry, eliminating chain transfer to solvent as a side reaction [7], and the low viscosity of sc-CO_2 removes issues of Tromsdorf effect and autoacceleration [7]. Therefore, as a reaction medium, CO_2 is ideally suited to free-radical polymerizations.

Additionally, statistical copolymers of fluorinated acrylates and common hydrocarbon monomers, such as methyl methacrylate (MMA), butyl acrylate (BA), styrene, and ethylene have been effectively synthesized using CO_2 as a reaction medium (Table 9.3) [28]. This is possible because of the high solubility of most low-molecular-weight liquid hydrocarbon monomers in sc-CO_2. Further, copolymers of FOA and either 2-(dimethylamino)ethyl acrylate or 4-vinyl pyridine have been successfully synthesized via homogeneous solution polymerization in sc-CO_2, producing CO_2-soluble polymeric amines [29]. These polymerizations were observed to be homogeneous throughout each reaction period even at high concentrations of comonomer. Furthermore, these materials were easily isolated from the reaction solution as a powder or gum by releasing pressure and venting with CO_2.

Kinetic studies of free-radical initiation in CO_2 were conducted using AIBN as the initiator [16, 29]. The study was undertaken using high-pressure UV spectroscopy. The results showed decomposition to be slower in CO_2 relative to that measured in benzene owing to the reduced dielectric strength of carbon dioxide. However, it was also determined that the initiator efficiency is much higher in CO_2 because of reduced viscosity and solvent "cage" effects. Initiator

Table 9.3 Statistical copolymers of FOA with hydrocarbon, vinyl monomers. Polymerizations were conducted at $(59.4\pm0.1)\,^\circ C$ and (34.8 ± 0.05) MPa for 48 h in CO_2. Intrinsic viscosity was determined in 1,1,2-trifluoro-trichloroethane (Freon-113) at $30\,^\circ C$.

Copolymer	Feed ratio	Incorporation	Intrinsic viscosity (dL g^{-1})
Poly(FOA-co-MMA)	0.47	0.57	0.10
Poly(FOA-co-styrene)	0.48	0.58	0.15
Poly(FOA-co-BA)	0.53	0.57	0.45
Poly(FOA-co-ethylene)	0.35	–	0.14

efficiency is defined as the fraction of radicals, formed in the initial decomposition step, which diffuse out of the solvent cage and are successful in initiating polymerization of monomer. This again underscores the earlier statement that CO_2 is an effective and ideal radical polymerization medium that is inert even in the presence of highly electrophilic hydrocarbon (or fluorocarbon) radicals.

9.4
Amphiphilic Poly(alkylacrylates)

As fluorinated poly(alkylacrylates) have proved to be highly soluble in compressed CO_2, they present themselves as possible components in the stabilization of heterogeneous reaction systems in CO_2. Indeed, there has been considerable effort in the development and synthesis of polymeric emulsifiers for lipophilic materials and stabilizers for hydrocarbon polymer dispersions in CO_2. De-Simone and coworkers demonstrated the feasibility of using fluorinated alkyl acrylate homopolymers, such as PFOA, as efficient amphiphiles, owing to the lipophilic acrylate backbone and the CO_2-philic fluorinated pendant chains [35, 36]. As was described earlier, these fluorinated alkyl acrylates are readily synthesized homogeneously in CO_2. The solution properties and phase behavior of PFOA in compressed CO_2 have been thoroughly examined and reported elsewhere [37–39].

Further, the ability to synthesize random copolymers with various hydrocarbon monomers allows the anchor-soluble balance to be tuned while maintaining solubility even with high incorporations of hydrocarbon comonomers [29]. Because of the amphiphilic nature of such copolymers, it was predicted that these materials would self-assemble into micelles consisting of a highly fluorinated corona segregating the lipophilic core from the compressed CO_2 continuous phase. Thus, PFOA-*b*-PS block copolymers were synthesized via controlled free-radical techniques (Fig. 9.3), and it was confirmed (by small-angle neutron scattering) that these copolymers spontaneously assemble into multimolecular micelles in solution [40]. In addition to amphiphilic materials, which physically adsorb to the surface of polymer particles in dispersion polymerizations, fluorinated acrylates can be utilized as polymerizable comonomers in the stabilization of CO_2-phobic polymer colloids [41].

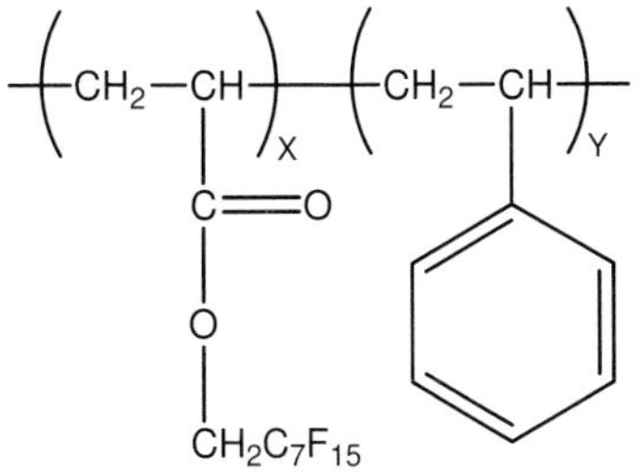

Fig. 9.3 Poly(1,1-dihydroperfluorooctyl-acrylate-*b*-styrene).

9.5
Photooxidation of Fluoroolefins in Liquid CO_2

As previously mentioned, another class of fluoropolymers known to be readily soluble in CO_2 are perfluoropolyethers (PFPEs) [1]. A unique class of fluoropolymers, PFPE polymers and copolymers have been established as high performance materials, exhibiting low surface energies and low moduli, as well as excellent thermal and chemical stabilities. PFPEs are primarily found in high-performance lubricant applications, e.g., for magnetic data storage media and as heat exchanger fluids. One of the main industrial processes for the production of PFPEs is photooxidation of fluoroolefins [41]. Currently, only TFE and HFP are used commercially in this process. Typically, HFP is photooxidized in bulk because of its very low reactivity, while TFE requires an inert diluent in order to prevent homopolymerization of the olefin.

DeSimone and coworkers recently reported the photooxidation of various concentrations of HFP in CO_2 in concert with parallel reactions carried out in bulk HFP and in perfluorocyclobutane for purposes of comparison [42]. Scheme 9.5 and Table 9.4 detail the reaction conditions and results. Based on the data collected in this study, it was demonstrated that there is a strong dependence of molecular weight and composition on HFP concentration (i.e. lower HFP concentrations gave lower peroxide content). Additionally, viscosity effects in liquid CO_2 appeared to significantly reduce the amount of peroxidic linkages in the final product, as seen from entries 4 and 5 in Table 9.4.

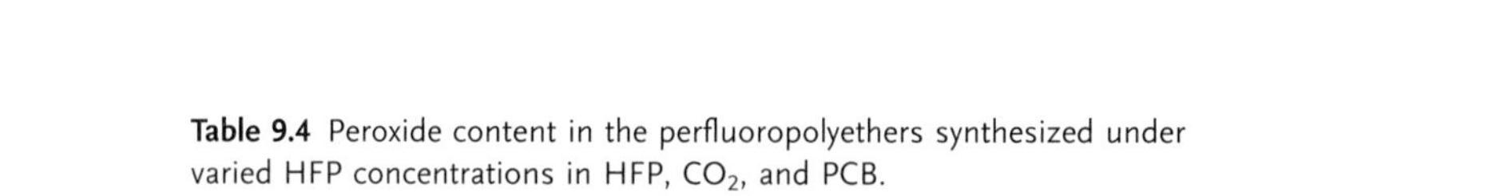

$$R_1, R_2 = C(O)F, CF_3 \text{ (Major)}, CF_2C(O)F \text{ (Minor)}$$

Scheme 9.5 Perfluoropolyether synthesis through photooxidation of hexafluoropropylene.

Table 9.4 Peroxide content in the perfluoropolyethers synthesized under varied HFP concentrations in HFP, CO_2, and PCB.

Entry	[HFP] (M)	Solvent	Peroxide content (wt.%)	Volume HFP (vol. %)
1	10.2	HFP	1.8	100
2	6.8	CO_2	1.1	67
3	6.5	PCB	1.2	64
4	3.9	CO_2	0.29	38
5	3.7	PCB	0.85	36

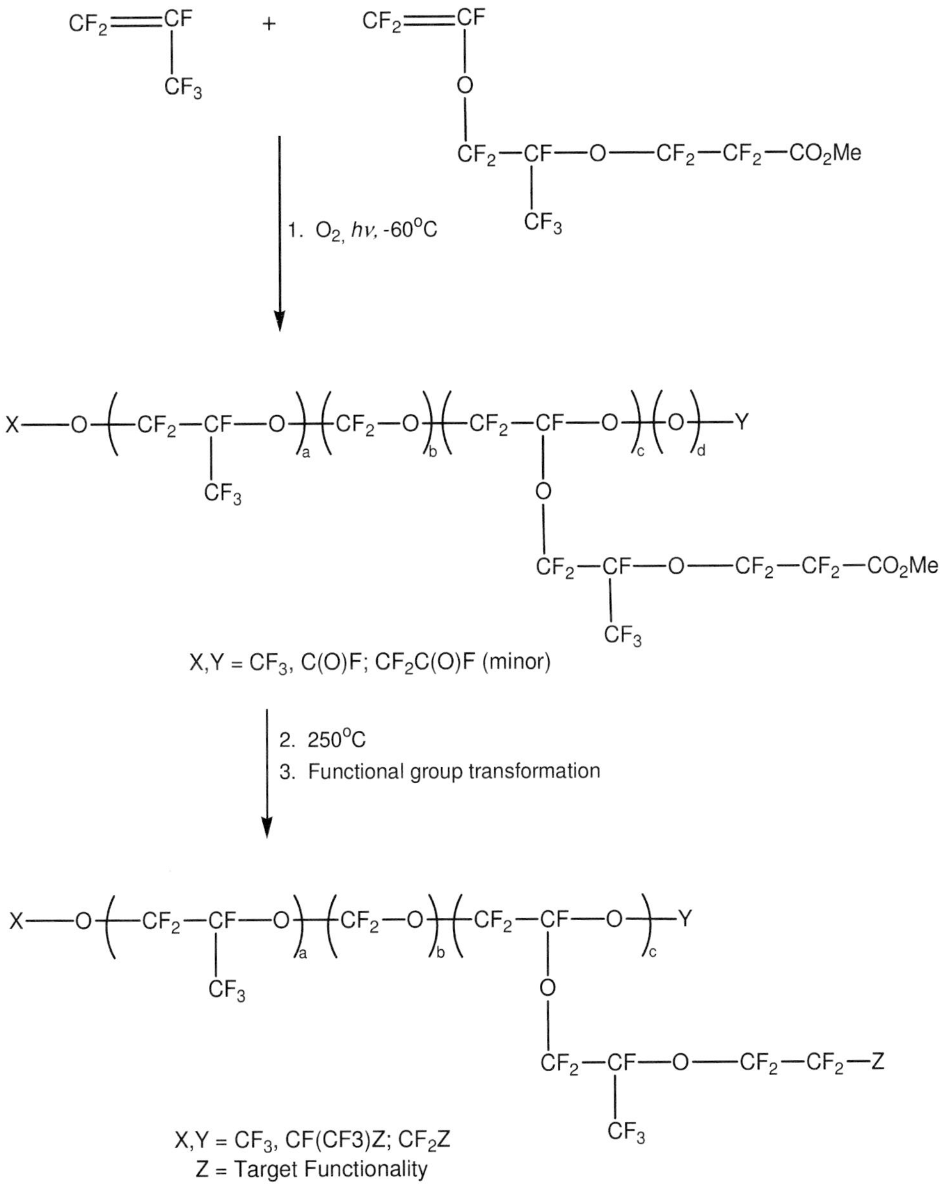

$CF_2\!\!=\!\!CF$ + $CF_2\!\!=\!\!CF$
  | |
  CF_3 O
                            |
                            CF_2—CF—O——CF_2—CF_2—CO_2Me
                                    |
                                    CF_3

1. O$_2$, hv, -60°C

X—O$\left(\!CF_2\text{—}CF\text{—O}\!\right)_a\!\!\left(\!CF_2\text{—O}\!\right)_b\!\!\left(\!CF_2\text{—}CF\text{—O}\!\right)_c\!\!\left(\!O\!\right)_d$—$Y$
         | |
         CF_3 O
                                                  |
                                                  CF_2—CF—O——CF_2—CF_2—CO_2Me
                                                          |
                                                          CF_3

X,Y = CF$_3$, C(O)F; CF$_2$C(O)F (minor)

2. 250°C
3. Functional group transformation

X—O$\left(\!CF_2\text{—}CF\text{—O}\!\right)_a\!\!\left(\!CF_2\text{—O}\!\right)_b\!\!\left(\!CF_2\text{—}CF\text{—O}\!\right)_c$—$Y$
         | |
         CF_3 O
                                                  |
                                                  CF_2—CF—O——CF_2—CF_2—Z
                                                          |
                                                          CF_3

X,Y = CF$_3$, CF(CF3)Z; CF$_2$Z
Z = Target Functionality

Scheme 9.6 Synthesis of multifunctional PFPEs.

Additionally, the photooxidation of fluoroolefins provides a method by which to synthesize PFPEs with varying degrees of functionality. DeSimone and co-workers have reported the synthesis of multifunctional PFPEs by cophotooxidation of HFP and an ester-functionalized vinyl ether (EVE), the details of which are illustrated in Scheme 9.6 [43]. This was the first report of the utilization of functional monomers by this method. Prior to this work, the anionic ring-opening polymerization and fluoroolefin photooxidation methods commonly used to synthesize functional PFPEs provided polymers with chain functionalities no greater than one and two, respectively. However, the polymers reported exhibited average chain functionalities as high as 7.3 as determined by repeat unit and endgroup composition (^{19}F NMR).

9.6
CO$_2$/Aqueous Hybrid Systems

PTFE is manufactured primarily by free-radical methods in aqueous media. De-Simone et al. developed a CO$_2$/aqueous hybrid process that allows for the safer handling of the TFE, resulting in high-molecular-weight PTFE resins [44]. This system represents a substantial deviation from traditional systems, as CO$_2$ and water exhibit low mutual solubilities, allowing for compartmentalization of monomer, polymer, and initiator based on their solubility characteristics.

9.7
Conclusions

In general, CO$_2$ is an attractive alternative solvent for the synthesis of fluoropolymers because it is "environmentally friendly", non-toxic, non-flammable, inexpensive, and readily available in high purity from a number of sources. Product isolation is straightforward because CO$_2$ is a gas under ambient conditions, removing the need for energy-intensive drying steps. Moreover, additional advantages are realized in the manufacture of fluoropolymers, such as increased safety in handling explosive monomers and enhanced polymer properties in many cases. As demonstrated, CO$_2$ is not only an adequate alternative, but in many cases a superior one. This is strikingly borne out by DuPont's recent $275 million investment in its CO$_2$-based Teflon^{TM} production process. Thus, bearing in mind the ever-increasing regulations concerning fluorinated surfactants, CO$_2$ may find even more widespread use for the synthesis of fluoropolymers.

References

1 M.A. McHugh, V. Krukonis, *Supercritical Fluid Extraction*, 2nd edn., Butterworth Heinemann, **1994**.

2 K.D. Bartle, A.A. Clifford, S.A. Jafar, G.F. Shilstone, *J. Phys. Chem.*, **1991** *Ref. Data* 20713–20778.

3 K.P. Johnson, Penninger (Eds.), *Supercritical Fluid Science and Technology*, American Chemical Society, Washington D.C. **1989**.

4 C.L. Phelps, N.G. Smart, C.M. Wai, *J. Chem. Educ.*, **1996**, *73*, 1163–1168.

5 J.L. Kendall, D.A. Canelas, J.L. Young. J.M. DeSimone, *Chem. Rev.*, **1999**, *99*, 543.

6 J.P. DeYoung, T.J. Romack, J.M. DeSimone, In *Fluoropolymers 1: Synthesis*; G. Hougham, P.E. Cassidy, K. Johns, T. Davidson, (Eds.), Kluwer Academic/Plenum Publishers, New York, **1999**, pp. 191–205.

7 T.A. Davidson, J.M. DeSimone, In *Chemical Synthesis Using Supercritical Fluids*; Jessop, P.G. Leitner, W. (Eds.), Wiley-VCH, Weinheim, **1999**, p. 297.

8 R.P. Resnick, W.H. Buck, In *Modern Fluoropolymers*; Scheirs, J., (Ed.), John Wiley & Sons, Chichester, **1997**, p. 397.

9 J.P. DeYoung, T.J. Romack, J.M. De Simone, *Polym. Prep. (Am. Chem. Soc., Div. Polym. Chem.)*, **1997**, *38*, 424.

10 C.A. Moody, J.A. Field, *Environ. Sci. Technol.*, **2000**, *34*, 3864–3870.

11 C. Hogue, *C & EN.*, August 30, **2004**, p. 17.

12 D.J. Van Bramer, M.B. Schiflett, A. Yokozeki, United States Patent 5,345,013, **1994**.

13 T.J Romack, J.M. DeSimone, *Macromolecules*, **1995**, *28*, 8429.

14 W.C Bunyard, J.F. Kadla, J.M. De Simone, *J. Am. Chem. Soc.*, **2001**, *123*, 7199.

15 A.E. Fiering, In *Organofluorine Chemistry: Principles and Commercial Applications*, R.E. Banks, B.E. Smart, J.C. Tatlow, (eds.), Plenum Press, New York, **1994**.

16 J.M. DeSimone, Z. Guan, C. Eisbernd, *Science*, **1992**, *257*, 945.

17 T.J. Romack, J.M DeSimone, *Macromolecules*, **1995**, *28*, 8429–8431.

18 G. Parkinson, *Chem. Eng.*, **1999**, *106*, 17.

19 J.M. DeSimone, *Science*, **2002**, *297*, 799.

20 T.J. Romack, Doctoral Dissertation, University of North Carolina at Chapel Hill, **1996**.

21 U. Michel, P. Resnick, B.E. Kipp, J.M. DeSimone, *Macromolecules*, **2003**, *36*, 7107–7113.

22 R.D. Lousenberg, M.S. Shoichet, *Macromolecules*, **2000**, *33*, 1682–1685.

23 B. Baradie, M.S. Shoichet, *Macromolecules*, **2002**, *35*, 3569–3575.

24 P.A. Charpentier, K.A. Kennedy, J.M. DeSimone, G.W. Roberts, *Macromolecules*, **1999**, *32*, 5973–5975.

25 P.A. Charpentier, J.M. DeSimone, G.W. Roberts, *Ind. Eng. Chem. Res.*, **2000**, *39*, 4588–4596.

26 M.K. Saraf, S. Gerrard, L.M. Wojcinski, P.A. Charpentier, J.M. DeSimone, G.W. Roberts, *Macromolecules*, **2002**, *35*, 7976–7985.

27 J.R. Combes, Z. Guan, J.M DeSimone, *Macromolecules*, **1994**, *27*, 865.

28 T.J. Romack, J.R. Combes, J.M. De Simone, *Macromolecules*, **1995**, *28*, 1724.

29 Z. Guan, C.S. Elsbernd, J.M. DeSimone, *Polym. Prepr. (Am. Chem. Soc., Div. Polym. Chem.)*, **1992**, *34*, 329.

30 Z. Guan, J.R. Combes, C.S. Elsbernd, J.M. DeSimone, *Polym. Prepr. (Am. Chem. Soc., Div. Polym. Chem.)*, **1993**, *34*, 446.

31 P. Ehrlich, *Chemtracts: Org. Chem.*, **1993**, *6*, 92.

32 J.M. DeSimone, US Pat. 5,496,901, **1996**.

33 J.M. DeSimone, US Pat. 5,688,879, **1997**.

34 J.M. DeSimone, US Pat. 5,739,223, **1998**.

35 J.M. DeSimone, E.E. Maury, Y.Z. Menceloglu, J.B. McClain, T.J. Romack, J.R. Combes, *Science*, **1994**, *265*, 356.

36 Z. Guan, J.M. DeSimone, *Macromolecules*, **1994**, *27*, 5527.

37 J.B. McClain, D.E. Betts, D.H. Canelas, E.T. Samulski, J.M. DeSimone, J.D. Londono, G.D. Wignall, *Polym. Mater. Sci. Eng.*, **1996**, *74*, 234.

38 J. B. McClain, J. D. Londono, J. R. Combes, T. J. Romack, D. A. Canelas, D. E. Betts, G. D. Wignall, E. T. Samulski, J. M. DeSimone, *J. Am. Chem. Soc.*, **1996**, *118*, 917.

39 Y. L. Hsiao, E. E. Maury, J. M. DeSimone, S. M. Mawson, K. P. Johnson, *Macromolecules*, **1995**, *28*, 8159.D.

40 D. Chillura-Martino, R. Triolo, J. B. McClain, J. R. Combes, D. E. Betts, D. A. Canelas, E. T. Samulski, J. M. DeSimone, H. D. Cochrane, J. D. Londono, G. D. Wignall, *Molec. Struct.*, **1996**, *383*, 3.

41 D. Sianesi, G. Marchionni, R. J. DePasquale, In *Organofluorine Chemistry: Principles and Commercial Applications*, R. E. Banks, B. E. Smart, J. C. Tatlow (Eds.), Plenum Press, New York, **1994**, p. 431.

42 W. C. Bunyard, T. J. Romack, J. M. DeSimone, *Macromolecules*, **1999**, *32*, 8224.

43 W. C. Bunyard, J. M. DeSimone, *Polym. Prep (Am. Chem. Soc., Div. Polym. Chem.)*, **1999**, *40*, 827.

44 T. J. Romack, B. E. Kipp, J. M. DeSimone, *Macromolecules* **1995**, *28*, 8432.

10
Polymer Processing with Supercritical Fluids *

Oliver S. Fleming and Sergei G. Kazarian

10.1
Introduction

In this chapter we review some of the key issues associated with the use of supercritical fluids for polymer processing. The development of supercritical enhanced processing requires understanding of the phase behavior of polymer/supercritical fluid systems. Therefore, the first part of the chapter is focused on the solubility of CO_2 in polymers – a key factor that determines conditions and outcome of polymer processing with supercritical carbon dioxide. The associated phenomena, such as supercritical CO_2-induced plasticization, diffusion, crystallization, and foaming of polymers are also discussed and interrelated to CO_2 solubility in polymers. Spectroscopy played an important role in the elucidation of these phenomena and provided molecular-level insight, and thus, applications of spectroscopy to polymer processing are also discussed in this section and throughout the chapter. The rheological behavior of polymers subjected to high-pressure CO_2 is reviewed in the next section of the chapter, and implications of CO_2-induced viscosity reduction for polymer processing are discussed. These effects open up a whole new dimension for polymeric materials that are difficult to process because of their high viscosity. The following section focuses on effects of supercritical CO_2 on polymer blends – mixing and phase separation. A novel spectroscopic imaging approach for the analysis of polymer blends subjected to high-pressure CO_2 is also discussed in this section. The plasticizing effect of supercritical CO_2 on polymeric materials is related to the processes of supercritical fluid impregnation for dyeing applications and the preparation of advanced optical materials and biomaterials. Finally, we assess the outlook for new opportunities and the further use of supercritical fluids in polymer processing for sustainable technology.

---

* The symbols used in this chapter are listed at the end of the text, under "Notation".

Supercritical Carbon Dioxide: in Polymer Reaction Engineering
Edited by Maartje F. Kemmere and Thierry Meyer
Copyright © 2005 WILEY-VCH Verlag GmbH & Co. KGaA, Weinheim
ISBN: 3-527-31092-4

10.2
**Phase Behavior of CO_2/Polymer Systems and the Effect of CO_2
on Polymerson Polymers**

10.2.1
Solubility of CO_2 in Polymers

Supercritical carbon dioxide has proved its applicability in the area of polymer
processing due to its unique properties as a temporary plasticizer. One of the
differences between traditional plasticizers and CO_2 is that CO_2 is a gas under
normal conditions and is therefore easy to remove. The solubility of high-pressure CO_2 in a polymeric matrix is the fundamental concept that facilitates all
forms of processing discussed in this chapter. In contrast to polymer synthesis
with CO_2, polymer processing involves CO_2 as the relatively minor phase dissolved within the polymeric matrix. Typically the equilibrium fraction of CO_2 in
common polymers such as poly(methylmethacrylate) (PMMA) reaches ca. 10%
(35 °C and 4.1 MPa), although it has been shown to reach levels of 21% in
high-molecular-weight poly(dimethylsiloxane) (PDMS) (50 °C and 9.09 MPa) [1].
The solubility of CO_2 in polymers is largely based on its ability to weakly interact with the basic sites in polymers.

The importance of polar groups in polymers to CO_2 solubility has been recognized for almost two decades [2], with enhancement in CO_2-philicity observed
for an increase in the number of electron-donating groups for a given polymer
density [3, 4]. The accessibility of the polar group is also an important factor [5];
for example, CO_2 solubility is greater in PMMA than in poly(ethylene terephthalate) glycol-modified (PETG) because of the access of CO_2 to the main-chain ester group in PETG being more hindered than in the case of PMMA, which has
a side-chain ester functionality [6]. The morphology of the polymer matrix also
affects CO_2 solubility; thus, glassy polymers [6] usually exhibit higher CO_2 solubility compared to semi-crystalline polymers [7].

Many experimental [3, 6–12] and theoretical [13–16] methods have been applied
to investigate the solubility of CO_2 in polymers. Of all the methods used for solubility determination, IR spectroscopy stands out as the key technique that has directly specified the interaction responsible for the phenomena. We obtained the
first spectroscopic evidence for the interaction between CO_2 and functional groups
in polymers (such as carbonyl groups or phenyl rings) [17] which has proven to
have impact on and implications for polymer modifications and synthesis, membrane technology, and polymer processing. *In situ* FTIR spectroscopy was applied
to probe the molecular level interaction of CO_2 with a wide range of polymers.
Specifically, interactions were assessed by the splitting of the v_2 degenerate bend-
ing mode of the CO_2, which indicates a Lewis acid–base interaction. This was
further substantiated by Nelson and Borkman [13], who applied *ab initio* molecular orbital calculations to study CO_2 interacting with PET, revealing specific structural information that was absent from the spectroscopic work. They predicted
that the CO_2 molecule is slightly bent when bound to the carbonyl oxygen of

PET, indicating a transfer of electron density to CO_2, which acts as a Lewis acid. Furthermore, preferential interaction was found to occur at the carbonyl group over the electron-donating ring system in PET. The interaction of CO_2 with polymers containing fluorinated groups was investigated by Fied and Hu [14] using computational chemistry. Quadrupole-dipole interactions were predicted to make a substantial contribution to the total energy of the interaction, estimated to occur at a close (4.3 Å) approach distance.

The dependence of CO_2 solubility with temperature [18] is a feature that is exploited in polymer processing. Temperature control provides an effective way to alter the conditions within the matrix. A good example of this is the application of CO_2-induced foaming, which can be initiated by rapid temperature ramping from a supersaturated state. At low temperatures the solubility of CO_2 in the polymer matrix is high, but with increasing pressure the solubility is lowered, which dictates careful temperature control during processing. A thorough understanding of CO_2 solubility in polymers is required to facilitate process optimization. For a concise summary of CO_2 solubility data in polymers and detailed information regarding the measurement techniques, the reader is directed to a recent review article complied by Tomasko et al. [19].

10.2.2
CO_2-Induced Plasticization of Polymers

The plasticization of polymers involves an increase in the inter-chain distances and an enhanced degree of chain and segmental mobility. Polymer plasticization is accompanied by changes in mechanical and physical properties, and thus a knowledge of this phenomenon is vital for processing applications. The plasticization facilitates the processes described in this chapter and and has the important characteristic that following processing there is no residual CO_2 remaining in the matrix. Plasticization of the polymer matrix occurs at the glass transition temperature (T_g) and may be induced thermally or by the use of an organic solvent which acts to depress the T_g. The use of an organic solvent has obvious drawbacks due to the presence of solvent residues in the matrix that may affect the final material and limit the applicability (i.e. applications such as food, biotechnology, and pharmaceutical technology where the presence of the residual solvent may be harmful).

The presence of dissolved CO_2 molecules in a polymer results in the plasticization of the amorphous component of the matrix. In this respect CO_2 mimics the effect of heat but with the important distinction that the T_g is depressed. The extent of the T_g depression is dependent on the wt% of CO_2 in the matrix. As previously mentioned, one of the characteristics of plasticization is the enhancement of segmental motion, which has been observed spectroscopically for the ester groups of PMMA [20] and the phenyl rings of polystyrene [21]. The consequential increase in free volume of the matrix has been studied by methods such as laser dilatometry [22], *in situ* FTIR spectroscopy [20], high-pressure partition chromatography [23], and inverse gas chromatography [24].

This T_g depression enables the processing of polymeric systems that were previously not possible by conventional methods. If one wanted to impregnate a polymer with an active component, the selection of active components would be dictated by the T_g of the host matrix in terms of a requirement for thermal stability of the component at temperatures of at least the T_g. With the use of high-pressure CO_2 and the consequential reduction of the T_g, the selection of active components is broadened to include components that would have suffered from thermal degradation via traditional melt processing. This ability of high-pressure or supercritical CO_2 to plasticize the matrix is at the heart of all applications described in the following sections.

10.2.3
CO$_2$-Induced Crystallization of Polymers

CO_2-induced plasticization of the amorphous domains in semi-crystalline polymers has significant morphological implications due to the formation of crystallites. CO_2-induced crystallization stems from free-energy considerations in the CO_2/polymer system. The CO_2-induced mobility of the polymer chains permits reconfiguration to occur, both in terms of conformation and 3-D arrangement (intra and inter) into the lowest-energy configuration and thus results in the formation of crystallites. Following depressurization of the system, CO_2 escapes quickly from the matrix and leaves the induced morphological alterations "frozen", shown schematically in Fig. 10.1. In this manner CO_2 can be viewed as a "temporary morphological modification tool", as once the desired morphology has been induced CO_2 can be completely removed from the polymer. Table 10.1 summarizes polymers that have been crystallized with CO_2 and includes the methods used to detect the crystallization process.

The CO_2-induced crystallization of poly(ethylene terephthalate) (PET) has received much attention because of the industrial relevance of this polymer [25]. The rate of the crystallization has been shown to increase with increasing pressure [26] and follow the Avrami equation [27]. As expected, FTIR and Raman spectroscopy have proved to be highly valuable techniques to study the CO_2-induced crystallization of PET [28], which enabled Brantley et al. [18] to study the crystallization process by *in situ* IR spectroscopy. That work established that at $0\,°C$ crystallization was suppressed even at a CO_2 pressure of 175 bar and thus demonstrated the possibility to process PET without inducing morphological alterations. Recently, Fleming and Kazarian [29] applied confocal Raman microscopy (CRM) to evaluate scCO$_2$-induced morphological alterations in PET film. That work revealed that crystallinity does not occur uniformly through the polymer but results in the formation of a crystallinity gradient along the line normal to the film surface as a result of the relative concentration of CO_2 as a function of depth. Fleming et al. [30] complemented that work with an FTIR imaging study of thin-film sections exposed to scCO$_2$ for a range of exposure times. That approach permitted the progression of crystallinity through the polymer film to be visualized as a function of time, as shown in Fig. 10.2. The

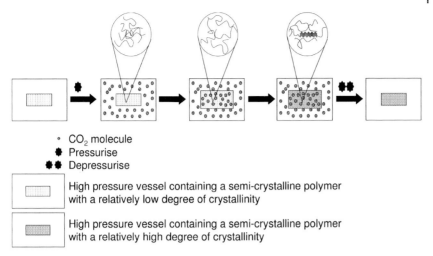

- • CO_2 molecule
- ✿ Pressurise
- ✿✿ Depressurise

High pressure vessel containing a semi-crystalline polymer with a relatively low degree of crystallinity

High pressure vessel containing a semi-crystalline polymer with a relatively high degree of crystallinity

Fig. 10.1 Schematic diagram of CO_2-induced crystallization.

Table 10.1 Summary of polymers crystallized by CO_2.

Polymer	Technique	Reference
Poly(ethylene terephthalate) (PET)	DSC	[168]
Poly(ethylene terephthalate) (PET)	X-ray diffraction, infrared spectroscopy and density measurements	[25]
Poly(ethylene terephthalate) (PET)	Quartz spring gravimetric measurements	[26]
Poly(ethylene terephthalate) (PET)	Infrared and FT-Raman spectroscopy	[28]
Poly(ethylene terephthalate) (PET)	DSC	[27]
Poly(ethylene terephthalate) (PET)	Confocal Raman microscopy	[29]
Poly(ethylene terephthalate) (PET)	FT-IR imaging	[30]
Poly(ethylene terephthalate) (PET)	DSC	[169]
Poly(ethylene terephthalate) (PET)	DSC and X-ray diffraction	[170]
Poly(ethylene terephthalate) (PET)	*In situ* FTIR spectroscopy	[171]
Bisphenol a polycarbonate	DSC	[33]
Methyl-substituted poly(aryl ether ether ketone) (MePEEK)	DSC and density measurements	[35]
Poly(ether ether ketone) PEEK	DSC	[172]
Syndiotactic polystyrene	DSC and X-ray diffraction	[31]
Syndiotactic polystyrene	*In situ* FT-IR spectroscopy	[32]
Poly(L-lactide)	High-pressure DSC and X-ray diffraction	[36]
tert-Butyl poly(ether ether ketone)	DSC	[34]

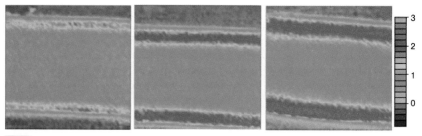

50 μm

Fig. 10.2 Transmission FTIR images (266×266 μm²) of cross-sections of PET film. The images are based on the distribution of the integrated absorbance of the band at 1340 cm⁻¹ after exposure to supercritical CO_2 for 15, 20 and 30 min (from left to right at 40 °C and 100 bar.

(Reprinted from Vibrational Spectroscopy, Vol 35, pp. 3–7, O. S. Fleming, K. L. A. Chan and S. G. Kazarian, FT-IR imaging and Raman microscopic study of poly(ethylene terephthalate) film processed with supercritical CO_2, Copyright (2004), with permission from Elsevier).

FTIR imaging technique, a novel and powerful method for polymer characterization, is discussed in detail in Section 3.3.

The facet of "morphological selectivity" is also demonstrated by CO_2 in its ability to induce a specific polymorph of syndiotactic polystyrene (sPS). The morphology of sPS is rather complex in that it may exhibit a gamut of polymorphs; α and β which possess planer structures and two helical forms denoted as the γ and δ phases. Handa and colleagues [31] investigated the effect of compressed CO_2 on the morphology of sPS by applying high-pressure DSC and X-ray diffraction. They observed a transition from glassy sPS to form the α-phase, and the β-phase was formed via a transition from the γ-phase under compressed CO_2. Kazarian et al. [32] applied *in situ* FTIR spectroscopy to investigate CO_2-induced morphology under supercritical conditions, which proved to be a route to the formation of the helical, nano-porous δ-phase. The use of scCO₂ is particularly attractive alternative to the traditional solvent-processing route, where the formation of the δ-phase is only possible via a two-stage process that requires extraction of the solvent from all clathrate forms.

CO_2-induced crystallization has also been observed for bisphenol A polycarbonate [33], *tert*-butyl poly(ether ether ketone) [34], and methyl substituted poly (aryl ether ether ketone) [35]. This CO_2-induced enhanced crystallization rate with increasing pressure has been explained in terms of the depression of T_g being far greater than the depression of the T_m. However, in contrast to the aforementioned polymers, it has been shown that exposure of poly(L-lactide) [36] and isotactic polypropylene [37] to CO_2 suppresses the crystallization rate. This observation was suggested to be due to the T_m being depressed to the same extent as the T_g, which prevents the formation of critical size nuclei. It has recently been shown by Hu et al. [38] that the crystallization of polycarbonate may be initiated under scCO₂ conditions using nano-scale clays.

CO_2-induced crystallization of polymers presents a unique, tunable degree of morphological control of the polymeric matrix, as revealed by CRM. Control over the operating conditions enables control of the diffusivity of CO_2 combined with control over the kinetics of crystallization. This degree of control enables the distribution of morphological changes in the polymer to be tailored, which may facilitate the formation of novel materials with a skin-core morphology with possible unique mechanical properties.

10.2.4
Interfacial Tension in CO_2/Polymer Systems

Interfacial tension in CO_2/polymer systems is an important parameter in applications such as the preparation of composite fibers [39], nucleation of polymer or void phases, and growth and coalescence of dispersed phases [40]. CO_2 acts to lower the interfacial tension in polymer systems by "shielding" unfavorable contacts at the interfaces between the system phases. The reduction of interfacial tension in a binary polymer/CO_2 system with increasing CO_2 density has been succinctly described by the "free energy densities of the two fluids becoming more alike" [40]. Interfacial tension is directly related to miscibility and responsible for (as well as viscosity) droplet breakup and coalescence during blending. However, the reduction of domain size in immiscible polymer blends due to exposure to high-pressure CO_2 has generally been interpreted in terms of viscosity reduction of one of the phases as a result of preferential affinity of CO_2 to one of the polymer phases.

Interfacial tension measurement techniques can be divided into two categories: equilibrium and transient methods [41]. The pendent-drop method is the most commonly applied method to measure interfacial tension under pressure and involves the measurement of density differences between two fluids and the equilibrium drop profile shape. In the following section, examples of interfacial tension reduction are presented for binary polymer/CO_2 systems and for polymer blends.

Harrison and colleagues [40] measured the interfacial tension between scCO_2 and polyethylene glycol (MW 600) using a tandem variable-volume pendent drop tensiometer, as shown schematically in Fig. 10.3. At 45 °C the interfacial tension between the PEG-CO_2 rich phase and the scCO_2 phase was reduced from 6.9 dyn cm^{-1} at 100 bar to 3.08 dyn cm^{-1} at 300 bar. Experimental observations were accurately predicted using a gradient model which utilized the lattice equation of state. In another piece of work, the effect of surfactants on polymer/CO_2 interfacial tension was addressed [42].

Li et al. [43] applied the pendent-drop method to study the interfacial tension of a PS/CO_2 system. A linear decrease in interfacial tension was observed in the temperature range 190–240 °C. Interfacial tension reduced from 24 to 12 dyn cm^{-1} up to a pressure of 70 atm, but at higher pressure the rate of reduction was suppressed. These observations were explained by two competing effects: interfacial tension reduction with temperature increase combined with a reduc-

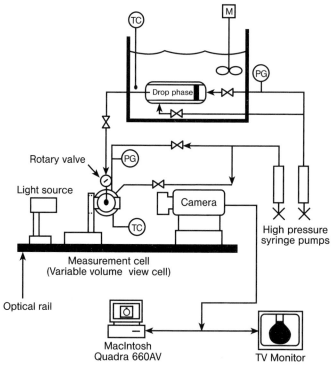

Fig. 10.3 Schematic diagram of a tandem variable-volume pendent drop tensiometer. (Reprinted with permission from Langmuir, Vol. 12, K. L. Harrison, K. P. Johnston and I. C. Sanchez, Effects of surfactants on the interfacial tension between supercritical carbon dioxide and polyethylene glycol, pp. 2637–2644, Copyright (1996) American Chemical Society).

tion in CO_2 solubility with increasing temperature. In the same study the interfacial tension in an immiscible blend of polystyrene and polypropylene saturated with CO_2 was measured. Experiments were recorded at 220 °C and pressures up to 165 atm for two molecular weights of polypropylene. In both cases, the interfacial tension was found to decrease by more than 20%. The data was successfully modeled at high pressure using the density gradient theory, but at low pressure the interfacial tension was severely underestimated.

The importance of interfacial tension reduction in polymer blends by CO_2 has been addressed by Xue et al. [44] The pendent-drop method was utilized to investigate the interfacial tension between PS/LDPE saturated with $scCO_2$ and compared to the same system in the absence of CO_2. At 200 °C the interfacial tension decreased from 6.62 mN m^{-1} at 0.1 MPa to 4.69 mN m^{-1} at CO_2 pressure of 9.2 MPa, corresponding to a 30% absolute reduction. This decreased interfacial tension was explained by the presence of dissolved CO_2 at the interface of the polymers reducing unfavorable interactions between the two phases and thereby enhancing the miscibility.

10.2.5

Diffusion of CO_2 in Polymers and Solutes in Polymers Subjected to CO_2

Above the T_g, the rate of polymer chain relaxation is faster than the diffusion of CO_2, and hence Fickian diffusion is to be expected. The diffusion of CO_2 is believed to occur within the amorphous domains of the polymer matrix, and for this reason the diffusion in semi-crystalline polymers may be more complex than it in the case for glassy polymers. In the case of semi-crystalline polymers, CO_2 is not soluble in the crystalline domains, and therefore the degree of crystallinity and hence the amorphous fraction available for CO_2 molecules may influence the diffusion characteristics. Furthermore, CO_2-induced crystallization is likely to lead to an increase in the tortuosity factor, and thus the diffusion path length may increase as a function of time. Syndiotactic polystyrene and poly(4-methyl-1-pentene) [45] are semi-crystalline polymers which have crystalline phases (helical in the case of sPS) with lower densities than that of the amorphous phase and are exceptions, as CO_2 access is not restricted to the amorphous domains, in fact CO_2 diffuses faster in the helical sPS than in the amorphous polymer [46].

The techniques available to study the diffusion of CO_2 through polymeric material include the barometric method, gravimetric method [47], and *in situ* FTIR spectroscopy [18]. Visual inspection of the polymer via an optical view cell enables the swelling characteristics to be assessed during the diffusion analysis [48]. It should be noted that diffusion coefficient determinations from these measurements rely on information from the total mass uptake of the polymer and hence do not permit region-specific determination, since the measurements correspond to the overall changes occurring throughout the total polymer sample. The diffusion coefficients of CO_2 for a range of solid and liquid polymers are summarized in Table 10.2.

The diffusion of CO_2 in PET has received much attention because of the industrially viable technology of $scCO_2$ dyeing, as described in detail in Section 10.4.1. Schnitzler and Eggers [49] implemented a gravimetric approach to study the mass transport of CO_2 in PET. Diffusion coefficient range was estimated at 3.5×10^{-10} to 5.3×10^{-9} m^2 s^{-1} with a temperature increase from 80 to 120 °C at various pressures. The sorption isotherm exhibited an S-bend shape, which is characteristic of a change in mass transport mechanism from dual mode sorption at low pressure and Fickian diffusion at higher pressures, also observed by Tang et al. for CO_2 in polycarbonate [50]. The sorption of CO_2 into PET was measure using *in situ* FTIR spectroscopy by Brantley et al. [18], and the diffusion followed Fickian behavior.

The plasticizing and swelling effect of CO_2 dissolved in the polymeric matrix can accelerate the diffusion of additives into the matrix. The "molecular lubrication" [20] provided by dissolved CO_2 molecules within the polymer matrix enables the diffusion of solutes to proceed in a less hindered manner than that in the absence of CO_2. The ease of diffusion of the solute through the matrix is due to the increase in the free volume as a result of the polymer being in a

Table 10.2 Diffusion coefficients for CO_2 in a range of polymers.

Polymer	Measurement method	Temperature (K)	Pressure (MPa)	Diffusion coefficient (cm^2/s)	Reference
Poly(butylenes succinate) (PBS)	Magnetic suspension balance	393.15 453.15 453.15	12.341 2.466 8.304	$1.23 \cdot 10^{-5}$ $2.04 \cdot 10^{-5}$ $2.68 \cdot 10^{-5}$	[173]
Poly(butylenes succinate-co-adipate) (PBSA)	Magnetic suspension balance	393.15 453.15 453.15	12.229 2.34 8.616	$0.95 \cdot 10^{-5}$ $2.06 \cdot 10^{-5}$ $2.05 \cdot 10^{-5}$	[173]
Gelatinized starch	Pressure decay	343 343 343	2.6 9.2 11.8	$7.5 \cdot 10^{-6}$ $1.9 \cdot 10^{-6}$ $0.9 \cdot 10^{-6}$	[174]
Polyamide 11	Swelling observation from a high-pressure view cell	488 488	10.3 37.9	$5.29 \cdot 10^{-5}$ $2.29 \cdot 10^{-5}$	[48]
Poly(vinyl acetate) (PVAc)	Magnetic suspension balance	313.15 313.15 313.15	0.329 3.039 6.626	$2.15 \cdot 10^{-8}$ $60.9 \cdot 10^{-8}$ $554 \cdot 10^{-8}$	[175]
Polystyrene (PS)	Magnetic suspension balance	373.15 423.15 473.15	8.320 8.319 8.420	$1.67 \cdot 10^{-6}$ $5.33 \cdot 10^{-6}$ $9.90 \cdot 10^{-6}$	[175]
High density polyethylene (HDPE)	Pressure decay method	453.2 453.2 453.2	3.965 9.743 13.98	$9.2 \cdot 10^{-5}$ $9.1 \cdot 10^{-5}$ $16.43 \cdot 10^{-5}$	[176]
Poly(methylmethacrylate) (PMMA)	Time absorption	313	10.5	$1.04 \cdot 10^{-6}$	[10]
Poly(chlorotrifluoroethylene) (PCTFE)	Time absorption	313	10.5	$7.08 \cdot 10^{-8}$	[10]
Cross-linked poly (dimethylsiloxane) (PDMS)	Time absorption	313	10.5	$7.08 \cdot 10^{-10}$	[10]
Low density polyethylene (LDPE)	Volumetric sorption technique based on pressure decay	423.15 423.15 423.15	0.654 1.564 2.997	$4.4 \cdot 10^{-5}$ $3.9 \cdot 10^{-5}$ $4.6 \cdot 10^{-5}$	[177]
PET	*In situ* FTIR spectroscopy	301.15 301.15 322.15 322.15	17.75 5.49 17.4 6.2	$8.2 \cdot 10^{-8}$ $5.4 \cdot 10^{-8}$ $2.2 \cdot 10^{-7}$ $1.8 \cdot 10^{-7}$	[18]

plasticized state combined with the possibility of the CO_2 molecules solvating the solute molecules and thus transporting them through the matrix. It should be noted that the ease of the diffusion is dependent on the extent of "molecular friction" occurring between functionalities of the diffusing solute and the polymer chains [20].

The diffusion mechanism by which the solute may diffuse into the polymer matrix is important and is dependent on the solubility of the additive in the CO_2 phase. In the case of the additive being soluble in the CO_2 phase, it is simply carried into the matrix, and hence deposition upon depressurization of the system is the mechanism which enables successful impregnation to occur. In the case of a solute having a low solubility in the CO_2 phase, successful impregnation may occur because of the affinity of the solute to the polymer, and hence the impregnation process occurs via partitioning of solute between polymer and CO_2 phases.

The CO_2 content in the matrix may be varied up to the limit of the equilibrium weight fraction at a certain pressure. This enables one to manipulate the polymer free volume and thus the diffusion characteristics of the solute may be "tuned" through temperature and pressure control. At low concentrations of dissolved CO_2 the polymer inter-chain distances are less than they are at higher concentrations, and thus the diffusion rate is lower. An increase in the weight fraction of CO_2 in PS at 60 °C from 4 to 11% enhances the diffusion of pyrene from 10^{-14} cm^2 s^{-1} to 10^{-10} cm^2 s^{-1}, revealed by steady state fluorescent measurements [51]. Similarly for the case of the diffusion of the relatively large molecules of decacyclene in PS, an increase in CO_2 weight fraction from 7 to 11% results in the diffusion coefficient increasing from 6.6×10^{-14} to 1.4×10^{-12} cm^2 s^{-1} [52]. This typical diffusion enhancement has been observed for phenol in poly(bisphenol A carbonate) [53], azo-dye/PMMA [54, 55], and azo-dye /PET [49, 56, 57].

The ability to control the diffusion rate simply by adjusting the operating conditions makes it possible to tailor the properties of the material, enabling the preparation of novel materials with a high degree of accuracy and customization. The control over solute diffusivity lends itself to the preparation of a surface-restricted distribution, as seen in examples of grafting and blending which are discussed in a later section.

10.2.6
Foaming

Polymeric foams have a closed-shell structure approximately 10 μm in diameter, with a cell density between 10^9 and 10^{15} cells cm^{-3}. These foamed materials have superior properties to their unfoamed counterparts in terms of enhanced ratio of flexural modulus to density and impact strength [58]. High-pressure CO_2 offers a particularly attractive alternative to traditional blowing agents because of its environmentally benign nature and relatively low cost. The phasing out of ozone-depleting substances associated with the Montreal Protocol acts as a substantial driver toward the acceptance of this novel technology. Polymers that have been foamed using CO_2 include PMMA [59], PVDF [60], PS [61], PS nanocomposite foams [62], PET [63], poly(ethylene-co-vinyl acetate) [64], styrene-co-acrylonitrile (SAN) [65], conductive polypyrrole/polyurethane foams [66], conductive elastomeric foams [67], and biodegradable macroporous scaffolds [68].

In the preparation of polymeric foams the polymer is saturated with CO_2 and hence the matrix is in a plasticized state. Rapid temperature ramping or depressurization results in CO_2 escaping from the material, which can cause nucleation, and as the T_g rises the foamed structure is "frozen". The processing route to these microcellular materials can be achieved in a continuous [61, 62] or discontinuous manner [69–73]. Rodeheaver and Colton [74] developed a model to predict the conditions required for the formation of open-cell microcellular foams in batch processes. Knowledge of the relationship of the T_g depression to pressure is vital in this application as it dictates the conditions required for cell nucleation and growth to occur [75].

Microcellular foams prepared via the discontinuous method involve a 3-stage process: polymer saturation with CO_2 at a temperature below the T_g, a transfer stage, and finally a rapid heating stage above the T_g. A characteristic feature of the foams prepared in this manner is the formation of a dense unfoamed skin layer [63, 72]. The appearance of the dense skin layer is due to CO_2 desorption from the film surface during the transfer stage. The thickness of this layer may be controlled by the time of the transfer period and the morphology of the polymer [63]. Krause et al. [70] studied the foaming of thin films of polysulfone (PSU), poly(ether sulfone) (PES), and cyclic olefin copolymer (COC) in a discontinuous manner. Control of the cell size distribution was influenced by the saturation pressure, with higher pressures corresponding to a reduction in cell size diameter and an increase in the cell size distribution. Furthermore, foaming was shown to occur within a finite temperature window in which the cell density passes through a maximum. The reason for the temperature dependence on the foam architecture is due to two competing processes: nucleation and growth of cells and the CO_2 solubility. The resultant foaming temperature window is depicted by the series of SEM micrographs presented in Fig. 10.4, which shows PSU film foamed at 180 °C and 200 °C and clearly depicts distinct differences in the foam architecture. The sample foamed at 200 °C shows a decrease in the cell density and an increase in the cell wall thickness due to the loss of CO_2 from the thin film before nucleation and cell growth can occur. The CO_2 content in the polymer may be used to manipulate the foam architecture, as shown with polysulfone/polyimide blends [71]. Closed-cell microcellular structures were formed at low concentration, and at high concentrations an open nano-porous foam was formed, as confirmed by gas permeation measurements.

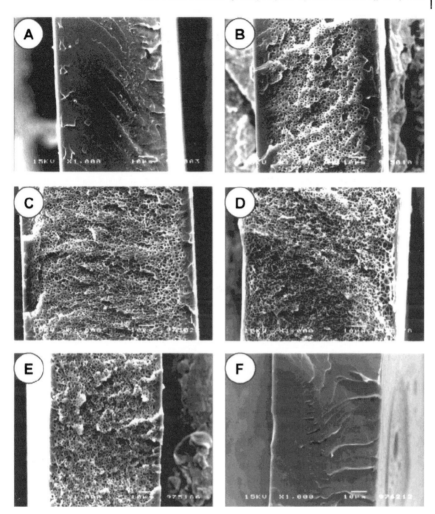

Fig. 10.4 SEM micrographs of PSU films satured at 50 bar and foamed at 70 (A), 100 (B), 130 (C), 160 (D), 190 (E), and 210 °C (F). The white horizontal bar indicates 10 μm. (Reprinted with permission from Macromolecules, Vol. 34, B. Krause, R. Mettinkhof, N. F. A. van der Vegt, and M. Wessling, Microcellular foaming of amorphous high-T_g polymers using carbon dioxide, pp. 874–884, Copyright (2001) American Chemical Society).

10.3
Rheology of Polymers Under High-Pressure CO$_2$

10.3.1
Methods for the Measurements of Polymer Viscosity Under High-Pressure CO$_2$

Viscosity measurement methods for polymers under high-pressure CO$_2$ may be divided into four classes: pressure driven, falling body, rotational devices, and vibrating wire. Accurate viscosity measurements require a homogeneous solution to be prepared, thus ensuring that phase separation does not occur [19, 76].

Pressure-driven devices include capillary viscometers and slit-die viscometers, in both of which the flow is driven by pressure. In the case of the capillary viscometer the pressure is generated by an upstream piston, and in the case of the slit-die viscometer flow is generated by an extruder. In both cases, measurements of pressure drop and flow rate are used to determine the viscosity. Both techniques have the inherent problem of pressure drop, which may result in phase separation. For this reason, the techniques are suitable for low-pressure measurements, which may mean that the polymer has not reached equilibrated CO$_2$ concentrations.

Capillary viscometers have the advantage of a relatively simple flow field and a small amount of sample required for the measurement. The technique yields an average steady-shear viscosity in a precise and reproducible manner. However, the technique is limited by the need for corrections for end effects (Bagley correction), and is only suitable for the study of relatively high shear rates. A further problem is that the pressure drop across the capillary affect the viscosity, and the capillary flow model equations assume a viscosity independent of pressure. However, the advantages of capillary viscometers are also associated with the slit-die rheometer, which involves the measurement of viscosity at a fixed shear rate. The technique has been widely utilized to study polymer/CO$_2$ systems at high shear rates [77–82], although a significant disadvantage of this technique is the pressure drop across the die which may severely affect the viscosity.

Falling weight viscometers and rolling ball viscometers are types of falling body devices. The common feature of these techniques is that a known body falls through a fluid under the influence of gravity. The viscosity of the fluid can be determined by the time required for the body to fall a known distance through the fluid. Falling body techniques are suited to low-viscosity systems because of the time required for the body to fall through a high-viscosity fluid (temporal stability of the measurement). Falling weight viscometers enable viscosity to be calculated from a single measurement. The application of falling weight viscometers has been limited to the study of very low-viscosity polymer/CO$_2$ systems [83, 84]. However, the simple design and ease of operation of the device combined with its high-pressure capability make the falling weight viscometer advantageous for the study of Newtonian fluids. The falling ball viscometer has been used to measure the viscosity reduction of PDMS in the pressure range of 1–3 MPa [85].

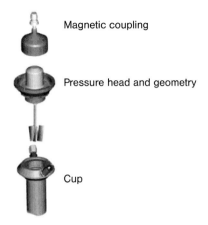

Magnetic coupling

Pressure head and geometry

Cup

Fig. 10.5 High-pressure attachment for the rheometer, consisting of a magnetic coupling, a pressure head, a rotating vane geometry, and a cup. (Reprinted with permission from Industrial and Engineering Chemistry Research, Vol. 42(25), N. M. B. Flichy, C. J. Lawrence, and S. G. Kazarian, Rheology of poly(propylene glycol) and suspensions of fumed silica in poly(propylene glycol) under high-pressure CO_2, pp. 6310–6319, Copyright (2003) American chemical Society).

Rotational viscometers are the most widely used devices for rheological determination. Continuous measurements are made at a given shear rate or shear stress for extended periods of time. A specific benefit with rotational viscometers is the ability to perform subsequent measurements on the same sample under different conditions. Specifically, the time dependency of shear rate may be determined. A particular feature of rotational viscometers is the approximately uniformly distributed shear rate throughout the fluid specimen, which gives good control of the fluid deformation, not possible with capillary rheometers. A specific issue relating to the operation and design of high-pressure devices is the seal required for the spinning shaft and the determination of the torque. Flichy et al. [86] adapted a concentric cylindrical rheometer for use with scCO₂. The accessory is capable of pressures up to 150 bar and temperatures up to 300 °C, and uses a magnetic coupling to transfer the torque from the instrument motor to the measurement system, as shown in Fig. 10.5. The elimination of wall slip effects was ensured by the incorporation of vane geometries which also provided good and fast mixing.

Royer et al. [87] developed a magnetically levitated sphere rheometer (MLSR) based on a falling sphere rheometer, and thus measurements are conducted at constant pressure. In essence, magnetic levitation is used to hold the sphere at a fixed point, and the cylindrical sample is chamber moved vertically with a motor, thus generating shear flow around the sphere. The change in magnetic force required to maintain the sphere in a stationary position is related to the shear stress. Specifically, the effects of pressure and concentration of CO_2 can be decoupled without the need for theoretical modeling.

10.3.2
Viscosity of Polymer Melts Subjected to CO_2

An important implication of the plasticizing nature of CO_2 is the associated viscosity reduction of the polymer bulk. An increase in the CO_2 concentration results in a reduction of viscosity due to the associated increase in free volume of the polymer and the "molecular lubrication" offered by CO_2. The ability to control the viscosity of the polymer with CO_2 concentration is well demonstrated by PS, for which the viscosity can be reduced between 25 and 80%, depending on the operating conditions [81]. The T_g depression enables polymers to be processed at milder temperature conditions compared to conventional melt processing. This provides a route to processing high-melt viscosity polymers [88] and enables thermally sensitive polymers to be manipulated without thermal degradation being a limiting factor. The processing temperature needs to be optimized because of its consequential reduction in CO_2 solubility and hence in viscosity [81]. The reduced operating temperature enables polymers such as acrylonitrile copolymer (65 wt% AN) to be processed while avoiding unfavorable crystallization and cross-linking that is associated with traditional melt processing [89].

CO_2-assisted viscosity reduction has been observed for the following polymer systems: PMMA [82], polypropylene [82], poly(vinylidene fluoride) [82], poly(dimethylsiloxane) [1, 85, 87, 90, 91], poly(ethylene glycol) [92–96], poly(ethylene glycol) nonphenyl ether [97], acronitrile copolymer (65 wt% AN) [89], polyamide 11 [48], low-density polyethylene (LDPE) [82, 98], poly(propylene glycol) and suspensions of fumed silica in poly(propylene glycol) [86], polystyrene [80, 81, 99], binary mixtures of polystyrene and toluene [84, 100, 101], and biomaterials [101]. Viscosity reduction has also been observed for the following blends: polyethylene and polystyrene [79, 102], polystyrene and PMMA [103], PMMA/rubber and polystyrene/rubber [104].

Bortner and Baird [89] utlilized a pressurized capillary rheometer to investigate the viscosity reduction of an acrylonitrile copolymer (65 wt% AN) under CO_2. The system was modified by the addition of a sealed chamber at the capillary exit to allow a static pressure to be applied, preventing CO_2 loss and thus ensuring that phase separation did not occur. Control of the static pressure was maintained with an adjustable pressure relief valve. At 6.7 wt% CO_2, the T_g was depressed by 31 °C (confirmed by DSC), with an associated viscosity reduction of 60%.

10.3.3
Implications for Processing: Extrusion

In the continuous method of extrusion, CO_2 is fed into the polymer melt and nucleation (and hence foaming) is initiated at the exit die. Pressure and temperature conditions at the exit die are controlled to result in supersaturation of the polymer. The inter-relationships between the key variables to control cell nucleation and growth in the continuous extrusion foaming process were summarized by Tomasko et al. [19]. Extrusion of polymers under CO_2 pressure enables operation to occur at reduced temperatures, facilitates the blending of polymer blends, and provides an environment for reactions to occur (reactive extrusion) [105].

Martinache et al. investigated the foaming of polyamide 11 using a slit-die extruder, producing a homogenous foam featuring well-distributed spherical closed shells with an average diameter of ca. 30 μm [48]. Garcia-Leiner and

Lesser [88] studied the continuous extrusion of a range of high-melt-viscosity polymers using a single screw extruder with a temperature-controlled die. Control of the foam morphology (cell density and cell size distribution) of polytetrafluoroethylene (PTFE), fluorinated ethylene propylene copolymer (FEP), and syndiotactic polystyrene (sPS) was possible through the control of the die temperature. Enhancement of nucleation was found to occur at reduced die temperatures, and coalescence of the foamed structure was evident at higher temperatures. Polystyrene has been continuously foamed in a two-stage single screw extruder by Han et al. [106]. CO_2 concentration was used to restrict the nucleation point within the die and to control the cell morphology (size and density). Han et al. [62] demonstrated that an introduction of ca. 5 wt% intercalated nanoclay was found to reduce the cell size from 25.3 to 11.1 μm and the density from 2.7×10^7 to 2.8×10^8 cells cm^{-3}, yielding a foam with an enhanced tensile modulus and improved barrier properties.

Siripurapu et al. [60] studied the continuous foaming of highly crystalline poly (vinylidene fluoride) (PVDF) at 175 °C with 2 wt% of scCO₂, which resulted in a foam structure with a heterogeneous cell size distribution and a low cell density, attributed to the low solubility of CO_2 in the polymer and therefore little CO_2 available for bubble formation. A further restriction is the semi-crystalline structure of PVDF. As the temperature falls below the T_m, crystallization occurs, which expels CO_2 into the amorphous domains and may lead to cracking of the cell walls and therefore formation of discrete particles. Improvements in the foam characteris-

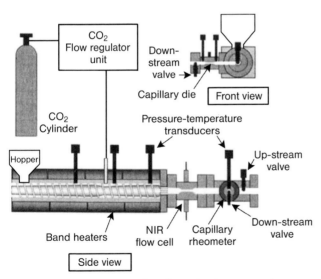

Fig. 10.6 Schematic diagram of capillary tube rheometer with a CO_2 injection system, a foaming extruder and an NIR sensor.
(Reprinted with permission of John Wiley & Sons, Inc. from measurement and prediction of LDPE/CO_2 solution viscosity, S. Areerat, T. Nagata and M. Oshima, Copyright (2002) Polymer Engineering and Science).

tics were achieved by blending PVDF with PMMA. The solubility of CO_2 in the blend was increased by 4% compared with that of neat PVDF because of the affinity of CO_2 to PMMA. A significant improvement in the foam morphology was observed with an increasing content of PMMA in the blend.

Areerat et al. [98] developed a capillary rheometer equipped with a foam extruder to measure the viscosity of a LDPE/CO_2 system. The design included an NIR sensor to monitor the CO_2 concentration within the polymer during processing. A schematic diagram of the equipment is shown in Fig. 10.6. Viscosity reduction was caused by an increase in CO_2 concentration, with 5% CO_2 corresponding to a viscosity reduction of 30% compared to that of the neat polymer.

10.4
Polymer Blends and CO_2

10.4.1
CO_2-Assisted Blending of Polymers

The role of CO_2 in assisting the blending of polymers may be viewed in one of the following ways:

- A transport medium used to impregnate a scCO_2-swollen matrix with an initiator and monomer, thus facilitating *in situ* polymerization (which may occur in the presence of CO_2 or following venting).
- A viscosity-reducing agent to facilitate the blending of high-melt-viscosity polymers.
- Route to the formation of polymer blends with a unique morphology, as demonstrated by Vega-González. et al. [107], who produced fibrous networks of PMMA/poly(ε-caprolactone) via a semi-continuous SAS process.

In situ polymerization to prepare immiscible blends was pioneered by Watkins and McCarthy [108], stimulating other researchers to apply this methodology to prepare novel polymer blends [109–112], fiber-reinforced composite materials[39], and electrically conducting composites [66, 67, 113–116]. Polymer blends produced in this manner include polystyrene/poly(vinyl chloride) [117, 118], polystyrene/PET [119], nanometer-dispersed polypropylene/polystyrene interpenetrating networks [120], polypropylene/polystyrene [121] and polyethylene/polystyrene [122]. The resultant polymer blend may have a unique morphology compared to the traditionally prepared counterpart (if it is feasible to prepare such a blend via conventional procedures) and therefore demands a thorough investigation.

Kung and colleagues [123] prepared a polystyrene/polyethylene blend via the radical polymerization of styrene within a scCO_2-swollen high-density polyethylene matrix. Polystyrene was localized within polyethylene spherulites, which served as reinforcement in a "scaffolds" type manner. The resultant blend exhibited a significant increase in modulus and strength, but had the associated disadvantage of a loss in fracture toughness.

The diffusion of the monomer and initiator within the $scCO_2$-swollen matrix determines the extent to which the polymerization and hence the blending of the polymers may occur, i.e. surface modification may occur whilst the bulk remains unchanged. This surface-specific blending of polymers was demonstrated by Muth and coworkers [112], who impregnated and polymerized methacrylic acid within a $scCO_2$-swollen PVC matrix to a depth of ca. 180 µm. The specific modification of a polymer surface is obtainable through grafting [119, 122, 124–127], which involves the surface modification of the bulk polymer with functional groups.

The ability of CO_2 to induce morphological alterations in polymers requires examination of the morphology of the blends prepared using CO_2. Zhang et al. [128] investigated the morphology of a (53/47 w/w) ultra-high-molecular-weight polyethylene (UHMWPE)/PMMA blend, utilizing *tert*-butyl peroxybenzoate as the initiator. Tapping mode atomic force microscopy (TMAFM) was used to investigate the morphological structure of the resultant blend, revealing the presence of three distinct phases: crystalline UHMWPE, amorphous UHMWPE, and domains of amorphous PMMA, which were estimated to range in size between 10 and 100 nm. Furthermore, the CO_2-assisted polymerization was found to increase the crystallinity of UHMWPE from 50.4% in the starting material to 59.2% following the polymerization blending. Busby et al. [129] applied a combination of FTIR spectroscopy, DSC, and AFM to characterize nano-structured blends of polymethacrylates in UHMWPE, prepared by *in situ* polymerization. Methacrylates were found to exist in dispersed nano-scale domains, which have never previously been prepared via conventional routes. An increase in methacrylate fraction resulted in the fragmentation of the UHMWPE lamellae into low T_m crystallites. Moreover, the steric hindrance of the methacrylate side chain, depending on its length, was found to control the extent of disruption to the lamellae and ultimately the degree of fragmentation.

Shieh et al. [130] probed the CO_2-induced morphological alterations in a poly(ethylene oxide)/poly(methyl methacrylate) blend using small-angle X-ray scattering (SAXS). The crystal and amorphous layer thickness of PEO was found to increase following exposure to CO_2, and the extent of the layer thickening was found to be controlled by the fraction of PMMA in the blend. The increase in thickness of the lamellae was attributed to recrystallization of the PEO domains.

The preferential interaction of CO_2 with individual phases of polymer blends is dictated by the affinity of CO_2 to the individual components. Zhou et al. [131] applied atomic force microscopy (AFM) and phase contrast microscopy to study the effect of $scCO_2$ on the surface structure and phase morphology of PMMA/PS (50/50 w/w) thin films. AFM was used to monitor the return to thermodynamic equilibrium of the cast film as a function of temperature, pressure, and exposure time. Initially, PMMA appeared as a dispersed surface phase elevated from a continuous PS matrix. The series of AFM images in Fig. 10.7 indicate that the T_g of PMMA/PS blend at 20 MPa is <60 °C, and at a temperature of 70 °C the T_g occurs at a pressure greater than 20 MPa, as seen by the initially elevated PMMA phase falling into the PS phase.

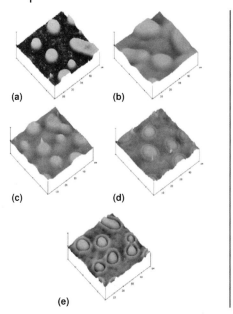 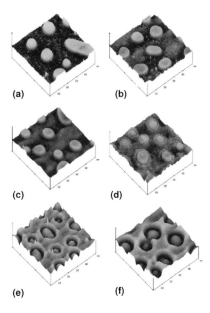

Fig. 10.7 Three-dimensional AFM images (50×50 μm^2) of PMMA/PS (50/50 w/w) blend thin films. Images on the left were treated for 2 h in scCO2 (20 MPa) at different temperatures: (a) as cast film, (b) 40 °C, (c) 50 °C, (d) 60 °C, (e) 70 °C. Images on the right were treated for 2 h in scCO$_2$ (70 °C) at different pressures: (a) as cast, (b) 5 MPa, (c) 10 MPa, (d) 20 MPa, (e) 30 MPa, (f) 40 MPa. (Reprinted from Journal of Supercritical Fluids, Vol. 26, H. Zhou, J. Fang, J. Yang and X. Xie, Effect of supercritical CO$_2$ on the surface structure of PMMA/PS blend thin films, pp. 137–145, Copyright (2004), with permission from Elsevier).

10.4.2
CO$_2$-Induced Phase Separation in Polymer Blends

Polymer blends may be characterized in terms of the temperature dependence of the Flury-Huggins interaction parameter (χ). In the case of an upper critical solution temperature (UCST) blend, χ decreases with temperature, and the blend remains miscible. For phase separation to occur in a UCST blend, the temperature must be lower than the critical solution temperature. In the case of a lower critical solution temperature (LCST) blend, χ increases with temperature, and thus phase separation occurs above the critical solution temperature. The ability of CO$_2$ to mimic heat means that miscibility is enhanced in the case of UCST blends, and for the case of LCST blends the miscibility is depressed. Ramachandrarao et al. [132] explained this phenomenon by postulating a dilation disparity occurring at higher CO$_2$ concentration as a result of the preferential affinity of CO$_2$ to one of the components of the blend, inducing free-volume and packing disparity.

The effect of CO$_2$ sorption on LCST blends is to depress the lower critical solution temperatures of the polymer system. Watkins et al. [133] observed phase sep-

40°C, 1 bar   40°C, 60 bar

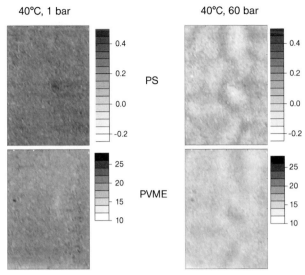

Fig. 10.8 ATR-IR images of PS and PVME blend before and after exposure to 60 bar CO$_2$. The top line of images corresponds to PS and the bottom images to PVME. The size of each image is 820×1140 µm^2. (Reprinted with permission from Macro-molecules, Vol. 37, S.G. Kazarian and K.L.A. Chan, FTIR imaging of polymeric materials under high-pressure carbon dioxide, pp. 579–584, Copyright (2004) American Chemical Society).

aration of homogeneous blends of poly(deuterated styrene)/poly(vinyl methyl ether) more than 115 °C below the ambient pressure LCST in the presence of 3.3 wt% CO$_2$. High-pressure fluorescence spectroscopy was applied by Ramachandrarao et al. [134] to study various compositions of polystyrene/poly(vinyl methyl ether). Phase separation was found to occur as much as 90 °C below the coexistence temperature for the corresponding binary blends under ambient conditions in the presence of ca. 3.5 wt% CO$_2$. The depression of LCSTs has also been observed for the blends of deuterated polystyrene/poly(n-butyl methacrylate) [133] and deuterated polybutadiene/polyisoprene [132]. Kazarian and Chan [135] applied *in situ* FTIR imaging to visualize the phase separation of a (50/50 w/w) homogeneous LCST PS/PVME polymer blend, observing a domain size of ca. 200 µm. Fig. 10.8 shows the resultant images before and after phase separation; details of the imaging technique employed can be found in Section 3.3.

The presence of CO$_2$ in UCST polymer blends has been shown to enhance blend miscibility. Walker et al. [136] observed a depression of the cloud point temperature for low-molecular-weight blends of polystyrene/polyisoprene using a combination of visual inspection, small-angle neutron scattering, and spectrophotometry. In the presence of 13.8 MPa of CO$_2$, a reduction of the cloud point was observed compared to the same system in the absence of scCO$_2$. This demonstrates the ability of CO$_2$ to promote miscibility in UCST blends, and consequently an increase in the processing window is available. Walker et al. [137]

performed a thermodynamic analysis of the UCST blend of PDMS/PEMS in the presence of high-pressure CO_2.

10.4.3
Imaging of Polymeric Materials Subjected to High-Pressure CO_2

This section introduces a novel application of IR spectroscopy, namely IR imaging, and the specific sampling technique of attenuated total reflectance (ATR). FTIR imaging in ATR mode allows one to visualize the spatial distribution of different components in polymeric materials and to study directly the effect of high-pressure CO_2 on this distribution. This novel approach should benefit polymer scientists studying polymer blends and their processing with scCO$_2$.

The ATR measurement mode is achieved by placing the sample onto a crystal that is IR-transparent (e.g. diamond) with a high refractive index and to allow the IR light to contact the crystal at an angle above the critical angle, resulting in total internal reflection. At the point of total internal reflection, the IR light penetrates the sample with an evanescent wave which decays exponentially into the sample. The depth of penetration is on the order of a few micrometers and is dependent on the refractive indices of the crystal and the sample and the wavelength of the light. Because of the relatively shallow depth of penetration, ATR is a very versatile technique that allows the sampling of most liquid and solid samples. The accessory used in the imaging under pressure application is the Golden Gate (Specac), which has a diamond as the ATR crystal and has been adapted to enable *in situ* measurements under supercritical conditions. This accessory has been used to study phenomena such as polymer swelling [138] and the decomposition of PET subjected to near-critical water [139].

FTIR imaging with the focal plane array (FPA) detector enables one to record spatially resolved IR spectra. The FPA records thousands of IR spectra simultaneously, making the study of a dynamic system possible. Chemical images that represent the distribution of different components can be generated based on the chemical specificity of different functional groups in the mid-infrared region. FTIR imaging using the Golden Gate accessory, which is not specially designed for imaging applications, was developed at Imperial College London by Chan and Kazarian [140]. This allows the combination of the high-pressure cell and FTIR imaging which makes possible a qualitative chemical imaging analysis of polymer systems exposed to high-pressure and supercritical environments [135]. A schematic diagram of the high-pressure cell can be seen in Fig. 10.9.

This accessory was used to image, *in situ*, the phase separation of a homogeneous LCST PS/PVME (50/50 w/w) polymer blend. A homogeneous mixture was cast directly onto the diamond, which was the ATR crystal used for the measurements. The resultant FTIR images are presented in Fig. 10.8 (Section 3.2), showing the distribution of both PS and PVME, before and after exposure to 60 bar of CO_2. Following exposure to CO_2 it can be seen that phase separation occurs, resulting in domains of ca. 200 μm [135].

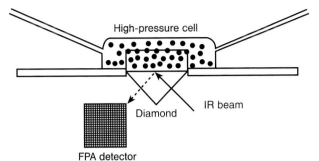

Fig. 10.9 Schematic diagram of the high-pressure ATR-FTIR spectroscopic accessory used to measure polymers under high-pressure and scCO_2. (Reprinted with permission from Macromolecules, Vol. 37,

S.G. Kazarian and K.L.A. Chan, FTIR imaging of polymeric materials under high-pressure carbon dioxide, pp. 579–584, Copyright (2004) American Chemical Society).

This approach was also applied to study the simultaneous sorption of CO_2 into PMMA and PEO under identical conditions. ATR-FTIR spectroscopy enables the extent of polymer swelling to be qualitatively assessed, the measurement probing a finite depth within the sample (wavelength dependent). Thus, a reduction in absorbance can be used to interpret the extent of swelling. A sharp interface between the two polymers was created by first casting PMMA from solution onto half of the diamond; removal of the solvent was then achieved by heating the sample at 60 °C and was confirmed by the elimination of spectral bands associated with the solvent from the spectra. PEO was then melted onto the remaining half of the diamond at 60 °C. The characteristic IR signatures associated with the individual components in the ternary system allow their distribution to be visualized simultaneously, as shown in Fig. 10.10. The images show that sorption of CO_2 at a pressure of 50 bar is greater in PMMA than in PEO, which shows a smaller amount of CO_2. The associated swelling of PMMA by CO_2 is interpreted from the reduction in absorbance of PMMA. Following the temperature increase to 50 °C the solubility of CO_2 decreases and consequently the extent of swelling is reduced, as seen by the increase in PMMA absorbance and the reduction in CO_2 absorbance. PEO exhibited characteristic spectral features associated with the molten state, indicating that the dissolved CO_2 molecules act to reduce the T_m. After 15 min exposure the CO_2 sorption was seen to increase in PEO, especially at close proximity to the interface. After 45 min exposure at 50 bar and 50 °C, CO_2 has equilibrated and is shown to have the greatest concentration in PEO [135]. This example shows that in situ imaging allows one to analyze simultaneously several phenomena occurring in two polymer samples, such as CO_2 sorption, polymer swelling, and melting. This points to the possibility of high-throughput chemical analysis of many samples under high-pressure CO_2.

This imaging technique has tremendous potential for the measurement of polymer systems under the influence of CO_2. The ability to visualize the indi-

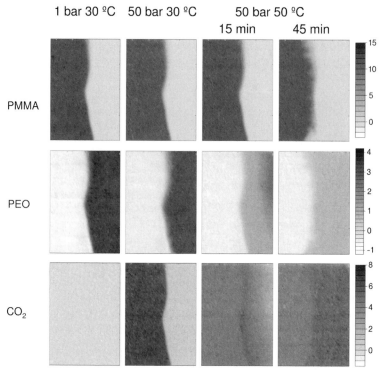

Fig. 10.10 ATR-IR images of PMMA/PEO system. The top row shows images based on the spectral band of PMMA, the middle row shows images based on the spectral bands of PEO, and the bottom row shows images based on CO_2 dissolved in the polymers. (Reprinted with permission from Macromolecules, Vol. 37, S. G. Kazarian and K. L. A. Chan, FTIR imaging of polymeric materials under high-pressure carbon dioxide, pp. 579–584, Copyright (2004) American Chemical Society).

vidual components within a dynamic system provides a means of studying the effects of processing conditions on polymer systems. Furthermore, this technique gives the possibility of studying diffusion processes in several polymer samples simultaneously under pressure, and ensures the reliability of a comparative analysis of these samples.

10.5
Supercritical Impregnation of Polymeric Materials

Polymer impregnation involves the introduction of a guest doping solute into a host polymer matrix. The SCF impregnation of polymeric materials offers a particularly attractive processing route due to the "solvent free" nature of the technology. Once the system is depressurized, CO_2 quickly escapes from the polymer and the guest solute remains "trapped" in the matrix without the presence of CO_2 in the finished product. Supercritical carbon dioxide has been successfully used to impregnate various dyes, drugs, and metal complexes into polymer hosts.

10.5.1
Dyeing of Polymeric Materials

Traditionally, the textile industry dyes hydrophobic polymer fibers from an aqueous medium that includes surfactants and dispersing agents to assist the dyeing process. Not only does this method require a clean water source, but the effluent water stream needs post processing treatment to remove the liquor constituents. CO_2 is able to swell hydrophobic fibers by penetrating into them and thus allow the dye to penetrate the polymer matrix. The use of $scCO_2$ technology as a replacement for the traditional use of water as the dyeing medium offers substantial environmental advantages [141], and has been the subject of recent reviews [141–143]. Structural modifications as a result of the dyeing process have been investigated and showed that the migration of oligomers to the surface of PET fibers and the shrinkage of polyolefin fibers (history dependant) occurs [144]. The use of supercritical CO_2 technology as a replacement for the traditional aqueous method of dyeing polyesters is a viable alternative that has grown in international interest since 1995 [142] and operates under the principles demonstrated by Berens [145]. An obvious driver for $scCO_2$ dyeing is the environmental benefit that has been an underlying feature of all of the applications described in this chapter. Additionally, the use of CO_2 results in a great gain in the rate of diffusion of dye through the matrix [55, 56].

Initially it was deemed that high dye solubility in the CO_2 phase was required to enable impregnation to proceed via a deposition mechanism, which stimulated investigation of solubility characteristics [146, 147]. However, successful dyeing has been demonstrated for solubility as low as 10^{-6} M as a result of the affinity of the dye to the polymer, thus facilitating impregnation via partitioning [20, 55]. This impregnation mechanism is beneficial for the dyeing of polymeric materials, since at any one time there is a very low concentration of dye in the CO_2 phase and hence dye loss is minimized.

The diffusion mechanism under supercritical conditions has been studied by methods including the film roll technique [56, 57] and the gravimetric approach [49], demonstrating that the diffusivity is dye-specific and may be "tuned" with temperature and pressure control. West et al. [55] revealed that the functionalities on the diffusing dye molecules need to be taken into account because of so-called "molecular friction" (intermolecular interactions such as H-bonding) that can occur between the functional groups of the dye and the polymer matrix. This was demonstrated by the diffusion rate of 4,4'-(dimethylamino)nitroazobenzene (DENAB) in PMMA, which is greater than that of Disperse Red1 (DR1) because of the hydroxyl groups in DR1 interacting with the carbonyl functionalities of PMMA. Dye diffusion in PMMA has also been monitored by *in situ* UV-vis [20].

The spatial distribution of the dye within the polymer provides information required to model the dyeing process. Confocal Raman microscopy (using an oil immersion objective) has emerged as a powerful technique to obtain accurate profiles of the dye distribution as a function of depth [148]. The technique is specifically suited to this application because of the high Raman activity of azo-dyes used in the dyeing process and thus enables the dye to be detected at low concentrations. Depth profiling of polymers dyed from a supercritical solution have also been achieved with the use of photoacoustic (PA) spectroscopy [149].

10.5.2
Preparation of Materials for Optical Application

Materials with non-linear optical (NLO) properties are of particular interest to the telecommunications industry and specific to the technology of photonics [150]. The impregnation of polymeric materials with organic, azobenzene dyes is a possible route to the preparation of these materials. Poling techniques are required to avoid the polar molecules aligning antiparallel to each other, which avoids centro-symmetric arrangements [151]. A thorough review of the preparation and assessment of polymeric materials with azo-dyes for non-linear optics has recently been prepared by Yesodha et al. [152].

The use of CO_2 to prepare such materials involves the plasticization of the polymer matrix followed by the impregnation of the chromophore. The application of an electric field results in the alignment of the dye molecules, and this is followed by depressurization of the system, which results in the rapid escape of CO_2 and the consequential "freezing" of the dye alignment in the matrix. A specific benefit of this method is that the rapid release of CO_2 enables the matrix to be "frozen", and hence the dye molecules do not have time to reorientate [76]. A general problem with the polymeric guest-host system is the instability of the system due to polymer chain relaxation, which can result in the loss of the necessary alignment. Supercritical fluid treatment of such materials allows one to process them at lower temperatures because of plasticization and thus possibly to achieve better orientation of the dyes in the polymer matrix.

10.5.3
Preparation of Biomaterials and Pharmaceutical Formulations

The "solvent free" nature of CO_2 makes it an ideal choice for the processing of pharmaceutical materials, which require stringent control over product integrity and the absence of harmful solvent residues. Retention of the bioactivity of the active component in the formulation is a prerequisite for the processing route of pharmaceutical formulations. The ability to plasticize and reduce the melting temperature [153–155] of polymers enables the thermally labile active components to be processed, thereby avoiding the possibility of thermal degradation and enabling the bioactivity to remain following release from the formulation.

The reduction of polymer viscosity also facilitates the mixing of the polymer with an active component [156]. The ability of CO_2 to foam polymers provides a convenient method to prepare scaffolds for the application of tissue engineering, providing an environment able to promote the differentiation and growth of tissue. Particle formation [157] provides a route to prepare formulations with controlled release characteristics via encapsulation. The use of CO_2 also provides a convenient method to deliver coatings to pharmaceutical formulations [158]. A detailed discussion of polymer processing for various pharmaceutical applications has recently been published [159].

One of the interesting applications where $scCO_2$-processing of polymeric materials is beneficial is the method of PGSS (Particles from Gas Saturated Solution). PGSS is a technique able to form polymer/active compound foams, solid particles, or droplets [160]. The principle of the technique is for CO_2 to form a gas-saturated solution/suspension which may then be foamed or passed through a nozzle to produce solid particles or droplets. The technique is suited to thermally labile components, since the process is undertaken at near ambient temperature. Control of particle size has been achieved by the introduction of N_2 back-pressure in the collection chamber, as demonstrated by Hao et al. [161] with poly(DL-lactic acid). The PGSS method has been shown to enhance the dissolution characteristics of nifedipine from PEG 400 [162, 163].

Another use of $scCO_2$-processed polymers is in tissue engineering. 3-D scaffolds for tissue engineering applications require an environment that is conducive to cell attachment and cell growth, an interconnected/vascularized architecture to facilitate the transport of the necessary components into and out of the scaffold, and the mechanical compatibility for the specific tissue. The foaming of polymeric materials with $scCO_2$ is an effective, "clean" processing route to the formation of 3-D scaffolds. Successful applications involve the preparation of scaffold materials for the delivery of specific factors such as chondrocytes [164], providing the required environment to promote osteogenesis [165] and the ability to deliver a range of growth factors [166]. The cellular structure of polymers foamed using CO_2 often results in closed-shell morphology. This is disadvantageous for 3-D scaffolds as it does not fulfill the requirement of interconnectivity, and thus the diffusion of the required factors into and out of the scaffold is restricted. This limitation may be overcome with the additional step of particulate leaching [167]. This method also gives control over porosity and mechanical strength by adjustment of acid-base gas-evolving reaction between the two salts.

The formation of a solid dispersion may be achieved by the impregnation of an active component from a supercritical solution. The benefit of this approach is that it often results in the active component being molecularly dispersed within the host matrix [5], which is advantageous compared to the reduced bioavailability in the crystallized form which may be formed in the case of traditional methods. Furthermore, this study [5] proved that the impregnation of a drug from a supercritical solution reduces the water uptake into the formulation as a result of the interaction between the polymer and the drug, which helps to extend the shelf life of the product. The topic of supercritical fluid impregnation of polymers for drug delivery is discussed in more detail in the recent book mentioned above [159].

10.6
Conclusions and Outlook

The objective of this chapter was to provide an overview of polymer processing with supercritical fluids and to underline some of the key advantages of this type of processing compared to the conventional means of polymer processing. Bearing in mind the costs associated with processing under high pressures, these advantages must be capable of adding significant value to the processed products. Recent studies demonstrate that the potential of supercritical fluids in polymer processing is being realized. The key advantage of the use of $scCO_2$ as a "temporary" plasticizer in polymer processing lies in the ability of CO_2 to weakly interact with the functional groups in polymers. This leads to a reduction of the glass transition temperature, reduction of melting temperature in some polymers, polymer swelling, and the facilitation of solute mass transport within polymer matrices. The number of applications discussed in this chapter is by no means exhaustive, as there are numerous recent articles and reviews dedicated to specific issues in $scCO_2$ polymer processing. It is hoped that this chapter will rather provide a summary of some key recent developments which have exploited advantages of the use of $scCO_2$.

In 2000 we attempted to predict some future developments in this field and suggested a number of possible breakthroughs [76]. Interestingly, while most of these breakthroughs have now been achieved, the realization of some of these speculations has actually exceeded expectations. Thus, although general development of novel *in situ* characterization techniques was expected, the ability of "chemical photography" using spectroscopic imaging of polymeric materials under pressure has emerged as a powerful innovative method with great potential. In particular, the ability to obtain "chemical snapshots" of a number of different polymers subjected to high-pressure CO_2 promises to improve the reliability of analysis, accelerate laborious high-pressure investigations, and contribute to studies that were not previously possible. The proposed development of new surfactants for polymer processing in $scCO_2$ has recently seen a significant increase in activity with several exciting new results. New speculative developments in $scCO_2$-assisted extrusion of polymeric materials have been realized in part, although $scCO_2$-assisted extrusion of ceramic pastes and many biodegradable polymers has yet to be realized. Given the benign solvent nature of $scCO_2$, it is not surprising that the expected pharmaceutical and biomedical applications of this technology have resulted in a number of interesting and useful studies and developments. This chapter has highlighted some of these recent developments, and it is hoped that this book will facilitate further application of supercritical fluids in polymer processing.

Acknowledgements

We thank EPSRC for support and Dr. K. L. A. Chan for help and advice.

Notation

Abbreviations

PMMA	Poly(methylmethacrylate)
PDMS	Poly(dimethylsiloxane)
PETG	Poly(ethylene terephthalate) glycol modified
PES	Poly(ether sulfone)
COC	Cyclic olefin copolymer
SAN	Styrene-co-acrylonitrile
PTFE	Polytetrafluoroethylene
FEP	Fluorinated ethylene propylene copolymer
PVDF	Poly(vinylidene fluoride)
LDPE	Low density polyethylene
sPS	Syndiotactic polystyrene
PS	Polystyrene
PSU	Polysulfone
UHMWPE	Ultra-high-molecular-weight polyethylene
NIR	Near infrared
Å	Angstrom
FTIR	Fourier transform infrared spectroscopy
wt%	Weight percent
atm	Atmosphere
DSC	Differential scanning calorimetry
SEM	Scanning electron micoscopy
MW	Molecular weight
sc	Supercritical
SAS	Supercritical antisolvent
PVC	Poly(vinylchloride)
TMAFM	Tapping mode atomic force microscopy
SAXS	Small-angle X-ray scattering
AFM	Atomic force microscopy
Pa	Pascal
CRM	Confocal Raman microscopy
PGSS	Particles from gas-saturated solution

Symbols

T_g	Glass transition temperature
T_m	Melting temperature

References

1 L.J. Gerhardt, C.W. Manke, and E. Gulari, *J. Polym. Sci., Part B: Polym. Phys.* (**1997**), 35, 523–534.

2 W.J. Koros, *J. Polym. Sci., Part B: Polym. Phys.* (**1985**), 23, 1611.

3 J.H. Aubert, *J. Supercrit. Fluids* (**1998**), 11, 163–172.

4 Z. Shen, M.A. McHugh, J. Xu, J. Belardi, S. Kilic, A. Mesiano, S. Bane, C. Karnikas, E. Beckman, and R. Enick, *Polymer* (**2003**), 44, 1491–1498.

5 S.G. Kazarian and G.G. Martirosyan, *Int. J. Pharm.* (**2002**), 232, 81–90.

6 Y.T. Shieh, J.T. Su, G. Manivannan, P.H.C. Lee, S.P. Sawan, and W.D. Spall, *J. Appl. Polym. Sci.* (**1996**), 59, 707–717.

7 Y.T. Shieh, J.H. Su, G. Manivannan, P.H.C. Lee, S.P. Sawan, and W.D. Spall, *J. Appl. Polym. Sci.* (**1996**), 59, 695–705.

8 Y. Zhang, B.P. Cappleman, M. Defibaugh-Chavez, and D.H. Weinkauf, *J. Appl.Polym. Sci: Part B: Polym. Phys.* (**2003**), 41, 2109–2118.

9 Y. Zhang, K.K. Gangwani, and R.M. Lemert, *J. Supercrit. Fluids* (**1997**), 11, 115–134.

10 K.F. Webb and A.S. Teja, *Fluid Phase Equilib.* (**1999**), 158–160, 1029–1034.

11 B. Wong, Z. Zhang, and P. Handa, *J. Polym. Sci., Part B: Polym. Phys.* (**1998**), 36, 2025–2032.

12 N. von Solms, J.K. Nielsen, O. Hassager, A. Rubin, A.Y. Dandekar, S.I. Andersen, and E.H. Stenby, *J. Appl. Polym. Sci.* (**2004**), 91, 1476–1488.

13 M.R. Nelson and R.F. Borkman, *J. Phys. Chem. A* (**1998**), 102, 7860–7863.

14 J.R. Fried and N. Hu, *Polymer* (**2003**), 44, 4363–4372.

15 I. Kikic, M. Lora, L.A. Cortesi, and P. Sist, *Fluid Phase Equilib.* (**1999**), 158–160, 913–921.

16 M. Hamedi, V. Muralidharan, B.C. Lee, and R.P. Danner, *Fluid Phase Equilib.* (**2003**), 204, 41–53.

17 S.G. Kazarian, M.F. Vincent, F. Bright, V., C.L. Liotta, and C.A. Eckert, *J. Am. Chem. Soc.* (**1996**), 118, 1729–1736.

18 N.H. Brantley, S.G. Kazarian, and C.A. Eckert, *J. Appl. Polym. Sci.* (**2000**), 77, 764–775.

19 D.L. Tomasko, H. Li, D. Liu, X. Han, M.J. Wingert, L.J. Lee, and K.W. Koelling, *Ind. Eng. Res.* (**2003**), 42, 6431–6456.

20 S.G. Kazarian, N.H. Brantley, B.L. West, M.F. Vincent, and C.A. Eckert, *Appl. Spectrosc.* (**1997**), 51, 491–494.

21 T. Miyoshi, K. Takegoshi, and T. Terao, *Macromolecules* (**1997**), 30, 6582–6585.

22 T.L. Sproule, J.A. Lee, H. Li, J.J. Lannutti, and D. Tomasko, *J. Supercrit. Fluids* (**2004**), 28, 241–248.

23 P. Alessi, A. Cortesi, I. Kikic, and F. Vecchione, *J. Appl. Polym. Sci.* (**2003**), 88.

24 I. Kikic, F. Vecchione, P. Alessi, A. Cortesi, and F. Eva, *Ind. Eng. Chem. Res.* (**2003**), 42, 3022–3029.

25 K. Mizoguchi, T. Hirose, Y. Naito, and Y. Kamiya, *Polymer* (**1987**), 28, 1298–1231.

26 S.M. Lambert and M.E. Paulaitis, *J. Supercrit. Fluids* (**1991**), 4, 15–23.

27 M. Takada and M. Oshima, *Polym. Engin. Sci.* (**2003**), 43, 479–489.

28 S.G. Kazarian, N.H. Brantley, and C.A. Eckert, *Vib. Spectrosc.* (**1999**), 19, 277–283.

29 O.S. Fleming and S.G. Kazarian, *Appl. Spectrosc.* (**2004**), 58, 391–394.

30 O.S. Fleming, K.L.A. Chan, and S.G. Kazarian, *Vib. Spectrosc.* (**2004**), 35, 3.

31 Y.P. Handa, Z. Zhang, and B. Wong, *Macromolecules* (**1997**), 30, 8499–8509.

32 S.G. Kazarian, C.J. Lawrence, and B.J. Briscoe, *Proceedings SPIE* (**2000**), 4060, 210.

33 E. Beckman and R.S. Porter, *J. Polym. Sci., Part B: Polym. Phys.* (**1987**), 25, 1511–1517.

34 P. Handa, Z. Zhang, and J. Roovers, *J. Polym. Sci., Part B: Polym. Phys.* (**2001**), 39, 1501–1512.

35 Y.P. Handa, J. Roovers, and F. Wang, *Macromolecules* (**1994**), 27, 5511–5516.

36 M. Takada, S. Hasegawa, and M. Oshima, *Polym. Eng. Sci.* (**2004**), 44, 186–196.

37 M. Takada, M. Tanigaki, and M. Oshima, *Polym. Eng. Sci.* (**2001**), 41, 1938–1946.

38 X. Hu and A. J. Lesser, *Polymer* (**2004**), 45, 2333–2340.

39 T. C. Caskey, A. J. Lesser, and T. J. McCarthy, *J. Appl. Polym. Sci.* (**2003**), 88, 1600–1607.

40 K. L. Harrison, K. P. Johnston, and I. C. Sanchez, *Langmuir* (**1996**), 12, 2637–2644.

41 P. Xing, M. Bousmina, and D. Rodrigue, *Macromolecules* (**2000**), 33, 8020–8034.

42 K. L. Harrison, S. R. P. da Rocha, M. Z. Yates, and K. P. Johnston, *Langmuir* (**1998**), 14, 6855–6863.

43 H. Li, J. Lee, and D. Tomasko, *Ind. Eng. Chem. Res.* (**2004**), 43, 509–514.

44 A. Xue, C. Tzoganakis, and P. Chen, *Polym. Eng. Sci.* (**2004**), 44, 18–27.

45 M. Hendenqvist and U. W. Gedde, *Prog. Polym. Sci.* (**1996**), 21, 299–333.

46 D. Larobina, L. Sanguigno, V. Venditto, G. Guerra, and G. Mensitieri, *Polymer* (**2004**), 45, 429–436.

47 O. Muth, T. Hirth, and H. Vogel, *J. Supercrit. Fluids* (**2001**), 19, 299–306.

48 J. D. Martinache, J. R. Royer, S. Siripurapu, F. E. Hénon, J. Genzer, S. A. Khan, and R. G. Carbonell, *Ind. Eng. Chem. Res.* (**2001**), 40, 5570–5577.

49 J. Schnitzler and R. Eggers, in *J. Supercrit. Fluids* (**1999**), Vol. 16, pp. 81–92.

50 M. Tang, T. Du, and Y. Chen, *J. Supercrit. Fluids* (**2004**), 28, 207–218.

51 T. Cao, K. P. Johnston, and S. E. Webber, *Macromolecules* (**2004**), 37, 1897–1902.

52 R. R. Gupta, V. S. Ramachandrarao, and J. J. Watkins, *Macromolecules* (**2003**), 36, 1295–1303.

53 C. Shi, G. W. Roberts, and D. J. Kizerrow, *J. Polym. Sci., Part B: Polym. Phys.* (**2003**), 41, 1143–1156.

54 T. T. Ngo, C. L. Liotta, C. A. Eckert, and S. G. Kazarian, *J. Supercrit. Fluids* (**2003**), 27, 215–221.

55 B. L. West, S. G. Kazarian, M. F. Vincent, N. H. Brantley, and C. A. Eckert, *J. Appl. Polym. Sci.* (**1998**), 69, 911–919.

56 S. Sicardi, L. Manna, and M. Banchero, *Ind. Eng. Chem. Res.* (**2000**), 39, 4707–4713.

57 S. Sicardi, L. Manna, and M. Banchero, *J. Supercrit. Fluids* (**2000**), 17, 187–194.

58 F. Rodriguez, *Principles of polymer systems*, 4th Edn., Taylor & Francis, **1996**.

59 R. Gendron and P. Moulinie, *J. Cell. Plast.* (**2004**), 40, 111–130.

60 S. Siripurapu, Y. J. Gay, J. R. Royer, J. M. DeSimone, R. J. Spontak, and S. A. Khan, *Polymer* (**2002**), 43, 5511–5520.

61 X. Han, K. W. Koelling, D. L. Tomasko, and L. J. Lee, *Polym. Eng. Sci.* (**2002**), 42, 2094–2106.

62 X. Han, C. Zeng, K. W. Koelling, D. L. Tomasko, and L. J. Lee, *Polym. Eng. Sci.* (**2003**), 43, 1261–1275.

63 V. Kumar and O. S. Gebizlioglu, *ANTEC'91* (**1991**), 1297–1299.

64 M. A. Jacobs, M. F. Kemmere, and J. T. F. Keurentjes, *Polymer* (**2004**), 45, 7539–7547.

65 K. N. Lee, H. J. Lee, and J. H. Kim, *Polym. Int.* (**2000**), 49, 712–718.

66 Y. P. Fu, D. R. Palo, C. Erkly, and R. A. Weiss, *Macromolecules* (**1997**), 30, 7611.

67 S. L. Shenoy, P. Kaya, C. Erkly, and R. A. Weiss, *Synth. Methods* (**2001**), 123, 509.

68 J. J. Yoon and T. P. Park, *J. Biomed. Mater. Res.* (**2001**), 55, 401–408.

69 P. Handa and Z. Zhang, *Macromolecules* (**2000**), 38, 716–725.

70 B. Krause, R. Mettinkhof, N. F. A. van der Vegt, and M. Wessling, *Macromolecules* (**2001**), 34, 874–884.

71 B. Krause, K. Diekmann, N. F. A. van der Vegt, and M. Wessling, *Macromolecules* (**2002**), 35, 1738–1745.

72 V. Kumar and J. E. Weller, *ANTEC'91* (**1991**), 1401–1405.

73 V. Kumar and M. M. Van der Wel, *ANTEC'91* (**1991**), 1406–1410.

74 B. A. Rodeheaver and J. S. Colton, *Polym. Eng. Sci.* (**2001**), 41, 380–400.

75 Y. D. Hwang and S. W. Cha, *Polym. Test.* (**2002**), 21, 269.

76 S. G. Kazarian, *Polym. Sci. Ser. C.* (**2000**), 42, 78–101.

77 C. D. Han and C. Y. Ma, *J. Appl. Polym. Sci.* (**1983**), 28, 2961–2982.

78 D. Kropp and W. Michaeli, *J. Cell. Plast.* (**1998**).

79 M. Lee, C. B. Park, and C. Tzoganakis, *Polym. Eng. Sci.* (**1998**), 38, 1112–1120.

80 M. Lee and C. B. Park, *Polym. Eng. Sci.* (**1999**), 39, 99–109.

81 J. R. Royer and Y. J. Gay, *J. Polym. Sci., Part B: Polym. Phys.* (**2000**), 38, 3168–3180.

82 J.R. Royer and J.M. DeSimone, *J. Polym. Sci., Part B: Polym. Phys.* (**2001**), 39, 3055–3066.

83 Y. Xiong and E. Kiran, *Polymer* (**1995**), 36, 4817–4826.

84 S.D. Yeo and E. Kiran, *Macromolecules* (**1999**), 37, 7325–7328.

85 Y.C. Bae and E. Gulari, *J. Appl. Polym. Sci.* (**1997**), 63, 459–466.

86 N.M.B. Flichy, C.J. Lawrence, and S.G. Kazarian, *Ind. Eng. Chem. Res.* (**2003**), 42, 6310–6319.

87 J.R. Royer, Y.J. Gay, M. Adam, J.M. DeSimone, and S. A. Khan, *Polymer* (**2002**), 43, 2375–2383.

88 M. Garcia-Leiner and A.J. Lesser, *J. Appl. Polym. Sci.* (**2004**), 93, 1501–1511.

89 M.J. Bortner and D.G. Baird, *Polymer* (**2004**), 45, 3399–3412.

90 Y. Xiong and K. Erdogan, *Polymer* (**1995**), 36, 4817–4826.

91 R. Mertsch and B.A. Wolf, *Macromolecules* (**1994**), 27, 3289–3294.

92 E. Weidner, V. Wiesmet, Z. Knez, and M. Skerget, *J. Supercrit. Fluids* (**1997**), 10, 139–147.

93 J.A. Lopes, D. Gourgouillion, P.J. Pereira, A.M. Ramos, and M. Nunes da Ponte, *J. Supercrit. Fluids* (**2000**), 16, 261–267.

94 D. Gourgouillion and M. Nunes da Ponte, *Phys. Chem. Chem. Phys.* (**1999**), 1, 5369.

95 D. Gourgouillion, H.M.N.T. Avelino, J.M.N.A. Fareleira, and M. Nunes da Ponte, *J. Supercrit. Fluids* (**1998**), 13, 177–185.

96 M. Danesshaver, S. Kim, and E. Gulari, *J. Phys. Chem.* (**1990**), 94, 2124.

97 K. Dimitrov, L. Boyadzhiev, and K. Tufeu, *Macromol. Chem. Phys.* (**1999**), 200, 1626.

98 S. Areerat, T. Nagata, and M. Oshima, *Polym. Eng. Sci.* (**2002**), 42, 2234–2245.

99 C. Kwag, C.W. Manke, and E. Gulari, *J. Polym. Sci., Part B: Polym. Phys.* (**1999**), 37, 2771–2781.

100 S.D. Yeo and E. Kiran, *J. Appl. Polym. Sci.* (**2000**), 75, 306–315.

101 D.Q. Tuan, J.A. Zollweg, P. Harriot, and S.S.H. Rizvi, *Ind. Eng. Chem. Res.* (**1999**), 38, 2129–2136.

102 M. Lee and C. Tzoganakis, *Adv. Polym. Tech.* (**2000**), 19, 300–311.

103 M.D. Elkovitch, D. Tomasko, and J. Lee, *Polym. Eng. Sci.* (**1999**), 39, 2075–2084.

104 M.D. Elkovitch, L.J. Lee, and D. Tomasko, *Polym. Eng. Sci.* (**2001**), 41, 2108–2125.

105 B.M. Dorscht and C. Tzoganakis, *J. Appl. Polym. Sci.* (**2003**), 87, 1116–1122.

106 X. Han, K.W. Koelling, D. Tomasko, and L.J. Lee, *Polym. Eng. Sci.* (**2002**), 42, 2094–2106.

107 A. Vega-Gonzalez, C. Domingo, C. Elvira, and P. Subra, *J. Appl. Polym. Sci.* (**2004**), 91, 2422–2426.

108 J.J. Watkins and T.J. McCarthy, *Macromolecules* (**1994**), 27, 4845–4847.

109 E. Kung, A.J. Lesser, and T.J. McCarthy, *Macromolecules* (**2000**), 33, 8192–8199.

110 P. Rajagopalan and T.J. McCarthy, *Macromolecules* (**1998**), 31, 4791–4797.

111 K.A. Arora, A.J. Lesser, and T.J. McCarthy, *Macromolecules* (**1999**), 32, 2562–2568.

112 O. Muth, T. Hirth, and H. Vogel, *J. Supercrit. Fluids* (**2000**), 17, 65–72.

113 M. Tang, T.Y. Wen, T.B. Du, and Y.P. Chen, *Eur. Polym. J.* (**2003**), 39, 151.

114 M. Tang, T.Y. Wen, T.B. Du, and Y.P. Chen, *Eur. Polym. J.* (**2003**), 39, 143–149.

115 K.F. Abbet, A.S. Teja, J. Kowalik, and L. Tolbert, *Macromolecules* (**2003**), 46.

116 Y. Tominaga, Y. Izumi, G.H. Kwak, S. Asai, and M. Sumita, *Macromolecules* (**2003**), 36, 8766.

117 D. Li and B. Han, *Macromolecules* (**2000**), 33, 4555–4560.

118 X. Dai, D. Liu, B. Han, G. Yang, X. Zhang, J. He, J. Xu, and M. Yao, *Macromol. Rapid Commun.* (**2002**), 23, 626–629.

119 D. Li, B. Han, and D. Zhao, *Polymer* (**2001**), 42, 2331–2337.

120 D. Li, Z. Liu, B. han, L. Song, G. Yang, and T. Jiang, *Polymer* (**2002**), 43, 5363–5367.

121 Z. Liu, Z. Dong, B. Han, J. Wang, J. He, and G. Yang, *Chem. Mater.* (**2002**), 14, 4619–4623.

122 D. Li and B. Han, *Ind. Eng. Chem. Res.* (**2000**), 39, 4506–4509.

123 E. Kung, A. J. Lesser, and T. J. McCarthy, *Macromolecules* (**1998**), 31, 4160–4169.

124 H. J. Hayes and T. J. McCarthy, *Macromolecules* (**1998**), 31, 4813–4819.

125 Z. Liu, L. Song, X. Dai, G. Yang, B. Han, and J. Xu, *Polymer* (**2002**), 43, 1183–1188.

126 D. Li, B. Han, and Z. Liu, *Macrol. Chem. Phys* (**2001**), 202, 2187–2194.

127 G. Spadaro, R. De Gregorio, A. Galia, A. Valenza, and G. Filardo, *Polymer* (**2000**), 41, 3491–3494.

128 J. Zhang, A. J. Busby, C. J. Roberts, X. Chen, M. C. Davies, S. J. B. Tendler, and S. M. Howdle, *Macromolecules* (**2002**), 35, 8869–8877.

129 A. J. Busby, J. Zhang, A. Naylor, C. J. Roberts, M. C. Davies, S. J. B. Tendler, and S. M. Howdle, *J. Mater. Chem.* (**2003**), 13, 2838–2844.

130 Y. T. Shieh, K. H. Liu, and T. T. Lin, *J. Supercrit. Fluids* (**2004**), 28, 101–112.

131 H. Zhou, J. Fang, J. Yang, and X. Xie, *J. Supercrit. Fluids* (**2003**), 26, 137–145.

132 V. S. Ramachandrarao, B. D. Vogt, R. R. Gupta, and J. J. Watkins, *J. Polym. Sci., Part B: Polym. Phys.* (**2003**), 41, 3114–3126.

133 J. J. Watkins, G. D. Brown, V. S. Ramachandrarao, M. A. Pollard, and T. P. Russel, *Macromolecules* (**1999**), 32, 7732–7740.

134 V. S. Ramachandrarao and J. J. Watkins, *Macromolecules* (**2000**), 33, 5143–5152.

135 S. G. Kazarian and K. L. A. Chan, *Macromolecules* (**2004**), 37, 579–584.

136 T. A. Walker, S. R. Raghaven, J. R. Royer, S. D. Smith, G. D. Wignall, Y. B. Melnichenko, S. A. Khan, and R. J. Spontak, *J. Phys. Chem. B* (**1999**), 103, 5472–5476.

137 T. A. Walker, C. M. Colina, K. E. Gubbins, and R. J. Spontak, *Macromolecules* (**2004**), 37, 2588–2595.

138 N. M. B. Flichy, S. G. Kazarian, C. J. Lawrence, and B. J. Briscoe, *J. Phys. Chem. B.* (**2002**), 106, 754–759.

139 S. G. Kazarian and G. G. Martirosyan, *Phys. Chem. Chem. Phys.* (**2002**), 4, 3759–3763.

140 K. L. A. Chan and S. G. Kazarian, *Appl. Spectrosc.* (**2003**), 57, 381–389.

141 S. G. Kazarian, N. H. Brantley, and C. A. Eckert, *CHEMTECH* (**1999**), 36–41.

142 E. Bach, E. Cleve, and E. Schollmeyer, *Rev. Prog. Color* (**2002**), 32, 88–102.

143 G. A. Montero, C. B. Smith, W. A. Hendrix, and D. L. Butcher, *Ind. Eng. Chem. Res.* (**2000**), 39, 4806–4812.

144 E. Bach, E. Cleve, and E. Chollmeyer, *The Journal of The Textile Institute. Part 1: Fiber Science and Textile Technology* (**1998**), 89, 647–656.

145 A. R. Berens, G. S. Huvard, R. W. Korsmeyer, and F. W. Kung, *J. Appl. Polym. Sci.* (**1992**), 46, 231–242.

146 B. Guzel and A. Akgerman, *J. Chem. Eng. Data* (**1999**), 44, 83–85.

147 F. Tessari, L. Devetta, G. B. Guarise, and A. Bertucco, presented at the CISF 99, 5th Conference on Supercritical Fluids and their Applications, Garda (Verona) Italy, 1999 (unpublished).

148 S. G. Kazarian and K. L. A. Chan, *Analyst* (**2003**), 128, 499–503.

149 L. Olenka, E. S. Nogueiran, A. N. Medina, M. L. Baesso, and A. C. Bento, *Rev. Sci. Inst.* (**2003**), 74, 328–330.

150 P. N. Prasad and D. J. Williams, *Introduction to nonlinear optical effects in molecules and polymers.* John Wiley & Sons, Inc., **1991**.

151 P. Yang, J. Huang, and J. Jou, *Proc. Natl. Sci. Counc.* (**2000**), 24, 310–315.

152 S. K. Yesodha, C. K. S. Pillai, and N. Tsutsumi, *Prog. Polym. Sci.* (**2004**), 29, 45–74.

153 Z. Zhang and P. Handa, *Macromolecules* (**1997**), 30, 8505–8507.

154 S. G. Kazarian, *Macromol. Symp.* (**2002**), 184, 215–228.

155 E. Weidner, V. Weismet, Z. Knez, and M. Skerget, *J. Supercrit. Fluids* (**1997**), 10, 139–147.

156 S. M. Howdle, M. S. Watson, M. J. Whitaker, V. K. Popov, M. C. Davies, F. S. Mandel, J. D. Wang, and K. M. Shakesheff, *Chem. Commun.* (**2001**), 109–110.

157 J. Jung and M. Perrut, *J. Supercrit. Fluids* (**2001**), 20, 179–219.

158 J. N. Hay and A. Khan, *J. Mater. Sci.* (**2002**), 37, 4734–4752.

159 S.G. Kazarian, in *Drug delivery and supercritcal fluid technology*, ed. York, Kompella, and Shekunov, Marcel Dekker, Inc., **2003**, pp. 343–366.

160 J. Fages, H. Lochard, J. Letourneau, M. Sauceau, and E. Rodier, *Powder Technol.* (**2004**), 141, 219–226.

161 J. Hao, M.J. Whitaker, B. Wong, G. Serhatkulu, K.M. Shakesheff, and S.M. Howdle, *J. Pharm. Sci.* (**2004**), 93, 1083–1090.

162 P. Sencar-Bozic, S. Srcic, Z. Knez, and J. Kerc, *Int. J. Pharm* (**1997**), 148, 123–130.

163 J. Kerc, S. Srcic, Z. Knez, and P. Sencar-Bozic, *Int. J. Pharm* (**1999**), 182, 33–39.

164 J.J.A. Barry, H.S. Gidda, C.A. Scotchford, and S.M. Howdle, *Biomaterials* (**2004**), 25, 2559–2568.

165 X.B. Yang, H.I. Roach, N.M.P. Clarke, S.M. Howdle, R. Quirk, K.M. Shakesheff, and R.O.C. Oreffo, *Bone* (**2001**), 19, 523–531.

166 T.P. Richardson, M.C. Peters, A.B. Ennett, and D.J. Mooney, *Nat. Biotechnol.* (**2001**), 19, 1029–1034.

167 M.H. Sheradin, L.D. Shea, M.C. Peters, and D.J. Mooney, *J. Controlled Release* (**2000**).

168 J.S. Chiou, J.W. Barlow, and D.R. Paul, *J. Appl. Polym. Sci.* (**1985**), 30, 3911–3924.

169 M. Takada and M. Ohshima, *Polym. Eng. Sci.* (**2003**), 43, 479–489.

170 Z. Zhong, S. Zheng, and Y. Mi, *Polymer* (**1999**), 40, 3829–3834.

171 N.H. Brantley, S.G. Kazarian, and C.A. Eckert, *J. Appl. Polym. Sci.* (**2000**), 77, 764–775.

172 Y.P. Handa, S. Capowski, and M. O'Neil, *Thermochem. Acta* (**1993**), 226, 177–185.

173 Y. Sato, T. Takikawa, A. Sorakubo, S. Takishima, H. Masuoka, and M. Imaizumi, *Ind. Eng. Chem. Res.* (**2000**), 39, 4813–4819.

174 B. Singh, S.S.H. Rizvi, and P. Harriott, *Ind. Eng. Chem. Res.* (**1996**), 35, 4457–4463.

175 Y. Sato, T. Takikawa, S. Takishima, and H. Masuoka, *J. Supercrit. Fluids* (**2001**), 19, 187–198.

176 Y. Sato, K. Fujiwara, T. Takikawa, Sumano., S. Takishima, and H. Masuoka, *Fluid Phase Equilib.* (**1999**), 162, 261–276.

177 P.K. Davis, G.D. Lundy, J.E. Palamara, J.L. Duda, and R.P. Danner, *Ind. Eng. Chem. Res.* (**2004**), 43, 1537–1542.

11
Synthesis of Advanced Materials Using Supercritical Fluids

Andrew I. Cooper

11.1
Introduction

Supercritical carbon dioxide ($scCO_2$) has attracted much interest as an alternative solvent for materials synthesis and processing, and a number of recent reviews have appeared on this subject [1–11]. Researchers have promoted CO_2 as a sustainable and "green" solvent because it is non-toxic, non-flammable, and naturally abundant. In fact, the economics of using dense CO_2 on an industrial scale are complex, and the benefits must be assessed in comparison with alternative technologies, on a case-by-case basis [12]. Issues such as the capital costs associated with high-pressure equipment and the energy requirements for compressing CO_2 into the dense state may prove prohibitive in many instances. Nevertheless, it is widely accepted that the advantages associated with this solvent are likely to lead to a number of new CO_2-based processes – particularly where unique benefits can be obtained in the properties and performance characteristics of the materials produced. The level of interest in supercritical fluid (SCF) technology can be gauged from the growing number of participating academic and industrial research groups worldwide. We highlight in this chapter four broad research areas of special interest in the field of advanced materials. We describe how the unusual physical properties of CO_2 – such as low toxicity, easy separations, variable density, low viscosity, low surface tension, and polymer plasticization – are exploited in each case.

11.2
Polymer Synthesis

There has been strong interest in the use of $scCO_2$ for polymer synthesis over the last 12 years, dating from seminal publications in this area by DeSimone and coworkers [13, 14]. This topic has been reviewed extensively [1, 15], and we will focus here on recent developments and new strategies for addressing challenges connected with industrial implementation.

Supercritical Carbon Dioxide: in Polymer Reaction Engineering
Edited by Maartje F. Kemmere and Thierry Meyer
Copyright © 2005 WILEY-VCH Verlag GmbH & Co. KGaA, Weinheim
ISBN: 3-527-31092-4

11.2.1
Reaction Pressure

A major issue associated with the use of $scCO_2$ as a solvent for polymerization is the reaction pressure; many of the processes published so far operate at pressures in the range 20.0–40.0 MPa [1, 15], which has significant implications for capital equipment and running costs. This problem has been approached in a number of ways. For example, DeSimone and coworkers have developed methods for the *continuous* precipitation polymerization of fluoropolymers [16]. In general, continuous SCF processes are likely to be more readily implemented than batch processes, and a continuous approach has also been applied to the processing of porous PVDF foams [17].

An alternative approach to the pressure issue has been to explore other supercritical fluids or "compressed fluid" solvents in the liquid state. For example, we have shown that liquid 1,1,1,2-tetrafluoroethane (R134a) is a good solvent for dispersion polymerization at pressures as low as 1.0–2.0 MPa [18, 19]. R134a is non-toxic and non-flammable and is widely regarded as having zero ozone depletion potential. The global warming potential for R134a is estimated to be 1300 times that of CO_2, but a widely held view is that HFCs will have a very small impact on overall climate change, which will arise mostly from the accumulation of CO_2 in the atmosphere from the burning of fossil fuels [20, 21]. R134a is more expensive than CO_2, and any HFC-based process would likely require effective recycling of the solvent. Energy-efficient recycling of R134a may be practical since it was developed originally as a refrigerant. An important chemical difference between R134a and CO_2 is the degree of polarity: CO_2 is symmetrical and has no permanent dipole moment (although it does possess a substantial quadrupole moment), while R134a is moderately polar and has a significant dipole moment (2.1 D).

We have shown that liquid R134a is an excellent solvent for the dispersion polymerization of styrene at low pressures (<2.0 MPa) using inexpensive hydrocarbon stabilizers [19]. We have also developed an entirely new process that we have termed "compressed fluid sedimentation polymerization" [22]. This process operates at moderate pressures (<5.0 MPa) and uses dense R134a–CO_2 mixtures as a sedimentation medium for polymerization in order to produced large (>1 mm) beads of cross-linked hydrogel materials in the absence of any organic solvents or surfactants (Fig. 11.1). A key advantage is that the sedimentation rate can be fine-tuned by varying the pressure – and therefore density and viscosity – of the compressed sedimentation medium.

11.2.2
Inexpensive Surfactants

A significant limitation of using $scCO_2$ in materials applications is that it is a feeble solvent for most polymers [23]. Certain amorphous fluoropolymers are exceptions to this rule [14], and these materials have been used, for example, as

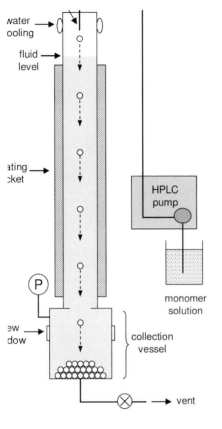

water
cooling

fluid
level

ating
cket

P

ew
dow

HPLC
pump

monomer
solution

collection
vessel

vent

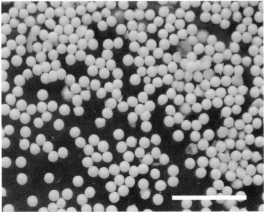

Fig. 11.1 Schematic layout for equipment used in compressed fluid sedimentation polymer process (top) and cross-linked polymer beads produced by this process (bottom, scale bar=10 mm).
(Adapted from [22]).

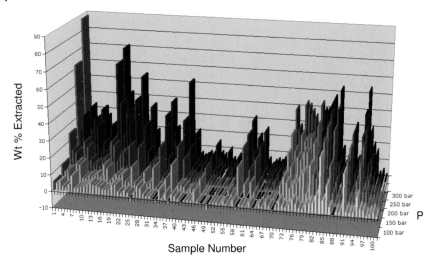

Fig. 11.2 High-throughput discovery of CO_2-soluble polymers: the graph shows cumulative weight % extracted in CO_2 for a library of 100 polymers at five different extraction pressures (1.0–30.0 MPa). Red bar = low molar mass internal standard material (AIBN); blue bars = aliphatic poly(ether ester)s (PE); yellow bars = aliphatic poly(ether carbonate)s (PEC); green bars = poly(vinyl acetate) (PVAc). (Reproduced from [29]).

241 for dispersion polymerization in $scCO_2$ [13]. It is even possible to produce CO_2-soluble conducting polythiophenes by extensive incorporation of fluoro-substituents, despite the fact that rigid polymers of this type are notoriously hard to dissolve, even in "good" solvents [24]. Fluorinated polymers exhibit a unique blend of properties such as solvent resistance, water repellency, and UV transparency and are therefore the materials of choice for certain advanced applications [25]. Conversely, fluoropolymers are relatively expensive and have poor environmental degradability, which may prohibit their use in other processes. There is thus a real need to consider alternative "CO_2-philes" that are less expensive and, ideally, environmentally benign. It is difficult to design such materials because of the small number of known examples and the embryonic "design rules" that exist [26]. Nonetheless, a number of groups have made recent progress in this area. For example, Beckman and coworkers have reported that inexpensive aliphatic poly(ethercarbonate) materials are soluble in CO_2, perhaps even more soluble than amorphous fluoropolymers of an equivalent degree of polymerization [27, 28]. The development of inexpensive functional polymeric surfactants and ligands based on this approach could underpin a number of new CO_2-based applications. Recently, we have adopted high-throughput screening approaches in order to discover more soluble polymers and ligands for use in conjunction with $scCO_2$ (Fig. 11.2) [29].

11.3
Porous Materials

This area has been reviewed recently in detail [9], and we will focus here mainly on new approaches that have appeared in the literature since this review was published. There are several specific reasons to consider SCFs as alternative solvents for the synthesis and processing of porous materials:

1. The production of porous materials is often solvent intensive, so that more sustainable alternatives could offer significant environmental benefits.
2. Drying steps can be energy intensive. With the exception of water, most SCF solvents studied so far are gases under ambient conditions.
3. Pore collapse can occur in certain materials (e.g., aerogels) when removing conventional liquid solvents. This can be avoided by the use of SCF solvents, which do not give rise to a liquid–vapour interface.
4. Porous structures are important in biomedical applications (e.g., tissue engineering), where there are strict limits on the amounts of residual organic solvent that may remain in the materials. This provides a strong driving force to seek non-toxic solvent alternatives.
5. Surface modification of porous materials frequently requires the use of solvents that will wet the pore structure efficiently. Supercritical fluids (and certain liquefied gases, such as CO_2) are extremely versatile wetting agents because of their low surface tensions (e.g., liquid CO_2 will wet Teflon).
6. Surface modification or templating of nanoporous materials presents special problems because organic solvents are often too viscous to fill such small pores. Even gaseous species (when below the critical temperature) can condense within small pores, thus forming a relatively viscous liquid "plug" that blocks the pore, preventing further penetration. SCF solvents have much lower viscosities than organic liquids and cannot condense into the liquid state. Moreover, mass transfer rates in SCF solvents tend to be high because of low solvent viscosity.
7. As a result of their compressed state, SCF solvents are highly suited to the generation of polymer foams. Moreover, polymer foaming requires that the material is either melted or highly plasticized, and many SCF solvents are excellent plasticizing agents (while being non-solvents) for a wide range of polymers.

In many applications, more than one these considerations is important. Therefore, the synthesis and processing of porous materials is a particularly fertile area for SCF research.

11.3.1
Porous Materials by SCF Processing

The use of SCFs for polymer foaming is quite well established [1, 9], but a number of novel applications have arisen recently. For example, ultraporous thin polymer films have been generated using physical constraints imposed by external surfaces (hard plates) and internal surfaces (hard nanoparticles and soft micelles) via foaming with supercritical CO_2 [30]. The constraints serve as diffusion barriers and/or heterogeneous nucleation sites, and the use of CO_2-philic copolymers further enables microcellular foaming at reduced pressures using liquid CO_2. Similarly, ultralow-k dielectric materials have been produced by foaming polyimides such as Matrimid using scCO$_2$ [31]. This approach allows the formation of both microcellular and bicontinuous, nanoporous structures with dielectric constants as low as $k = 1.77$. A process was also developed for the production of porous polyetherimide monofilaments by semicontinuous solid-state foaming using a modified "pressure cell" technique [32]. Dense, CO_2-saturated fibers were spun at rates up to 1 m s^{-1}, the porosity being introduced at the spinning head. The process was designed to allow the production of closed microcellular as well as open nanoporous filaments.

SCF solvents have also been used recently for the chemical modification of prefabricated porous materials. For example, direct synthesis (co-condensation-reaction) and post-synthesis reaction (grafting) were combined for the first time to produce bifunctionalized ordered mesoporous materials (OMM) [33]. Ethylenediamine-containing OMMs (ED-MCM-41) were first synthesized via direct synthesis and then further modified by the phenyl (PH) group in an SCF medium via a grafting reaction, resulting in OMMs with ED and PH groups (PH-ED-MCM-41). SCF solvents have also been quite widely used for the removal of guests from porous substrates [34].

It is clearly desirable to use non-toxic solvents for the synthesis or processing of biocomposite materials, for example, in tissue engineering [10]. Carbon dioxide is an obvious choice for such applications, although SCF alkanes (ethane, propane) and certain hydrofluorocarbons (R134a [20, 21]) could in principle fulfil similar requirements from a toxicological perspective. A major challenge in this area is to incorporate biologically active guest species into polymer hosts without loss of activity. For example, there are well-documented problems in maintaining protein activity under conventional processing methods because of either the presence of an organic–aqueous interface (double emulsion techniques), elevated temperatures (polymer melt processing), or vigorous mechanical agitation. A further challenge is to control the morphology of the composites, i.e. to generate porosity that optimizes release characteristics or allows cell infiltration into a scaffold. SCF mixing can be used to overcome many of these limitations in a single processing step. For example, CO_2-induced plasticization has been exploited to lower the viscosity of biodegradable polymers such as poly(D,L-lactide) (PLA), poly(lactide-*co*-glycolide) (PLGA), and polycaprolactone to such an extent that bioactive guests could be mixed into the polymer at temperatures close to ambient (e.g., 35 °C, 20.0 MPa) [35]. Foaming occurred upon venting the CO_2, which introduced a high degree of porosity into the composite materials. Biocomposites were formed encapsulating enzymes (e.g., ribonuclease A, catalase, β-D-galactosidase), and it was found that the enzyme activity was retained.

11.3.2
Porous Materials by Chemical Synthesis

We have developed two methods for the chemical preparation of porous organic polymers using SCF solvents. The first involves the formation of permanently porous cross-linked poly(acrylate) and poly(methacrylate) monoliths or beads using $scCO_2$ as the porogenic solvent [36–39]. No organic solvents are used, either in synthesis or in purification. It is possible to synthesize the monoliths in a variety of containment vessels, including chromatography columns and narrow-bore capillary tubing. Moreover, we have exploited the variable density associated with SCF solvents in order to "fine-tune" the polymer morphology. The average pore size and surface area in the materials can be tuned continuously over a considerable range (BET surface area = 90–320 m^2 g^{-1}) just by varying the SCF solvent density (Fig. 11.3). This can be rationalized by considering the variation in solvent quality as a function of CO_2 density and the resulting influence on the mechanism of nucleation, phase separation, aggregation, monomer partitioning, and pore formation [38]. We have applied the same concept to the synthesis of well-defined porous, cross-linked poly(methacrylate) beads (diameters = 100–200 µm) by suspension polymerization, again without the use of any organic solvents [37]. The surface area of the beads can be tuned over a wide range (5–500 m^2 g^{-1}) by varying the CO_2 density. Both of these techniques demonstrate how the variable density associated with SCF solvents can be exploited to precisely control the structure of porous materials produced by reaction-

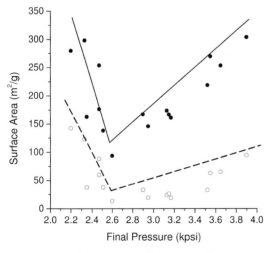

Fig. 11.3 Control over pore size for porous polymers prepared in $scCO_2$ can be achieved by adjusting the solvent density. The graph shows the variation in total BET surface area (closed circles) and micropore surface area (open circles) for a cross-linked poly(methacrylate) as a function of CO_2 pressue. (Reproduced from [38]).

induced phase separation. We have also shown that one can avoid high-pressure reaction conditions by using R134a as the porogenic solvent [39].

Our second general approach to the formation of porous materials by chemical synthesis has been templating of high-internal-phase CO_2-in-water (C/W) emulsions to generate highly porous materials in the absence of any organic solvents, only water and CO_2 being present [40, 41]. Providing that the emulsions are sufficiently stable – which depends strongly on the surfactant system – it is possible to generate low-density materials (~ 0.1 g cm^{-3}) with relatively large pore volumes (up to 6 cm^3 g^{-1}) from water-soluble vinyl monomers such as acrylamide and hydroxyethyl acrylate (Fig. 11.4) [40, 41]. Templated cell densities in these materials were found to be in the range 0.5×10^8 to 5×10^8 cells cm^{-3}. Initially, we used low-molecular-weight ($M_w \sim 550$ g mol^{-1}) perfluoropolyether ammonium carboxylate surfactants to stabilize the C/W emulsions [40], but a significant practical disadvantage is that this surfactant is expensive and non-degradable. We have subsequently shown that it is possible to use inexpensive hydrocarbon surfactants to stabilize the C/W emulsions and that these emulsions can also be templated to yield low-density porous materials [41]. We also showed that reaction pressures can be kept to moderate levels (<7.0 MPa) by employing liquid (rather than SCF) CO_2 at room temperature by using redox coinitiation [41].

The controllable swelling properties of CO_2 have been used to prepare templated inorganic materials by chemical synthesis at high pressures. For example, well-ordered mesoporous silicate films were prepared by infusion and selective condensation of silicon alkoxides within microphase-separated block copolymer templates dilated with scCO$_2$ [42]. Confinement of metal oxide deposition to specific subdomains of the preorganized template yielded high-fidelity, three-dimensional replication of the copolymer morphology, enabling the preparation of struc-

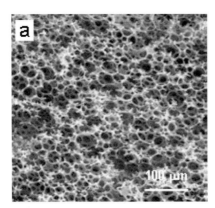 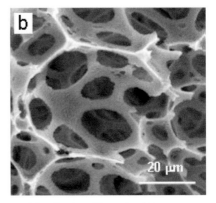

Fig. 11.4 Porous emulsion-templated polymer prepared by polymerization of a concentrated CO_2-in-water (C/W) emulsion. The "cells" that can be observed in the structure are direct replicas of the CO_2 emulsion droplets. These droplets are distorted into irregular polyhedra because the close packing in the concentrated emulsion (CO_2 internal phase volume= 80%). (Reproduced from [41]).

tures with multiscale order in a process that closely resembles biomineralization. The films survived the chemical-mechanical polishing step required for device manufacturing. Similarly, the controlled expansion of pores within mesoporous silicas has been achieved by exploiting the tunable density of $scCO_2$ to induce the swelling of triblock copolymer surfactant templating agents, P123 (PEO_{20}-PPO_{69}-PEO_{20}) and P85 (PEO_{26}-PPO_{39}-PEO_{26}) [43]. At pressures of approximately 48.2 MPa, pore diameters of up to 100 Å were achieved, representing a pore expansion of 54% compared to the conventionally-formed mesoporous silicas. The presence of $scCO_2$ was shown *not* to affect the hexagonal ordering of the silica, a distinct advantage over conventional pore-swelling techniques.

11.4
Nanoscale Materials and Nanocomposites

There has been a surge of interest in the last five years in the use of SCF solvents for the preparation of nanoscale materials and nanocomposites [5, 44]. Scientific drivers in this case include reduced solvent viscosity, pore wetting, and separation advantages.

11.4.1
Conformal Metal Films

SCF solvents are very versatile for the deposition of conformal metal films both on planar surfaces [46, 47] and within porous substrates [48]. For example, high-purity gold films were deposited onto metal, ceramic, and polymer substrates by the H_2-assisted reduction of dimethyl(acetylacetonate)gold(III) in $scCO_2$ at temperatures in the range 60–125 °C [49]. At 125 °C and 15.0 MPa, Au deposition proceeded readily on all surfaces studied, including SiO_2 and TiN films. This technique was used to generate continuous arrays of gold posts using etched Si wafers backfilled with SiO_2 as templates [48].

11.4.2
Synthesis of Nanoparticles

This is an area where fine control over solvent properties may offer distinct advantages. For example, Schiffrin and coworkers showed that the variable density associated with SCFs may be exploited in the size fractionation of functionalized metal nanoparticles [50]. Fulton and Wai have described a process for synthesizing and dispersing silver nanoparticles in a water-in-$scCO_2$ microemulsion [51]. A microemulsion of aqueous sodium nitrate solution was stabilized in $scCO_2$ by the addition of a mixed surfactant system involving sodium bis(2-ethylhexyl)-sulfosuccinate (AOT) and a perfluoropolyether-phosphate ether. Addition of a reducing agent [$NaBH(OAc)_3$] led to the formation of Ag(0) nanoparticles with an average diameter of about 4 nm, as estimated from UV-vis bandwidth analysis and TEM imaging. Both the precursor solution and the reduced Ag particle solution formed stable, optically clear microemulsions throughout the entire reaction sequence. More recently, Wai outlined a process for synthesizing silver halide nanoparticles (diameters 3–15 nm) by mixing two *different* water-in-scCO$_2$ microemulsions – one containing silver nitrate and the other containing a sodium halide (X=Cl, Br, I) [52]. In this case, nanoparticle formation involves several processes, such as droplet collision, intermicellar exchange, and reaction between silver ion and halide ion.

Titanium dioxide nanoparticles were produced by the controlled hydrolysis of titanium tetra*iso*propoxide (TTIP) in the presence of reverse micelles formed in CO$_2$ with the two fluorinated surfactants [53]. Based on dynamic light scattering measurements, the amorphous TiO$_2$ particles formed by injection of TTIP were larger than the reverse micelles, indicating surfactant reorganization. The size of the particles and the stability of dispersions in CO$_2$ were affected by the molar ratio of water-to-surfactant headgroup, (w_o), the precursor concentration, and the injection rate.

In general, the formation of nanoparticles in supercritical fluid solvents offers potential advantages, which include rapid solvent separation, accelerated reaction rates (due to high diffusivities), and the possibility of depositing particles *in situ* in porous materials, thereby taking advantage of the unique properties of the SCF phase. However, conventional approaches to nanoparticle synthesis and processing are not always directly transferable. For example, alkanethiol-capped nanoparticles have not been found to disperse readily in scCO$_2$ because of the very low van der Waals forces and polarizability exhibited by this solvent. Johnston and Korgel have addressed this problem by synthesizing robust fluorocarbon-capped silver nanocrystals, which may be dispersed in *pure* liquid or supercritical CO$_2$ [54]. The synthesis and processing of metal nanoparticles capped with fluorinated ligands (e.g., 1*H*,1*H*,2*H*,2*H*-perfluorodecanethiol) could prove important, for example, in memory storage applications where a low dielectric constant coating material may help to insulate the active charge-storing devices.

11.4.3
Synthesis of Nanowires

A number of research groups have investigated SCF solvents for the production of nanowires. For example, bulk quantities of defect-free silicon (Si) nanowires with nearly uniform diameters ranging from 40 to 50 Å were grown to a length of several micrometers with an SCF solution-phase approach [55]. Alkanethiol-coated gold nanocrystals (25 Å in diameter) were used as uniform seeds to direct one-dimensional Si crystallization in a solvent heated and pressurized above its critical point. The orientation of the Si nanowires produced with this method could be controlled with reaction pressure. Flow reactors can be employed to minimize the undesirable deposition of Si particulates formed in the bulk solution [56]. Scanning electron microscopy (SEM) images of Si nanowires grown at a series of temperatures, reactor residence times, and Si precursor concentrations revealed that the wire growth kinetics influences the nanowire morphology significantly and can be controlled effectively using such a system.

Recently, it was shown that ultrahigh densities (up to 10^{12} nanowires cm^{-2}) of ordered germanium nanowires can be produced on silicon and quartz substrates using $scCO_2^{57}$. The nanocomposites displayed room-temperature photoluminescence, the energy of which was dependent on the diameter of the encased nanowires. Metallic nanowires of cobalt, copper, and iron oxide magnetite (Fe_3O_4) have also been synthesized within the pores of mesoporous silica using a SCF inclusion technique [58]. The ability to synthesis ultrahigh-density arrays of semiconducting nanowires on-chip is a key step in future "bottom-up" fabrication of multilayered device architectures for nanoelectronic and optoelectronic devices.

Lastly, it is possible to "decorate" and elaborate nanowires using SCFs, for example by hydrogen reduction of metal precursors in $scCO_2$ [59]. Cu and Pd nanocrystals were deposited from different types of nanostructured materials including nanocrystal-nanowire, spherical aggregation-nanowire, shell-nanowire composites, and "mesoporous" metals supported by the framework of nanowires [59].

11.5
Lithography and Microelectronics

Microelectronic devices and materials now represent one of the largest manufacturing sectors in the world. Currently, water and solvent usage is required on a large scale for the precise fabrication of device elements for state-of-the-art microprocessors [12]. A typical microelectronics fabrication facility processing 5000 wafers per day will generate almost 5×10^6 L organic and aqueous solvent waste per year. For example, chemical mechanical planarization (CMP), an increasingly important operation in such facilities, uses ultra-purified water and supported silica slurries to polish wafer levels for improved manufacturability. A standard CMP tool will use 40 000 L ultra-pure water per day. In these circumstances, there is a strong drive to consider alternative solvents for such processes. A further challenge for the high end of the market is the identification of alternative process technologies that are not just "greener" but provide compelling technical advantages that could foster adaptation [11].

11.5.1
Spin Coating and Resist Deposition

Spin-coating processes involving CO_2 as the solvent have strong potential for the development of "dry lithography" technologies for the future. DeSimone and coworkers have developed a novel high-pressure CO_2 spin-coating apparatus to produce high-quality thin films of CO_2-soluble photoresists based on 1H,1H-perfluorooctyl methacrylate/*tert*-butyl methacrylate copolymers [60]. Film thicknesses were correlated to various process variables including rotational speed, solution viscosity, and evaporative driving force.

11.5.2
Lithographic Development and Photoresist Drying

In the supercritical state there is no liquid–vapour interface, so capillary stresses are suppressed. This has been used, for example, to avoid pore collapse by SCF drying of porous materials [9]. More recently, the same advantages have been exploited to allow the preservation of high-aspect ratio, nanometer-sized features in microlithography using $scCO_2$ as the developing solvent. Supercritical CO_2 has great potential for the development of photoresists, but only if the developed resist material has sufficient solubility in CO_2. Ober has developed diblock fluorinated copolymer resists for 193 nm wavelength lithography using $scCO_2$ as the developing solvent [61–63]. Lithographic resolutions as low as 0.2 μm can be achieved using these methods. This may be due in part to interfacial segregation behavior exhibited by the diblock copolymer resists, and it is suggested that $scCO_2$ development could play a key role in the fabrication of high-aspect-ratio features because of the absence of surface tension forces [61–63]. Similarly, $scCO_2$ containing hydrocarbon surfactants has been used to remove water from photoresists without pattern collapse due to capillary forces [64].

11.5.3
Etching Using SCF Solvents

Aqueous hydrofluoric acid (HF) solutions used in wet etching processes provide one of the most effective ways to etch films of silicon wafers. Traditionally, such solutions are used for the production of integrated circuits on silicon wafers and in surface micro-machining for the release of component parts in micro-electromechanical systems (MEMS) devices. However, using aqueous-based solutions often hinders the processes and/or poses environmental difficulties. For example, surface micro-machined MEMS devices often require the use of HF/water mixtures to etch sacrificial silicon dioxide (SiO_2) layers with high selectivity toward polycrystalline silicon structural layers. Capillary forces in subsequent drying of released wet-etched structures cause stiction: the released parts stick to adjacent surfaces. DeSimone and coworkers have reported an $scCO_2$-based nonaqueous HF etchant solution for "dry" etch processing of microelectronics devices involving the controlled dissolution of silicon dioxide (SiO_2) thin films [65]. Practical application of this etchant solution was demonstrated for cleaning microelectronic structures. Such nonaqueous etchant solutions could potentially contribute to replacing water- and solvent-based processes in microelectronics fabrication facilities (FABs), the ultimate goal being a totally "dry" FAB in the future.

11.5.4
"Dry" Chemical Mechanical Planarization

There is growing interest in using copper as the interconnect metal in microelectronic devices because of its superior electrical conductivity and better electromigration resistance than the current tungsten- or aluminum-based materials. Because of the difficulties involved in patterning copper by conventional dry-etch techniques, chemical mechanical planarization (CMP) has emerged as a new technology for implementing the use of copper in submicron semiconductor devices. During CMP, the chemical slurry removes copper from the wafer surface by a combination of chemical and mechanical means. In the chemical process, the metallic copper is oxidized and chelated, while in the mechanical process, the copper surface is rubbed against a polishing pad until global planarization is achieved. Most of the current processes use water as the solvent for the CMP slurry, leading to both technical and environmental difficulties. One particular technical drawback of the aqueous-based CMP processes is the incompatibility of porous low-k inorganic and organic interlayer dielectric materials with water. Inefficient water removal can result in higher than expected dielectric constants. Environmental issues include the generation of large quantities of contaminated aqueous waste (containing acids, buffers, abrasives, etc.), which cannot readily be recycled. Non-aqueous CMP processes suffer from similar environmental drawbacks, as many of these processes use highly undesirable organic solvents such as carbon tetrachloride. Thus, there exists a great demand for new CMP technologies which circumvent the technical and environmental drawbacks of the current aqueous and chlorinated organic solvents. DeSimone have developed a novel $scCO_2$-based CMT process that involves no organic or aqueous-based solvents [66]. Unlike water, condensed CO_2 has an extremely low viscosity and a low surface tension – characteristics desired for CMP processing. Additionally, the environmental difficulties associated with the use of organic solvents (e.g., solvent toxicity and oxidation) as well as the need to recycle the organic or aqueous solvents are avoided. Condensed CO_2 is easily separated from the other chemicals in the proposed CMP slurry by tuning the pressure and/or temperature.

11.6
Conclusion and Future Outlook

The use of SCF solvents for advanced materials synthesis and processing is a burgeoning field, as witnessed by the large number of breakthroughs in the last five years. For example, most of the research reviewed in this chapter was published in the three-year period 2002–2004. In general, there have been two main thrusts to this work:

1. the elimination of volatile organic solvents and polluted aqueous waste streams;
2. the generation of materials with new or improved properties.

We suggest that commercialization of this technology in the materials area will be brought about by the successful convergence of these two approaches.

There are a number of important questions that still remain to be addressed in terms of the "green credentials" for scCO$_2$-based processes. For example, high pressure-differentials (ΔP) translate to increased energy consumption and, ultimately, to an additional source of environmental pollution. Moreover, while CO$_2$ itself is cheap and non-toxic, many of the surfactants, ligands, and stabilizers that are used in conjunction with this solvent – particularly those based on fluorine – are not. There is, therefore, a real need to analyze the entire life cycle for any new process. Pressures can be reduced by working with liquid rather than supercritical CO$_2$ – that is, unless the process relies on a specific property of the SCF state. In the case of materials synthesis, this may require one to devise chemical methods which work at lower pressures – for example, redox initiation in the case of polymer chemistry. There is a strong driving force to move away from expensive fluorinated stabilizers and surfactants in many processes, although fluorinated materials do have unique performance benefits in some applications. The feeble solvent strength of CO$_2$ for most hydrocarbon polymers is a major obstacle in this respect.

A range of future opportunities exists for the exploitation of SCF solvents in various areas of materials science. The explosive growth of nanoscale materials applications in the last five years points to one such area. Another field that may hold great future promise is "SCF solvent engineering" in supramolecular chemistry; this has, as yet, been little exploited, despite the growing understanding that exists concerning micelle formation and self-assembly in SCFs.

Perhaps the best scientific "acid test" for materials produced using SCFs is whether or not the same materials could be produced using more conventional methods. In truth, it is hard to point to any materials produced so far that could not, in principle, be produced using more conventional liquid- or vapor-phase technologies. There are, however, a growing number of examples of advanced materials which can be produced very conveniently using SCFs which would be difficult or inconvenient to produce by other routes. To give a few examples:

1. Chemical fluid deposition (CFD) [47] has been used to deposit metal films from *non-volatile* precursors; equivalent volatile precursors for conventional CVD do exist, but the CFD approach broadens the range of available precursors and allows deposition at lower temperatures.

2. Concentrated CO$_2$-in-water emulsions (C/W) are useful for the preparation of highly porous templated structures [40, 41]; very similar materials can be generated using conventional O/W emulsions but the latter method is highly solvent intensive, and it is often hard to remove the organic solvent from the pore structure.

3. Lithographic patterns may be developed using a number of solvents, but "feature collapse" due to capillary forces is avoided conveniently by using SCF drying or development [12].

4. Acceptable residual solvent levels can be achieved in biomaterials processed using organic solvents; scCO$_2$ is completely non-toxic and solvent residues are simply not an issue [35].

To conclude, the use of SCF solvents for the preparation of advanced materials is a vibrant field, and a number of future opportunities exist in terms of both academic and commercial research.

References

1 Cooper, A. I. *J. Mater. Chem.* **2000**, *10*, 207–234.

2 Cooper, A. I. *Adv. Mater.* **2001**, *13*, 1111–1114.

3 Hakuta, Y.; Hayashi, H.; Arai, K. *Curr. Opin. Solid State Mater. Sci.* **2003**, *7*, 341–351.

4 Kikic, I.; Vecchione, F. *Curr. Opin. Solid State Mater. Sci.* **2003**, *7*, 399–405.

5 King, J. W.; Williams, L. L. *Curr. Opin. Solid State Mater. Sci.* **2003**, *7*, 413–424.

6 Shariati, A.; Peters, C. J. *Curr. Opin. Solid State Mater. Sci.* **2003**, *7*, 371–383.

7 Tomasko, D. L.; Li, H. B.; Liu, D. H.; Han, X. M.; Wingert, M. J.; Lee, L. J.; Koelling, K. W. *Ind. Eng. Chem. Res.* **2003**, *42*, 6431–6456.

8 Cansell, F.; Aymonier, C.; Loppinet-Serani, A. *Curr. Opin. Solid State Mater. Sci.* **2003**, *7*, 331–340.

9 Cooper, A. I. *Adv. Mater.* **2003**, *15*, 1049–1059.

10 Woods, H. M.; Silva, M.; Nouvel, C.; Shakesheff, K. M.; Howdle, S. M. *J. Mater. Chem.* **2004**, *14*, 1663–1678.

11 O'Neil, A.; Watkins, J. J. *Green Chem.* **2004**, *6*, 363–368.

12 DeSimone, J. M. *Science* **2002**, *297*, 799–803.

13 DeSimone, J. M.; Maury, E. E.; Menceloglu, Y. Z.; McClain, J. B.; Romack, T. J.; Combes, J. R. *Science* **1994**, *265*, 356–359.

14 DeSimone, J. M.; Guan, Z.; Elsbernd, C. S. *Science* **1992**, *257*, 945–947.

15 Kendall, J. L.; Canelas, D. A.; Young, J. L.; DeSimone, J. M. *Chem. Rev.* **1999**, *99*, 543–563.

16 Saraf, M. K.; Gerard, S.; Wojcinski, L. M.; Charpentier, P. A.; DeSimone, J. M.; Roberts, G. W. *Macromolecules* **2002**, *35*, 7976–7985.

17 Siripurapu, S.; Gay, Y. J.; Royer, J. R.; DeSimone, J. M.; Spontak, R. J.; Khan, S. A. *Polymer* **2002**, *43*, 5511–5520.

18 Wood, C. D.; Senoo, K.; Martin, C.; Cuellar, J.; Cooper, A. I. *Macromolecules* **2002**, *35*, 6743–6746.

19 Wood, C. D.; Cooper, A. I. *Macromolecules* **2003**, *36*, 7534–7542.

20 McCulloch, A. *J Fluorine Chem.* **1999**, *100*, 163–173.

21 Powell, R. L. *J. Fluorine Chem.* **2002**, *114*, 237–250.

22 Zhang, H.; Cooper, A. I. *Macromolecules* **2003**, *36*, 5061–5064.

23 Kirby, C. F.; McHugh, M. A. *Chem. Rev.* **1999**, *99*, 565–602.

24 Li, L.; Counts, K. E.; Kurosawa, S.; Teja, A. S.; Collard, D. M. *Adv. Mater.* **2004**, *16*, 180–183.

25 Rolland, J. P.; Van Dam, R. M.; Schorzman, D. A.; Quake, S. R.; DeSimone, J. M. *J. Am. Chem. Soc.* **2004**, *126*, 2322–2323.

26 Beckman, E. J. *Chem. Commun.* **2004**, 1885–1888.

27 Sarbu, T.; Styranec, T.; Beckman, E. J. *Nature* **2000**, *405*, 165–168.

28 Sarbu, T.; Styranec, T. J.; Beckman, E. J. *Ind. Eng. Chem. Res.* **2000**, *39*, 4678–4683.

29 Bray, C. L.; Tan, B.; Wood, C. D.; Cooper, A. I. *J. Mater. Chem.* **2005**, *15*, 456–459.

30 Siripurapu, S.; DeSimone, J. M.; Khan, S. A.; Spontak, R. J. *Adv. Mater.* **2004**, *16*, 989.

31 Krause, B.; Koops, G. H.; van der Vegt, N. F. A.; Wessling, M.; Wubbenhorst, M.; van Turnhout, J. *Adv. Mater.* **2002**, *14*, 1041.

32 Krause, B.; Kloth, M.; van der Vegt, N. F. A.; Wessling, M. *Ind. Eng. Chem. Res.* **2002**, *41*, 1195–1204.

33 Zhang, W. H.; Lu, X. B.; Xiu, J. H.; Hua, Z. L.; Zhang, L. X.; Robertson, M.; Shi, J. L.; Yan, D. S.; Holmes, J. D. *Adv. Funct. Mater.* **2004**, *14*, 544–552.

34 Patarin, J. *Angew. Chem. Int. Ed.* **2004**, *43*, 3878–3880.

35 Howdle, S. M.; Watson, M. S.; Whitaker, M. J.; Popov, V. K.; Davies, M. C.; Mandel, F. S.; Wang, J. D.; Shakesheff, K. M. *Chem. Commun.* **2001**, 109–110.

36 Cooper, A. I.; Holmes, A. B. *Adv. Mater.* **1999**, *11*, 1270–1274.

37 Wood, C. D.; Cooper, A. I. *Macromolecules* **2001**, *34*, 5–8.

38 Hebb, A. K.; Senoo, K.; Bhat, R.; Cooper, A. I. *Chem. Mater.* **2003**, *15*, 2061–2069.

39 Hebb, A. K.; Senoo, K.; Cooper, A. I. *Compos. Sci. Technol.* **2003**, *63*, 2379–2387.

40 Butler, R.; Davies, C. M.; Cooper, A. I. *Adv. Mater.* **2001**, *13*, 1459–1463.

41 Butler, R.; Hopkinson, I.; Cooper, A. I. *J. Am. Chem. Soc.* **2003**, *125*, 14473–14481.

42 Pai, R. A.; Humayun, R.; Schulberg, M. T.; Sengupta, A.; Sun, J. N.; Watkins, J. J. *Science* **2004**, *303*, 507–510.

43 Hanrahan, J. P.; Copley, M. P.; Ryan, K. M.; Spalding, T. R.; Morris, M. A.; Holmes, J. D. *Chem. Mater.* **2004**, *16*, 424–427.

44 Johnston, K. P.; Shah, P. S. *Science* **2004**, *303*, 482–483.

45 Shah, P. S.; Hanrath, T.; Johnston, K. P.; Korgel, B. A. *J. Phys. Chem. B* **2004**, *108*, 9574–9587.

46 Blackburn, J. M.; Long, D. P.; Watkins, J. J. *Chem. Mater.* **2000**, *12*, 2625–2631.

47 Long, D. P.; Blackburn, J. M.; Watkins, J. J. *Adv. Mater.* **2000**, *12*, 913–915.

48 Fernandes, N. E.; Fisher, S. M.; Poshusta, J. C.; Vlachos, D. G.; Tsapatsis, M.; Watkins, J. J. *Chem. Mater.* **2001**, *13*, 2023–2031.

49 Cabanas, A.; Long, D. P.; Watkins, J. J. *Chem. Mater.* **2004**, *16*, 2028–2033.

50 Clarke, N. Z.; Waters, C.; Johnson, K. A.; Satherley, J.; Schiffrin, D. J. *Langmuir* **2001**, *17*, 6048–6050.

51 Ji, M.; Chen, X. Y.; Wai, C. M.; Fulton, J. L. *J. Am. Chem. Soc.* **1999**, *121*, 2631–2632.

52 Ohde, H.; Rodriguez, J. M.; Xiang-Rong, Y.; Wai, C. M. *Chem. Commun.* **2000**, 2353–2354.

53 Lim, K. T.; Hwang, H. S.; Ryoo, W.; Johnston, K. P. *Langmuir* **2004**, *20*, 2466–2471.

54 Shah, P. G.; Holmes, J. D.; Doty, R. C.; Johnston, K. P.; Korgel, B. A. *J. Am. Chem. Soc.* **2000**, *122*, 4245–4246.

55 Holmes, J. D.; Johnston, K. P.; Doty, R. C.; Korgel, B. A. *Science* **2000**, *287*, 1471–1473.

56 Lu, X. M.; Hanrath, T.; Johnston, K. P.; Korgel, B. A. *Nano Lett.* **2003**, *3*, 93–99.

57 Ryan, K. M.; Erts, D.; Olin, H.; Morris, M. A.; Holmes, J. D. *J. Am. Chem. Soc.* **2003**, *125*, 6284–6288.

58 Crowley, T. A.; Ziegler, K. J.; Lyons, D. M.; Erts, D.; Olin, H.; Morris, M. A.; Holmes, J. D. *Chem. Mater.* **2003**, *15*, 3518–3522.

59 Ye, X. R.; Zhang, H. F.; Lin, Y. H.; Wang, L. S.; Wai, C. M. *J. Nanosci. Nanotechnol.* **2004**, *4*, 82–85.

60 Hoggan, E. N.; Flowers, D.; Wang, K.; DeSimone, J. M.; Carbonell, R. G. *Ind. Eng. Chem. Res.* **2004**, *43*, 2113–2122.

61 Ober, C. K.; Gabor, A. H.; GallagherWetmore, P.; Allen, R. D. *Adv. Mater.* **1997**, *9*, 1039–1045.

62 Yang, S.; Wang, J. G.; Ogino, K.; Valiyateettil, S.; Ober, C. K. *Chem. Mater.* **2000**, *12*, 33–40.

63 Sundararajan, N.; Yang, S.; Ogino, K.; Valiyaveetil, S.; Wang, J. G.; Zhou, X. Y.; Ober, C. K.; Obendorf, S. K.; Allen, R. D. *Chem. Mater.* **2000**, *12*, 41–48.

64 Zhang, X. G.; Pham, J. Q.; Ryza, N.; Green, P. F.; Johnston, K. P. *J. Vac. Sci. Technol. B* **2004**, *22*, 818–825.

65 Jones, C. A.; Yang, D. X.; Irene, E. A.; Gross, S. M.; Wagner, M.; DeYoung, J.; DeSimone, J. M. *Chem. Mater.* **2003**, *15*, 2867–2869.

66 Bessel, C. A.; Denison, G. M.; DeSimone, J. M.; DeYoung, J.; Gross, S.; Schauer, C. K.; Visintin, P. M. *J. Am. Chem. Soc.* **2003**, *125*, 4980–4981.

12
Polymer Extrusion with Supercritical Carbon Dioxide *

Leon P. B. M. Janssen and Sameer P. Nalawade

12.1
Introduction

The previous chapters emphasize that supercritical CO_2 can play an important role as solvent in both the synthesis and the processing of polymers. In various applications supercritical CO_2 has been adopted as a solvent for low-molecular-weight materials, which have a reasonable solubility in supercritical CO_2. Although very few polymers are significantly soluble in supercritical CO_2, the solubility of CO_2 in polymers can be substantial. This sorption of CO_2 in polymers has an important industrial potential. The dissolved CO_2 in a polymer acts as a plasticizer, thereby reducing the viscosity of the polymer. Two mechanisms that are mainly responsible for the reduction of the viscosity of molten polymers can be distinguished. The first is a decrease of the number of chain entanglements, while the second involves the creation of additional free volume, which increases the chain mobility. The reduction in the viscosity is mostly caused by the second mechanism [1]. Because of the viscosity reduction by plasticization, supercritical CO_2 has replaced volatile organic compounds (VOCs) and chlorofluorocarbons in various polymer synthesis and processing applications. Moreover, CO_2 technology allows the development of several new applications

Although supercritical CO_2 can induce a considerable viscosity decrease in polymer systems, an impeller is still a poor choice to ensure good mass and heat transfer in these systems. In general, an extruder is a much more suitable apparatus for handling materials with intermediate or high viscosities. For that reason, extruders have become inevitable items of apparatus for continuous operations in polymer processing and synthesis. They are used in a wide variety of applications such as polymer blending, foaming, and polymerization. In these applications, polymers or monomers are treated in a molten state. In the literature, various types of extruders and extrusion techniques are described [2, 3].

---

* The symbols used in this chapter are listed at the end of the text, under "Notation"

Supercritical Carbon Dioxide: in Polymer Reaction Engineering
Edited by Maartje F. Kemmere and Thierry Meyer
Copyright © 2005 WILEY-VCH Verlag GmbH & Co. KGaA, Weinheim
ISBN: 3-527-31092-4

In this chapter we mainly discuss some practical aspects of extrusion as well as the application of supercritical CO_2-assisted extrusion processes in particular. In supercritical CO_2-based extrusion processes, a modified conventional extruder can be used, and CO_2 gas is fed continuously into the molten polymer stream. Both single-screw and twin-screw extruders can be applied. The amount of CO_2 added varies between different polymers and is strongly dependent on the solubility of CO_2 in the polymer. The sorption of CO_2 in a polymer cannot be described as a classical physical phenomenon like the sorption of gas in a porous adsorbent, as it also depends on interactions between the molecules. Using Fourier transmission infrared (FT-IR) spectroscopy, these intermolecular interactions between CO_2 and polymers have been demonstrated [4–7]. The intermolecular interactions are very weak compared to the strong organic interactions (bonds). Nevertheless, these interactions are sufficient to provide guidance in the selection of polymer systems suitable for CO_2-assisted extrusion. The solubility of CO_2 in polymers containing perfluoro ether, siloxane, or acrylate groups is higher because of interactions between CO_2 and these groups. Poly-(methyl methacrylate) (PMMA) and polydimethylsiloxane (PDMS) are good examples of materials where the interactions occur because of the Lewis acidity of CO_2.

Supercritical CO_2-assisted extrusion applications mainly involve polymer blending, microcellular foaming, particle production, and reactive extrusion. Of course, supercritical CO_2 can also be used as an interfacial agent, foaming agent, or plasticizer in other applications.

12.2
Practical Background on Extrusion

Although a variety of different types of extruders exists, a main division can be made between single-screw and twin-screw extruders. The most important difference between those two types of machines is the transport mechanism. A single-screw extruder consists of one screw rotating in a closely fitting barrel, the transport mechanism being based on friction between the polymer and the walls of the channel. If the polymer slips at the barrel wall, it is easy to envisage that the material will rotate with the screw without being pushed forward. This makes these types of machines strongly dependent on the frictional forces at the wall and the properties of the material processed. Twin-screw extruders consist of two screws placed in an 8-shaped barrel. In the case of an intermeshing extruder, the flights of one screw extend into the channel of the other screw. Because of this, the polymer cannot rotate with the screw, irrespective of the rheological characteristics of the material. This is the most important advantage of intermeshing twin-screw extruders: the transport action depends on the characteristics of the material to a much lesser degree than is the case for a single-screw extruder. In twin-screw extruders the screws can intermesh or they can be tangential to each other. In the intermeshing configuration the screws can

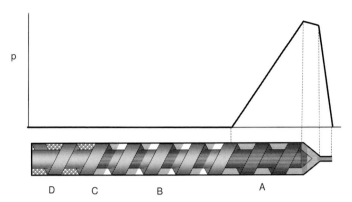

Fig. 12.1 Different zones in a twin-screw extruder: (A) fully filled zone (pump zone), (B) partially filled zone, (C) melting zone, (D) feed zone.

rotate in the same direction (co-rotating) or in opposite direction (counter-rotating). Although the designs of the different extruders can vary to a great extent, it can generally be stated that co-rotating extruders induce a larger shear on the polymer than counter-rotating machines and therefore provide better distributive mixing but also a higher viscous dissipation. Counter-rotating extruders, on the other hand, in general provide better sealing between the screw elements, which is better for preventing the escape of the supercritical component.

If a twin-screw extruder is stopped and opened, several zones can be distinguished clearly (Fig. 12.1). The channel near the feed hopper is more or less filled with solids. This material melts, and a zone with an only partly filled channel can be seen. At the end of the screw, where pressure has to be built up, the channel is completely filled with polymer. This can conveniently be used for supercritical processing. Screw elements that consume pressure are always preceded by a fully filled zone, which seals off the channel and prevents the supercritical material from leaking back.

12.3
Supercritical CO$_2$-Assisted Extrusion

Extrusion processes are generally carried out at elevated temperatures and pressures in order to process highly viscous polymers in a molten state. Elevated temperatures are not favored in a CO$_2$ dissolution process because the amount of CO$_2$ that dissolves into a polymer decreases with the temperature. On the other hand, this amount increases with pressure. However, a high pressure in an extruder causes various back leakage flows along the length of the screws, the amount of back leakage also being dependent on the viscosity of polymer. Since dissolved CO$_2$ strongly reduces the viscosity of polymer, extrusion with supercritical CO$_2$ can lead to large back leakages, resulting in long fully filled lengths and possible overfilling of the extruder compared to traditional extrusion processes.

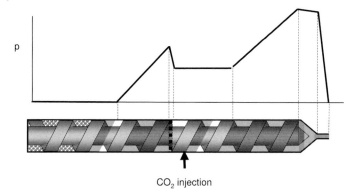

p

CO$_2$ injection

Fig. 12.2 Screw design and pressure profile for a counter-rotating twin-screw extruder with supercritical CO$_2$ injection.

A second effect associated with the decrease of viscosity is the sealing ability. If the supercritical CO$_2$ is added along the extruder, a fully filled zone has to be created between the hopper where the polymer is admitted and the part of the screw where the CO$_2$ is added. Prevention of leakage of the low-viscosity polymer/CO$_2$ mixture in a backward direction (towards the hopper) is a major challenge presented by the high-pressure extrusion process. Therefore, the extruder requires some modifications in its screw configuration to avoid leakage of CO$_2$. Understanding the pressure profile in an extruder is very important in this respect. An example of a screw profile with associated pressure profile when CO$_2$ is added into a twin-screw extruder is suggested in Fig. 12.2.

A melt seal is created prior to the CO$_2$ injection point. In this seal, a high pressure is built up which must be higher than the pressure in the CO$_2$ addition zone and therefore higher than the critical pressure of the supercritical substance. This seal has to be reliable and stable in order to prevent contact between CO$_2$ and the polymer melt prior to the melt seal. Back leakages into the seal have to be avoided at any cost, because the resulting decrease in viscosity will increase the back leakage and can lead to unstable operation and a blowout of CO$_2$ through the filling hopper.

The general schematic diagram of a supercritical CO$_2$-assisted extrusion setup is shown in Fig. 12.3 for a single-screw extruder.

After the zone where the polymer is molten the channel depth decreases. As a result, pressure is built up, and the channel is fully filled with melt. The screw part with the shallow channel provides the seal and should always be fully filled. After the addition of the CO$_2$ the screw geometry should be designed for a larger (volumetric) throughput. Therefore the channel depth is increased just after the melt seal. The extent of increase is also related to the amount of CO$_2$ that has to be added; if large amounts have to be accommodated a substantial increase is needed.

Although the principle of addition is the same for single-screw and twin-screw extruders the construction is different, mainly because in an intermesh-

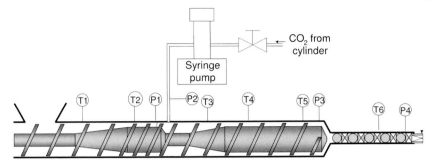

Fig. 12.3 Set-up of a supercritical CO_2-assisted extruder process with a single-screw extruder.

ing twin-screw extruder the channel depth is fixed and cannot be varied. In twin-screw extruders the seal can be provided by mixing elements or other types of restrictions in the screw profile. A good solution is the application of reversed pitch elements, which provide a very strong sealing action. An important consideration for every sealing is the existence of a fully filled zone prior to the restriction, where sufficient pressure is built up. This requires the use of screw elements with a narrow pitch that very closely intermesh. In this way the amount of leakage gaps increases per unit length and the resistance per gap increases too. Very small spacing between the flights and the barrel wall of an extruder also help to create the high pressure build-up. After the seal, flights with a higher pitch are used to accommodate the increase in volumetric flow. CO_2 is injected into a lower-pressure region created after the melt seal.

A syringe pump is used to inject the metered amount of CO_2 by carefully controlling the pressure and the volumetric flow rate. In order to process a molten polymer with supercritical CO_2, a sufficiently high pressure has to be maintained in the mixing zone. This high pressure is generated by the resistance provided by a small diameter nozzle, and is also dependent on the flow rate of the polymer melt and the viscosity of the CO_2-laden polymer. If the residence time of the polymer in the extruder is low, additional mixing can be obtained by a static mixer at the end of the screws. This static mixer improves the homogenization without increasing the distribution in residence times. Because the static mixer needs a certain pressure drop, care has to be taken that the pressure does not drop below the critical pressure before the end of the mixer.

12.4
Mixing and Homogenization

12.4.1
Dissolution of Gas into Polymer Melt

The amount of CO_2 absorbed in a molten polymer is one of the important factors that determine the morphology of the final product. This can be in the form of microcellular foams, polymer blends, or sub-micron polymer particles. Especially for foaming, the amount of CO_2 added should be less than the maximum amount of CO_2 that can be absorbed at the prevailing processing conditions. An excess of CO_2 causes pressure drops under supercritical conditions that lead to heterogeneous nucleation upon depressurization, leading to undesirable macro-void formations and the disappearance of micro-voids because of diffusion of CO_2 to the larger voids during expansion. This can be detrimental to the cell structure in microcellular foaming.

Fig. 12.4 is a representation of the classical dispersive mixing theory [8] that is also assumed to be valid for the initial mixing of a supercritical substance into a polymer [9]. This theory has been applied to the study of mixing of a polystyrene melt with a CO_2 solution in an extruder. In the solution, CO_2 is present as a dispersed minor component, while polymer is the continuous major component. In the gas-polymer melt solution, the minor component is dispersed into the major component in the form of bubbles. Because of the shear actions present in an extruder, the large-diameter bubbles are first stretched and then broken into smaller bubbles. The break-up of a large droplet into smaller droplets is a well-known phenomenon in emulsification studies and it can be described by the Weber number. The same theory is applicable to gas bubbles present in a polymer melt. The disintegration of a larger bubble into smaller bubbles takes place when a critical value of the Weber number is achieved depending on the viscosity ratio (dispersed to continuous) of two phases. Because of the high shear action in extrusion, stretching of the small bubbles occurs and eventually leads to dissolution of the gas by diffusion. In the breaking up

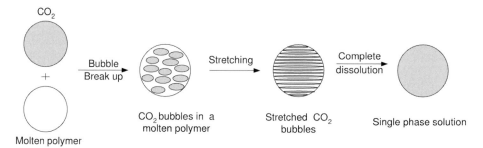

Fig. 12.4 The transformation of a two-phase solution into a single-phase solution.

process, the ratio between surface force and viscous force plays an important role. The surface force prevents the break-up while the shear force is responsible for the break-up of a droplet. Both forces are taken into consideration in the Weber number (We):

$$We = \frac{\dot{\gamma} d_b \eta_p f(\eta_g, \eta_p)}{2\sigma} \tag{1}$$

$$f(\eta_g, \eta_p) = \frac{19(\eta_g/\eta_p) + 16}{16(\eta_g/\eta_p) + 16} \tag{2}$$

Where, η_g, η_g, $\dot{\gamma}$, σ, and d_b are the dynamic viscosity of gas, the dynamic viscosity of polymer, the shear rate, the interfacial tension, and the diameter of the bubble, respectively. The above equations are used to determine the disintegrated bubble diameter for the given shear rate and viscosity ratio. When a bubble is stretched by shear forces, the thickness of the bubble is reduced. This enhances the mass transfer between the bubble and the polymer melt. The striation thickness is defined as the average distance between similar interfaces of a component in the mixture. The striation thickness as a function of stretching ratio (S) is expressed as

$$S = \frac{d_b}{\Phi_g \gamma} \tag{3}$$

$$\gamma = \frac{d_{max}}{d_b} \tag{4}$$

Where Φ_g and γ are the volume fraction of the minor component and the bubble stretching ratio. The bubble stretching ratio is defined as the maximum length of the elongated bubble under shear action, d_{max}, to the bubble diameter. The simple shear experiment can be carried out using a rotating disc and a stationary container to determine the bubble stretching ratio. Higher bubble stretching ratios are reported for the higher viscosity fluids at the same shear rates [9]. Different striation thicknesses can be expected in the various regions in an extruder depending on the different shear rates, a smaller striation thickness favoring a faster dissolution.

12.4.2
Diffusion into the Polymer Melt

The diffusivity (D) of gas or a supercritical component is generally lower in a polymer than in low-molecular-weight materials, and it increases with temperature. Good dissolution of the component in the polymer is enhanced by small striation thicknesses and by long residence times of the gas-loaded polymer melt in an extruder. Therefore, an estimation of the time required to completely dissolve the added component into a molten polymer to achieve a single-phase solution is necessary.

The required diffusion time (t_D) for the dissolution of gas into a molten polymer can be approximated from the striation thickness and the diffusivity of the gas, using the Fourier criterion, expressed as

$$Fo_m = 1 \text{ or } t_D = \frac{S^2}{D} \tag{5}$$

Park and Suh [9] have estimated the time required for the dissolution of CO_2 in a polystyrene melt using above expressions.

To ensure complete dissolution, the residence time of a mixture and the mixing can be enhanced using a static mixer at the exit of an extruder, as shown in Fig. 12.3. A static mixer consists of an array of stationary mixing elements placed in a pipe with a diameter approximately the same as that of the elements. The mixing is provided by division and rearrangement of fluid elements. Since the mixing elements do not move, the shear stresses in a static mixer are much lower than those in an extruder, and, although some dispersive mixing still occurs, the main reason for using a static mixer after the extruder is the increase in residence time.

12.5
Applications

12.5.1
Polymer Blending

The blending of polymers offers the opportunity to create materials with modified properties such as impact strength or rigidity without the necessity to synthesize a new polymer. However, because of entropic constraints, most polymer combinations are immiscible. A polymer blend can be produced by a non-reactive or reactive route.

In non-reactive blending, a two- (or multi-)phase mixture is formed when the immiscible polymers are physically mixed with each other. The minor phase, rich in B, is dispersed as droplets into a major phase rich in A. Apart from low interfacial tension, high shear rates and similar viscosities of both polymers are important for the size of the dipersed phase and therefore for the product quality. The reactive route follows the synthesis of a minor component via polymerization into a major component that acts as a host polymer. An alternative route for reactive blending is *in situ* formation of block co-polymers during the mixing process to decrease the interfacial tension. An extruder is the most commonly applied apparatus for the continuous production of polymer blends.

Until now, the use of supercritical CO_2 in blending has been restricted to the non-reactive process. In this process the main parameters that decide the morphology of the final blend are the concentration of the components and the dissolved CO_2, the viscosity of the polymers, and the interfacial tension between the polymers. Supercritical CO_2 acts as a solvent and influences the viscosity and the interfacial tension. The immiscibility of polymers is mainly due to a large interfacial tension present between the components. A compatibilizer is often used to decrease the interfacial tension and to enhance the favorable interactions between the polymers. Most compatibilizers are co-polymers that interact with the polymers used to produce a blend. In supercritical CO_2-assisted polymer blending, supercritical CO_2 plays the role of compatibilizer. The absence of surface tension in supercritical CO_2 helps to reduce the interfacial tension between the polymers. A second factor that is important for the final morphology – and therefore the product quality – of the blend is the viscosity ratio. The main theories about this influence can be summarized as: "The major component should have the highest viscosity and the viscosity difference should be as small as possible". To reduce the viscosity ratio, a particular amount of CO_2 can be dissolved in a high-viscosity material such that its viscosity is reduced to the viscosity of a low-viscosity material. The reduction in the viscosity of polymer due to the dissolved supercritical CO_2 can be explained in terms of enhancement in the free volume between the polymer molecules. The reduction in viscosity is different for different polymers under similar processing conditions (temperature, pressure, and shear rate), which allows the adjustment of the viscosity ratios. In addition, the inert nature (no chemical attack) is also an important factor, particularly for the selection of supercritical CO_2 as a solvent in this application.

Different polymer blends like PE (polyethylene)/PS (polystyrene) [10–11] and PMMA (polymethylmethacrylate)/PS [12–13] have been produced using supercritical CO_2-assisted extrusion. Fully intermeshing twin-screw extruders have been used in these studies. A decreased shear thinning behavior on dissolution of supercritical CO_2 into blends was observed. The obtained reduction in viscosity ratio resulted in a finer dispersion of the minor phase, which is desirable to create a good polymer blend. The effect of supercritical CO_2 on the dispersion of the minor phase for a PMMA/PS blend can be seen clearly in Fig. 12.5.

In these experiments [13], the viscosity of the discontinuous phase (PMMA) was higher than that of the continuous phase (PS). In the presence of CO_2, a fine dispersion of PMMA was observed. The size of the minor phase was 0.48 μm with supercritical CO_2 and 1.5 μm without CO_2. Also for different weight ratios of PMMA and PS (25/75, 50/50) the dispersion was better in the presence of CO_2, and finer dispersions of the PMMA phase were obtained. A completely different picture was observed in case of a 75/25 ratio blend. For this blend, PMMA is the major component while PS is a minor component, and the dispersion of PS into PMMA was not possible in the presence of supercritical CO_2. This result can be explained in terms of the viscosity of the major component being responsible for the stress transfer to the minor component. The reduction in viscosity of PMMA is higher because of the higher solubility of CO_2 in PMMA. The resulting low-viscosity major component results into a low stress transfer, and therefore it is not possible to disperse the minor phase into fine droplets.

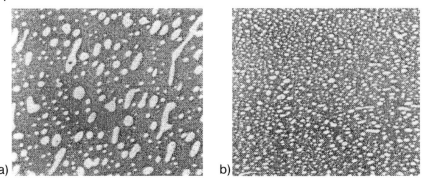

a)   b)

Fig. 12.5 TEM Micrographs of PMMA/PS (25/75) blend: (a) without
CO_2, and (b) with supercritical CO_2 [13] (reproduced with permission,
Copyright 2000, Society of Plastic Engineers).

Polymer blends can also be prepared using the phase inversion mechanism. Phase inversion is the transition of a dispersed phase into a continuous phase and vice versa. In blending with phase inversion, the minor component with a low melting temperature melts first. The low-melting component then surrounds the high melting component, the major component. Finally, when this major component starts melting it surrounds the minor component to form a continuous phase. The phase inversion depends on the volume ratio of the two polymers, the interfacial tension, and the viscosity ratio. As discussed earlier, supercritical CO_2 allows tuning of the interfacial tension and the viscosity ratio. In a twin-screw extruder, the compounding of both PMMA and PS with a rubber impact modifier (SP 2207) at a temperature of 473 K was successfully performed when only 2 wt% CO_2 was added [14]. A finely dispersed rubber phase was obtained for the two blends. In this study, the effect of CO_2 addition is measured in terms of the extruder length. The length of the extruder required for the phase inversion is much shorter in the presence of CO_2 than in its absence. The reduction in the glass transition temperature of the high-melting polymers due to dissolved CO_2 was responsible for the short length of the fully filled zone, and therefore the rest of the extruder length could be used effectively for reduction of the size of the dispersed phase. De-mixing of two phases after the CO_2 has been vented has been observed by Elkovitch et al. [12] for PMMA/PS blends. This undesirable situation can be overcome by using additives such as carbon black, calcium carbonate, and nano clay particles. Their use in a small amount (3 wt%) during blending prevents de-mixing after the CO_2 has been removed because the additives stabilize the interface between the phases.

The use of supercritical CO_2 can be beneficial in reactive blending also. Here, where a monomer polymerizes in a host polymer, supercritical CO_2 improves the infusion of reagents (monomer and initiator) into the host polymer, after which a polymerization reaction of the monomer is carried out. For this process it is necessary for the selected monomers and initiator to be soluble in supercritical CO_2. The dissolved CO_2 in the host polymer increases its free volume, causing it to swell, and, because of the resulting increase in intermolecular space, the diffusion of reagents in the polymer is enhanced. In this way, a high-molecular-weight polymer can be synthesized inside a host polymer. Pressure, temperature, swelling time, and monomer concentration are the important processing parameters for controlling the total mass uptake. The polymerization occurs both inside and outside the host polymer. However, the amount of polymer synthesized inside the host polymer is always much larger than the amount synthesized outside. The polymerization of styrene using supercritical CO_2 within the various host polymers has been carried out to produce different polystyrene blends [15–17]. A high mass uptake and a better distribution of the minor component over the major component are possible when using a reactive route.

It is expected that many novel and useful polymer blends will be obtained using supercritical CO_2 by applying both the reactive as well as the non-reactive route, and an extruder may well be a good choice of equipment for reactive blending in the continuous production of various polymer blends.

12.5.2
Microcellular Foaming

Continuous production of microcellular-foam plastic using supercritical CO_2-assisted extrusion is another emerging application. These materials are generally characterized by cell sizes less than 10 microns and have a large potential due to their unusual properties, such as a substantial reduction in weight, a high impact strength and toughness, and good resistance to fatigue. They are already used in various applications such as separation media, adsorbents, controlled release devices and catalyst supports [18–19]. Blowing agents that traditionally are applied in foaming processes include chlorofluorocarbons (CFC), hydrochlorofluorocarbons (HCFC), and volatile organic compounds (VOCs). Because of environmental regulations, plastic industries are now focusing on new blowing techniques to replaceme traditional blowing agents. The use of supercritical CO_2 leads to a clean and safe environment without severe emission constraints. A narrow cell size distribution, easy solvent recovery, good plasticizing ability, and high diffusivity are additional advantages of using supercritical CO_2 as a blowing agent.

The usual procedure to produce a foam plastic is to use a rapid pressure reduction or rapid heating to induce a liquid-to-gas transition. For microcellular foaming, supercritical CO_2 (or another supercritical component) is dissolved into the polymer at a certain pressure and temperature. Upon pressure release, the CO_2 transforms from the supercritical to the gas phase. This results in supersaturation, which gives the thermodynamic instability needed for nucleation. Both batch and continuous reactors can be used for the production of microcellular foam. The foaming temperature, the saturation pressure, and the depressurization rate (pressure drop per unit time) mainly determine the number of the cells, the cell size, and the cell size distribution in a product.

For many applications where complete saturation has not yet been reached, increasing the saturation time increases the amount of CO_2 dissolved in a polymer. The higher supersaturation that is achieved with longer saturation times will result in cells of smaller size. Semi-crystalline PMMA (a mixture of isotactic and syndiotactic PMMA) foams were produced by Mizumoto et al. [20] in batch experiments. The microcellular nucleation was caused by the supersaturation created by the depressurization of CO_2 in an isothermal condition.

The major drawback of a batch process is the amount of time that it requires. Low diffusivity of gas in a polymer in the absence of intense mixing is the main reason. This drawback can be overcome with a continuous microcellular foaming process using an extruder as the processing apparatus. An extruder provides intense mixing between the polymer melt and the CO_2, facilitating the formation of a single-phase solution. The formation of a good single-phase solution between the polymer and the supercritical CO_2 is very important in order to avoid the presence of excess voids together with the nucleated cells in the final product. In a two-phase solution, during nucleation the gas not only nucleates to form micro-cells but also diffuses to the already existing gas bubbles. The diffusion of nucleated gas to the already existing gas bubbles creates excess voids that may damage the structure of the microcellular foam, resulting in a product of lower strength.

An extruder is a suitable apparatus for handling high-viscosity polymer melts continuously. Mixing of a polymer melt and supercritical CO_2 takes place in an extruder because of the shear action and the convection mechanism. During this process, a single-phase solution forms, and the time needed is minimized because the shear actions in an extruder decreases the striation thickness and the diffusion distances.

The arrangement for producing microcellular foam consists of an extruder with a CO_2 injection system, a static mixer, and a nozzle. Mixing takes place in the extruder and continues in the static mixer. The nucleation of cells occurs in the nozzle attached to the exit of the static mixer. The following steps are involved in the continuous microcellular foaming process. A metered amount of CO_2 is continuously injected into the stream of polymer, which is plasticized prior to the injection point. At this point in the extruder a two-phase mixture is formed. The injected amount of CO_2 is always kept far below the solubility of CO_2 in a polymer under the processing conditions to ensure that a single-phase solution can be formed and that the CO_2 can dissolve completely into the polymer. The static mixer connected at the exit of the extruder enhances the rate of formation of single-phase solution and ensures complete mixing. The single-phase solution is then passed through the nozzle. The rapid depressurization provides the thermodynamic instability responsible for the nucleation of cells. Eventually, the nucleated cells (CO_2 bubbles) grow as micro-cells when the dissolved CO_2 diffuses to the nucleation sites. The nucleation of cells in a polymer melt can be expressed by the classical homogeneous nucleation theory [21].

$$N_{\text{hom}} = Cn \exp \left(\frac{-\Delta G_{\text{hom}}}{kT} \right) \tag{6}$$

$$\Delta G_{\text{hom}} = \frac{16\pi\sigma^3}{3\Delta P^2} \tag{7}$$

Where, N_{hom}, C, n, ΔG_{hom}, k, T, σ, and ΔP are the rate of homogeneous nucleation, the concentration of nucleation sites, the frequency factor for micro-cell nucleation, the change in Gibbs free energy of micro-cell nucleation, Boltzmann's constant, the temperature, the interfacial tension, and the pressure drop in the nozzle, respectively.

The residence time in the extruder and static mixer, the solubility of the supercritical component, and the supersaturation rate are the parameters that are responsible for the final structure of the microcellular foam. Single-phase solution formation depends on the residence time in the extruder and the static mixer and on the mixing prior to supersaturation. The residence time of a mixture should be higher than the diffusion time as calculated from the striation thicknesses to ensure homogeneous single-phase formation. The screw speed determines the flow rate of the solution, which in turn decides the residence time. A lower screw speed provides longer residence times for the dissolution of CO_2, but it also decreases the shear rate and increases the striation thickness. The time required to form a single-phase solution can be estimated from a convective diffusion theory. Pak and Suh [9] predicted that for a typical polymer viscosity of 200 Pa-s, single-phase solution formation is generally much faster than the industrial processing times. Effects of the other parameters on the microcellular foam production are explained below with some examples.

A continuous extrusion process for the production of various microcellular polymers such as high-impact PS, polypropylene (PP) and acrylonitrile butadiene styrene (ABS) using supercritical CO_2 as a blowing agent has been reported in detail by Baldwin et al. [22] and Park et al. [21, 23–24]. Attention has also been given to the process parameters such as the effect of pressure drop rate and the single-phase solution formation time. Park et al. [21] experimentally determined how a pressure drop rate affects the nucleation in a nozzle. Under identical processing conditions, including the pressure drop, different nozzle geometries were used to vary the depressurization rate. It was found that the cell density was increased by an increase in the pressure drop rate (Fig. 12.6).

A higher depressurization rate results in a higher supersaturation within a short period. Because of the instantaneous thermodynamic instability, the number of nuclei created is very large, and eventually the cell density is increased. A comprehensive theoretical explanation for the difference in nucleation at different pressure drop rates can be based on the competition between the micro-cell nucleation and the growth of cells [21].

Microcellular foams of PE/PS blends [10] and of poly(vinylidene fluoride) (PVDF) and its blends with PS and PMMA [25–26] can also be produced using an extruder. An increase in cell density and a decrease in cell size with increas-

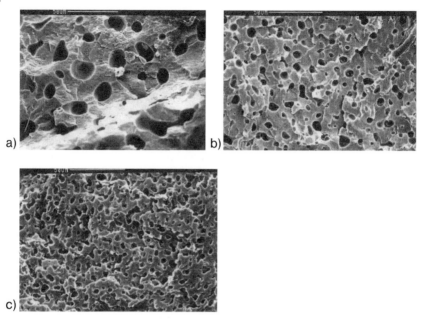

Fig. 12.6 SEM pictures of extruded HIPS for different pressure drop rates: (a) 0.076 Gpa s^{-1}, (b) 0.42 Gpa s^{-1}, (c) 3.5 Gpa s^{-1} [21]. (Reproduced with permission, Copyright 1995, Society of Plastic Engineers).

ing pressure and decreasing temperature have been reported for the process conditions used. This was mainly the effect of increased supersaturation due to a high concentration of dissolved CO_2 at higher pressures and lower temperatures. Recently, PS/nano clay composite foams were prepared using the extrusion foaming process with supercritical CO_2 [27]. Similar results were obtained in terms of pressure drop rate on cell nucleation to those described above [21].

12.5.3
Particle Production

Supercritical CO_2 has also emerged as a solvent in particle formation processes that can be used in the paint and pharmaceutical industries. Several advantages of the supercritical process can be considered, like a narrow particle size distribution, easy solvent recovery, and prevention of the emission of VOCs. The particle production using supercritical technologies utilizes the solubility of supercritical CO_2 in a polymer or vice versa. The mode of operation is either in batch or continuous. A rapid growth in research into particle production using supercritical CO_2 has been seen during the last decade. Various methods exist that use supercritical CO_2 as a solvent or anti-solvent. Recently, the various methods of particle production using supercritical CO_2 have been broadly reviewed [28].

These include rapid expansion of supercritical solutions (RESS), gas anti-solvent crystallization (GAS), supercritical anti-solvent precipitation (SAS), precipitation by compressed anti-solvent (PCA), solution-enhanced dispersion by supercritical fluids (SEDS), and particles from gas-saturated solutions (PGSS). Among all these processes the RESS and the PGSS process do not require any additional solvents. In the RESS process, a polymer is dissolved in supercritical CO_2, and the solution is then expanded via a nozzle. The RESS process is applicable to very few polymers because of the low solubility of most polymers in supercritical CO_2. The PGSS process is based on the solubility of supercritical CO_2 in various polymers and is much more widely applicable, as this solubility is very high in many polymers. The PGSS process can be carried out in a batch or continuous way. The PGSS process is mainly applicable to highly viscous, waxy polymers with a low glass transition temperature.

In the batch PGSS process, a polymer melt is saturated with a gas and then expanded though a small-diameter nozzle. The dissolved gas reduces the viscosity of molten polymer and enhances the atomization. The concentration of the dissolved supercritical CO_2 depends on the processing pressure and temperature, the residence time, and the amount of gas, which is generally higher than in microcellular foaming. During the expansion of the solution via a nozzle, very high supersaturation occurs and fine particles are formed. These particles are instantaneously solidified because of the cooling effect of the expanding gas. Moreover, the gas can easily be separated and recycled. The batch PGSS process has been patented by Weidner et al. [29]. However, because a batch process with highly viscous materials is not efficient in terms of mixing, the cycle times are very long. Therefore, an extruder again becomes a very good choice to handle a high-viscosity polymer melt in a continuous supercritical process. The continuous production of powder from highly viscous polymers using the PGSS process is still in its infancy. An extruder and a static mixer can be used to saturate the molten polymer with the gas. The continuous production of powder coatings from polyester resin using supercritical CO_2 in a twin-screw extruder and a static mixer has been patented [30], the patent claiming that the presence of CO_2 allows the processing temperature to be 80 °C lower than it would have been in its absence. This illustrates that if very good mixing of CO_2 and polymer can be achieved in an extruder, a low-viscosity polymer solution can be obtained.

12.5.4
Reactive Extrusion

High viscosities and low diffusion rates in polymers make polymerization or modification reactions significantly different from the reactions of smaller molecules. When a batch reactor is used for bulk polymerizations, mixing and heat transfer are only efficient during the initial reaction period. When the conversion increases the viscosity also increases following increase in the molecular weight of polymer during the polymerization reactions. The addition of solvents to a reaction mixture can overcome the problem associated with the viscosity enhancement during the reaction. The solvent reduces the viscosity of the reaction mixture and enhances heat and mass transfer. But the inevitable separation step at the end of the reaction is less economical.

An extruder can be used as a reactor for polymerizations. Its ability to handle high-viscosity materials, its good mixing ability, the continuous operation, and the easy volatilization of unreacted reactants are some of the advantages. The reactions performed in extruders include chain-growth polymerization, copolymerization, step-growth polymerization, modification reaction, grafting, and reactive blending [31]. Though an extruder can handle a high-viscosity material, a reduced viscosity can be advantageous because of lower energy consumption associated with lower heat dissipation and better mixing and diffusion.

The reduction in the viscosity of polymer melt due to dissolved CO_2 has led to its use as a solvent in reactive extrusion. Despite the high solubility of supercritical CO_2 in many polymers, only a few reactive extrusion studies (grafting and blending) using supercritical CO_2 as a solvent are to be found in the literature. Free-radical grafting of maleic anhydride (MAH) on polypropylene (PP) was observed to be improved in the presence of supercritical CO_2 by Dorscht and Tzoganakis [32]. The reactive blending of PE (polyethylene) functionalized with maleic anhydride (MAH) and polyamide-6 (PA-6) is another reaction in extruders where the use of CO_2 was beneficial, an increase in the conversion with increasing CO_2 concentration having been found [33]. These improved results are related to an increase in the free volume due to dissolved CO_2, which in turn improves the segmental chain mobility. The ultimate effect is a reduction in the viscosity of polymer melt. Therefore, the probability of contact between the reactive functional groups is also increased because of the lower diffusion resistance, enhancing the reaction rate and increasing the conversion.

12.6
Concluding Remarks

The high solubility of supercritical CO_2 in many polymers has led to its adoption as a solvent in the various applications for the processing and synthesis of polymers. For industrial scale applications the extruder plays an important role as a reactor or mixer in continuous production and in promoting better mixing of high-viscosity materials. Moreover, because of the high temperatures and pressures used in extruders, CO_2 is generally already in its supercritical state when it is inside an extruder. Before using an extruder for supercritical CO_2-assisted applications, it is necessary to understand the pressure profile in the extruder. This mainly depends on the screw configuration. The dissolved CO_2 causes a viscosity reduction of the polymer and allows extrusion at lower temperatures. Especially for degradation-sensitive specialty polymers, the addition of supercritical CO_2 is a simple way of obtaining an acceptable product. If mixing or residence times are not sufficient to obtain a single-phase mixture within the length of the extruder screws, the addition of a static mixer at the end of the extruder is a good way of improving the process. De-mixing of polymers in a blend after the CO_2 is released and the prevention of the formation of a two-phase solution prior to the supersaturation are the major concerns in reactive blending and in microcellular foaming, respectively. Despite the various applications of supercritical CO_2-assisted extrusion, not enough attention has been given to the different extruder types and screw configurations. Different results can be expected in terms of the product quality because of the different shear mixing regimes and residence times, depending on the extruder type and the screw configuration used. Among the applications discussed here, particle production and reactive extrusion are still in the early stages of the development.

Notation

We	Weber number	[–]
$\dot{\gamma}$	Shear rate	[s^{-1}]
η_g	Dynamic viscosity of gas	[Pa s]
η_p	Dynamic viscosity of polymer	[Pa s]
σ	Interfacial tension	[N m^{-1}]
d_b	Diameter of bubble	[m]
S	Stretching ratio	[–]
Φ_g	Volume fraction of gas	[–]
γ	Bubble stretching ratio	[–]
t_D	Diffusion time	[s]
D	Diffusivity of gas	[m^2 s^{-1}]
N_{hom}	Rate of homogeneous nucleation	[m^{-3} s^{-1}]
C	Concentration of nucleation sites	[m^{-3}]
n	Frequency factor for micro-cell nucleation	[s^{-1}]
ΔG_{hom}	Change in Gibbs free energy	[J]
k	Boltzmann's constant	[J K^{-1}]
T	Temperature	[K]
ΔP	Pressure drop in nozzle	[Pa]

References

1 L. Gerhardt, E. Gulari, C. Manke, *J. Polym. Sci.: Polym. Phys.* **1997**, *35*, 523–538.

2 L. P. B. M. Janssen, *Twin Screw Extrusion*, Elsevier, Amsterdam, 1978.

3 C. Rauwendaal, *Polymer Extrusion*, 3rd revised edition. Hanser Publishers, New York, 1994.

4 J. R. Fried, W. J. Li, *J. Appl. Polym. Sci.* **1990**, *41*, 1123–1131.

5 B. J. Briscoe, C. T. Kelly, *Polymer* **1995**, *36*, 3099–3102.

6 G. S. Kazarian, M. F. Vincent, F. V. Bright, C. L. Liotta, *J. Am. Chem. Soc.* **1996**, *118*, 1729–1769.

7 G. S. Kazarian, N. H. Brantley, B. L. West, M. F. Vincent, C. A. Eckert, *Appl. Spect.* **1997**, *51*, 491–494.

8 Z. Tadmore, C. G. Gogos, *Principles of Polymer Processing*, John Wiley and Sons, New York, 1979.

9 C. B. Park, N. P. Suh, *Polym. Eng. Sci.* **1996**, *36*, 1, 34–48.

10 M. Lee, C. B. Park, C. Tzoganakis, *Polym. Eng. Sci.* **1998**, *38*, 1112–1120.

11 M. Lee, C. Tzoganakis, C. B. Park, *Adv. Polym. Tech.* **2000**, *19*, 300–311.

12 M. D. Elkovitch, L. J. Lee, D. L. Tomasko, *Polym. Eng. Sci.* **1999**, *39*, 2075–2084.

13 M. D. Elkovitch, L. J. Lee, D. L. Tomasko, *Polym. Eng. Sci.* **2000**, *40*, 1850–1861.

14 M. D. Elkovitch, L. J. Lee, D. L. Tomasko, *Polym. Eng. Sci.* **2001**, *41*, 2108–2125.

15 J. J. Watkins, T. J. McCarthy, *Macromolecules* **1994**, *27*, 4845–4847.

16 J. J. Watkins, T. J. McCarthy, *Macromolecules* **1995**, *28*, 4067–4074.

17 X. Dai, Z. Liu, B. Han, G. Yang, X. Zhang, J. He, J. Xu, M. Yao, *Macromol. Rap. Comm.* **2002**, *23*, 626–629.

18 D. Klempner, K. C. Frisch, *Handbook of Polymeric Foams and Foam Technology*, Hanser Publishers, Munich, 1991.

19 F. A. Shutov, *Integral/ Structural Polymeric Foams: Technology, Properties and Applications*, Springer-Verlag, Berlin, Heidelberg, 1986.

20 T. Mizumoto, N. Sugimura, M. Moritani, *Macromolecules* **1994**, *33*, 6757–6763.

21 C. B. Park, D. F. Baldwin, N. P. Suh, *Polym. Eng. Sci.* **1995**, *35*, 432–440.

22 D. F. Baldwin, C. B. Park, N. P. Suh, *Polym. Eng. Sci.* **1996**, *36*, 1425–1435.

23 C. B. Park, N. P. Suh, *Polym. Eng. Sci.* **1996**, *36*, 34–48.

24 C. B. Park, N. P. Suh, *J. Manuf. Sci. Eng.* **1996**, *118*, 639–648.

25 S. Siripurapu, Y. J. Gay, J. R. Royer, J. M. DeSimone, S. A. Khan, R. J. Spontak, *Material Research Society Symposium Proceedings* **2000**, *629*, FF9.9.1-FF9.9.6.

26 S. Siripurapu, Y. J. Gay, J. R. Royer, J. M. DeSimone, R. J. Spontak, S. A. Khan, *Polymer* 2002, *43*, 5511–5520.

27 X. Han, C. Zeng, J. Lee, K. W. Koelling, D. L. Tomasko, *Polym. Eng. Sci.* **2003**, *43*, 1261–1275.

28 M. Perrut, J. Jung, *J. Supercrit. Fluids* 2001, *20*, 179.

29 E. Weidner, Z. Knez, Z. Novak, European Patent 0,744,992, 2000.

30 A. T. Daly, O. H. Decker, K. R. Wursthorn, F. R. Houda, US Patent 5,766,522, 1998.

31 L. P. B. M. Janssen, *Reactive Extrusion Systems*, Marcel Dekker, NewYork, 2004.

32 B. Dorscht, C. Tzoganakis, *59th Annual Technical Conference, SPE* **2001**, *1*, 298–302.

33 A. Xue, C. Tzoganakis, *59th Annual Technical Conference, SPE* **2003**, *1*, 2271–2275.

13

Chemical Modification of Polymers in Supercritical Carbon Dioxide *

Jesse M. de Gooijer and Cor E. Koning

13.1
Introduction

The use of supercritical fluids (SCFs) has recently received much attention in polymer science and technology. In Fig. 13.1 the potential interactions between polymers and supercritical fluids as well as possible applications, are outlined. Although CO_2 is a poor solvent for most polymers, it does have the capability to swell polymers up to several mass percentages, especially polymers with a non-polar character, since these have a better affinity with the non-polar CO_2. Besides bringing about swelling, $scCO_2$ can act as a plasticizer for a variety of polymers and thereby decrease the glass transition temperature (T_g) of the polymers [2, 3]. Thanks to the plasticizing effect of $scCO_2$ the mobility of the amorphous polymer segments will be increased. The swelling and plasticizing effects enable small molecules which are dissolved in the CO_2 to diffuse into the polymer. In this way, polymers can be modified not only on the surface of, for example, polymer sheets or polymer granules, but also closer to the core.

The extent of swelling and sorption depends on the type of polymer. Several scientists investigated the swelling and sorption of a variety of polymers [4]. FTIR spectroscopy is a frequently used technique to study the impregnation of SCFs and species dissolved therein [5, 6]. The small molecules which are dissolved in the CO_2 and impregnated into the polymer can, for example, be monomers of a different class, which can be polymerized in an additional step. In this way polymer blends can be obtained. The small molecules could also be a dye or a chemical reagent which could chemically modify polymers without melting the polymer first. Besides the impregnation of small molecules, the extraction of low-molecular-weight materials such as monomers, oligomers, plasticizers, and stabilizers has also been studied [7–9].

Comparing crystalline and amorphous polymers, the swelling of amorphous polymers, and therewith the diffusion of small molecules into the polymer, is

---

* The symbols used in this chapter are listed at the end of the text, under "Notation".

Supercritical Carbon Dioxide: in Polymer Reaction Engineering
Edited by Maartje F. Kemmere and Thierry Meyer
Copyright © 2005 WILEY-VCH Verlag GmbH & Co. KGaA, Weinheim
ISBN: 3-527-31092-4

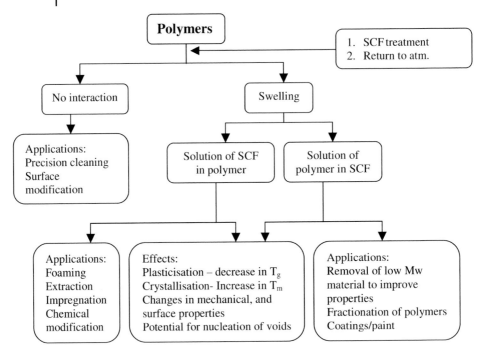

Fig. 13.1 Diagram of the interaction of SCFs with polymers [1].

much more pronounced than for crystalline polymers. Furthermore it is known that scCO$_2$ can induce, improve, and enhance the crystallization of several polymers, resulting in an increased melting temperature (T_m) and an increased melting enthalpy (ΔH_m), subsequently leading to a change in mechanical properties [1, 5, 10–14].

SCFs are considered to be of great interest in a variety of areas (see also Fig. 13.1), such as extraction [8, 15–19], chromatography [20–23], polymerization [24–36], polymer modification [37–42], impregnation [43–47], reactions [48–50] and catalysis [48, 51–55].

In the earlier days, SCFs were mainly used in extraction and chromatography applications. A well-known example of supercritical fluid extraction (SFE) is the extraction of caffeine from coffee [56]. Supercritical chromatography was frequently used to separate polar compounds [21, 23]. Nowadays, increasing interest is shown in SCFs for possible applications for polymerization, polymer modification, reactions, and catalysis [57]. Lora et al. [58] gave an overview of the use of SCFs in polymer processing. DeSimone et al. and Beckman et al. are among the pioneers using supercritical fluids as a polymerization medium [24, 27–35]. DeSimone, for example, described the synthesis of polyamides and poly(bisphenol A carbonate) in scCO$_2$ [24, 25]. Super and Beckman generated copolymers of CO$_2$ and cyclohexene oxide where CO$_2$ is both reactant and solvent [32].

Modification of polymers in supercritical fluids is an attractive alternative for more conventional approaches such as solution modification and melt modification. Neither conventional technique is economically or ecologically attractive because of the hazardous waste in the form of organic solvents, together with left-over monomer(s) and initiator(s). Furthermore, much energy is required to remove the solvents at the end of the solution process, or undesired side reactions may occur at elevated temperatures necessary for melt modification. Modification of polymers in supercritical fluids is therefore an attractive alternative, since very mild reaction conditions can be applied, and side reactions are expected to be limited, if not avoided. After modification is completed, the solvent can be easily released by reducing the pressure, and non-reacted chemicals can be supercritically extracted with pure supercritical fluid.

Supercritical modification of polymers was studied by several scientists to improve or change the properties of polymers. Polymers can either be chemically or physically modified. Examples of chemical modifications are the functionalization of polymers (grafting) or a chemical reaction of the functional groups of polymers to obtain new materials [38, 39]. Examples of physical modifications are the preparation of polymer blends, impregnation of polymers with additives [46], or foaming of polymers [59–61]. Another studied topic of polymer modification and impregnation is the supercritical dyeing of polymer fibers [40, 41].

The aim of this chapter is to present the scope and limitations of the chemical modification of (swollen) polymer particles in supercritical and subcritical CO_2. The supercritical modification technique will be critically evaluated as a promising alternative for the more conventional melt and solution modification techniques. In this respect, both effectiveness and economical and environmental issues play a role.

After a short review of the literature on this topic (Section 13.2), the modification of the end groups of polyamide 6 (PA-6) granules with a variety of blocking agents in supercritical or subcritical CO_2 is described in more detail (Section 13.3). In Section 13.4, the modification of the carboxylic acid end groups of poly(butylene terephthalate) (PBT) granules with 1,2-epoxybutane in supercritical or subcritical CO_2 is described.

Finally, the chapter ends with a short section on general conclusions, a technological assessment, and an outlook (Section 13.5).

13.2
Brief Review of the State of the Art

Few papers on the chemical modification of polymers in supercritical CO_2 have been published so far. In this short review we will mention the work on CO_2-assisted blending of polymers only very briefly. In the first place, this technique does not always imply a chemical modification of the supercritical carbon dioxide-swollen host polymer in which the monomer of the second blend component is polymerized, but, depending on the chosen polymerization conditions, merely a physical mixture of the two polymers may be generated. The second reason why we limit attention to this technology is that a detailed description of it is given elsewhere in this book.

The research group of McCarthy studied and described the formation of several polymer blends prepared in scCO$_2$. Sheets of polymers like poly(chlorotrifluoroethylene) (PCTFE), poly(4-methyl-1-pentene) (P-4-MP), poly(ethylene), nylon 6,6, poly(oxymethylene), and bisphenol-A polycarbonate were soaked with a homogeneous solution of styrene monomer and a free-radical initiator in scCO$_2$ for 4 h at a temperature below the decomposition temperature of the initiator, after which the temperature was raised to initiate the styrene polymerization inside the polymer host [62]. After polymerization and cooling, significant amounts of polystyrene proved to be incorporated into all the above-mentioned polymers. For two host polymers, namely P-4-MP and PCTFE, it was demonstrated by selective extraction procedures that graft polymerization, or chemical modification of the host polymer, had not occurred to a significant extent. For the PCTFE modification it was found that the polystyrene trapped inside the PCTFE matrix was present as discrete phase-segregated regions throughout the entire ca. 1500 μm thick film [44]. It was further demonstrated that the polystyrene content and its distribution can be controlled by adjusting the styrene concentration in the scCO$_2$ and/or the swelling time. By "filling" several scCO$_2$-swollen fluoropolymers with polystyrene, as described above, and subsequently sulfonating the styrene-rich surface at ambient temperature and pressure, the surface polarity of these chemically resistant polymers could be tailored [39]. In another paper, a controllable maleation procedure was described: by a free-radical grafting procedure, P-4-MP, swollen with scCO$_2$, was chemically grafted with maleic anhydride [63].

With the exception of the last example, the work of the McCarthy group can hardly be described as a chemical modification of a polymer. Muth et al. [64] applied the "McCarthy blending method" to the polymerization of the vinylic monomers styrene, methyl methacrylate, and methacrylic acid (MAA) inside scCO$_2$-swollen poly(vinylchloride), bisphenol-A polycarbonate, and poly(tetrafluoro ethylene). In the case of the polymerization of MAA inside PVC, the PMAA generated inside the PVC host could not be completely extracted, and chemical grafting of PMAA onto the PVC could not be completely excluded.

In 1993, Yalpani [38] published interesting work on the chemical modification of polymers in scCO$_2$. Mixtures of chitosan with glucose or malto-oligosaccharides were transformed into the corresponding imine-linked derivatives with high degrees of conversion. These modifications proved to be more facile and complete than equivalent reactions in conventional media. Amylose and poly(vinyl alcohol) were phosphorylated in scCO$_2$, yielding the corresponding phosphate ester derivatives. Poly(vinyl alcohol), starch, cellulose acetate, and other biopolymers were easily oxidized in mixtures of scCO$_2$ and oxygen.

The work by Catala's group is also convincing [65]. These researchers grafted isocyanato-isopropyl groups onto semi-crystalline poly(ethylene-co-vinyl alcohol) in scCO$_2$. Only the OH groups present in the amorphous phase were selectively modified. It proved to be possible to maintain the crystallinity of the copolymer while reducing its hydrophilicity. Indications were obtained that the grafting reaction predominantly occurred in the surface region of the material, which was ascribed to the poor solubility of the non-polar carbon dioxide in the polar polymer.

We would also like to mention the work by the group of Spadaro and Filardo, who used scCO$_2$ for the carboxylation (as well as the simultaneous crosslinking) of linear low-density polyethylene [66, 67] and high-density polyethylene [68] in the presence of γ-rays, and for the successful grafting of maleic anhydride onto polypropylene in the presence of both γ-rays and dicumylperoxide [69, 70].

Recently a series of papers was published in which the end-group modification of the step-growth polymers polyamide 6 [71–73] and poly(butylene terephthalate) [74] in scCO$_2$ was described. In some cases, the carbon dioxide, in which the end capping agent was dissolved, also contained a polar additive in order to enhance the sorption of the sc fluid by the relatively polar polymers. These papers illustrate the scope and limitations of the scCO$_2$ modification technique in a rather complete and subjective way, which will be discussed in detail in the rest of this chapter.

13.3
End-group Modification of Polyamide 6 in Supercritical CO$_2$

13.3.1
Background

Polyamides are widely used in a variety of applications. Polyamide 6 (PA-6) is used, for example, in the packaging industry. During film extrusion of PA-6, it is of great importance that the film is uniform and free of gel particles. These gel particles are caused by post-condensation of the polyamide, which results in highly viscous material. In PA-6 fiber and injection molding grades, chain stoppers like benzoic acid are added during polymerization to control the viscosity and prevent reaction of co-reactive end groups with one another. However, for film grade PA-6, chain stoppers cannot be used during polymerization, since the desired high viscosity will not be reached when these are added. After polymerization, an end group modification of the polymer is needed to avoid chain extension and gelation during film extrusion. Both amine and carboxylic acid end groups can be blocked, but blocking only the amine or only the carboxylic acid end groups can be sufficient. Usually, an attempt to modify the end groups of polyamides in the melt results in an incomplete reaction. Furthermore, if added chemicals at the high melt temperatures are not only reactive toward the end groups but also toward the amide bonds in the polymer chains, a change in molecular weight and molecular-weight distribution is observed, which will result in undesired changes in mechanical properties. Modification of these end groups in polymer particles swollen by supercritical fluids, especially by scCO$_2$, is an attractive alternative, since relatively mild reaction conditions can be applied and undesired side reactions can be suppressed. In order to enhance the swelling of relatively polar polyamide particles, and therewith the accessibility of the end groups, a polar additive can be added to the non-polar $scCO_2$. Based on extraction experience it is known that 5–10 mol% of a polar additive is usually sufficient [7, 8, 15, 20]. Since supercritical fluid extraction of low-molecular-weight materials from polymers is possible, it seems logical that the impregnation of polymer particles with low-molecular-weight (reactive) compounds, using the sc fluid as a carrier, should also be possible. Berens et al. [46] used supercritical CO_2 for the impregnation of additives into a number of glassy polymers, such as poly(vinyl chloride) (PVC) and polycarbonate (PC). Another interesting form of polymer modification by impregnation is supercritical dyeing of polymer fibers [40, 41].

Here, the selective blocking of end groups, present in PA-6 granules, applying supercritical and subcritical CO_2, with or without a small amount of a polar additive, is discussed. The fluids do not dissolve the polyamide but do have the possibility to diffuse into the granules and swell the amorphous phase of the polymer. When the PA-6 granules are swollen by a supercritical fluid, the blocking agent, which is dissolved in the fluid, is transported easily into the amorphous phase of the granules, where it can react with the end groups. In this way significant amounts of end groups can be blocked in a relatively short time. For blocking amine end groups of the polyamide chains, the highly reactive succinic anhydride is impregnated into the polyamide where it can react with the amine end groups, with the aim of enhancing the melt stability. Subsequently, the impregnation of 1,2-epoxybutane (1,2-EB) and two ketene derivatives will be described, and the results will be compared with the succinic anhydride (SA)-related results. An important difference between SA and the two other reactive blocking agents is the fact that SA only blocks the amine end groups, whereas both other end-cappers can react with both amine and carboxylic acid end groups. Moreover, the reactivity of the blocking agents is different, which will turn out to be crucial for optimizing the end-group modification of PA-6.

13.3.1.1 Sorption and Diffusion

It will be obvious that the transportation and diffusion of the blocking agent into the polymer granules is crucial to determining the extent of the chemical modification of the end groups of PA-6. Measurement of the sorption of the transporting medium $scCO_2$, with and without the blocking agent, by the polymer granules as a function of time gives an idea of the sorption kinetics and the diffusion coefficients, and tells us how much time is required to approach sorption equilibrium.

The amount of supercritical or subcritical (liquid or gaseous) fluid absorbed in the PA-6 granules is usually determined gravimetrically, e.g., by using a magnetic suspension balance [75]. In contrast to the conventional gravimetric equipment [46], where the balance is in direct contact with the sample, this balance is not. In this way, it is possible to weigh samples without direct contact under a wide range of conditions. Schnitzler et al. [76] discussed the fact that the sorption of penetrants into glassy polymers depends on the temperature. At temperatures well above the T_g, the polymer chains adjust so quickly to the presence of the penetrant that they do not cause diffusion anomalies [77]. Thus, it is appropriate to calculate diffusion coefficients by Fick's second law of diffusion in stagnant media.

A mathematical model to describe the diffusion in a cylinder, as presented by Crank [77], is used in this work to calculate diffusion coefficients. The model is based on the assumption that the direction of the diffusion is only radial. In the case of Fickian diffusion the following equation can be applied:

$$\frac{M_t}{M_\infty} = 1 - \frac{8}{\pi^2} \sum_{n=0}^{\infty} \frac{1}{(2n+1)^2} \exp\left(\frac{-D(2n+1)^2\pi^2 t}{r^2}\right) \qquad (1)$$

In this equation, M_t is the mass of absorbed gas or fluid, M_∞ the equilibrium amount absorbed, D the diffusion coefficient, r the radius of the initially non-swollen polymer particle, and t the sorption time. For short sorption times, Eq. (1) can be replaced by Eq. (2) [78].

$$\frac{M_t}{M_\infty} = \frac{4}{r} \sqrt{\frac{Dt}{\pi}} \qquad (2)$$

For Fickian diffusion, a plot of M_t/M_∞ versus the square root of sorption time t, according to Eq. (2), should be initially linear. The diffusion coefficient can be readily evaluated from the slope of this graph and the initial radius of the sample.

For longer sorption times and over 50% saturation, Eq. (1) can be approximated by Eq. (3). A plot of $\ln(1-M_t/M_\infty)$ vs time t results in a linear graph, from which the diffusion coefficient can be calculated.

$$\frac{M_1}{M_\infty} = 1 - \frac{8}{\pi^2} \exp\left(\frac{-D\pi^2 t}{r^2}\right) \qquad (3)$$

Where the equilibrium amount absorbed, M_∞, was not known because no equilibrium had yet been reached when the measurement was stopped, M_∞ was taken as a second-fit parameter in the calculation.

13.3.2
Amine End-Group Modification with Succinic Anhydride

The polar compound 1,4-dioxane is added to the CO_2 to enhance its polarity in order to raise the solubility of the polar blocking agent and to facilitate the penetration of the fluid into the polar PA-6. The chemical modification was performed in either sub- or supercritical mixtures of CO_2 with 10 mol% 1,4-diox-

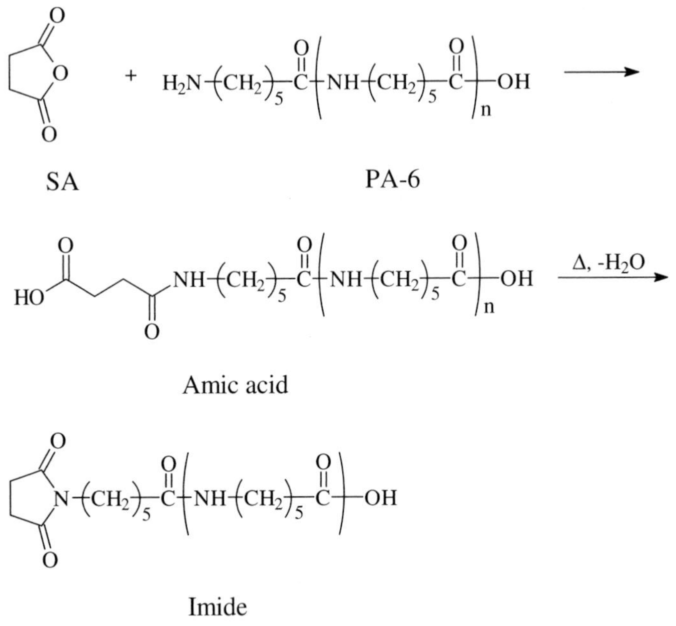

SA   PA-6

Amic acid

Imide

Fig. 13.2 Schematic representation of the amine end-group modification of PA-6 with succinic anhydride. (Reproduced with permission from Elsevier).

ane. Temperatures of 50–140 °C and pressures of 8 MPa were applied. By subcritical conditions we mean conditions below the critical point, i.e. homogeneous liquid media.

For pure CO_2 and 1,4-dioxane, the critical pressures (P_c) and temperatures (T_c) are 7.38 MPa and 31.1 °C, and 5.21 MPa and 313.9 °C, respectively. The critical pressures (P_c) and critical temperatures (T_c) for a CO_2/10 mol% 1,4-dioxane mixture, applied in the work described here, can be estimated using Kay's rule [79] with the following results: $P_c = 7.17$ MPa and $T_c = 59.3$ °C. In view of these calculations, modifications in supercritical fluids were performed at 100 or 140 °C and 8 MPa in CO_2/1,4-dioxane mixtures to ensure operation in the supercritical region. For reactions in the subcritical region in CO_2/10 mol% 1,4-dioxane mixtures, a temperature of 50 °C and a pressure of 8 MPa were applied. So, the sub- or supercritical character of the solvent mixtures was regulated by adjusting the temperature, while keeping the pressure at a constant level.

The chemistry taking place in the swollen polyamide 6 (PA-6) granules is given in Fig. 13.2. Most probably, at elevated temperatures the obtained amic acid is partially converted into the ring-closed imide with formation of a water molecule [80].

13.3.2.1 Sorption Measurements

In our work, sorption by PA-6 of CO_2 saturated with succinic anhydride (SA, around 5 wt% in CO_2) and of both supercritical and subcritical CO_2/10 mol% 1,4-dioxane mixtures saturated with SA was investigated in order to achieve a better idea of the amount of supercritical fluid absorbed in the amorphous region of the polymer. A magnetic suspension balance was used to determine the amount of fluid absorbed by the polymer. The mass increase of the polymer samples due to sorption of the fluid and the density of the fluid were simultaneously recorded.

The sorption of a fluid into a polymer is dependent on the crystallinity of the polymer. The diffusion of a supercritical fluid and therewith the diffusion of small molecules dissolved in this fluid into an amorphous polymer is much more pronounced than it is into a crystalline polymer because of better swelling of the amorphous polymer. As a consequence, the sorption by the PA-6 granules is not expected to be large, since PA-6 is a semi-crystalline polymer. Also, the fact that PA-6 has a polar character, due to the amide bonds in the polymer chain, does not have a positive effect on the amount of CO_2 absorbed by the polymer.

In Fig. 13.3 the sorption of the CO_2 saturated with SA as well as the sorption of CO_2/1,4-dioxane saturated with SA by the PA-6 granules at 50 and 140°C is plotted against time.

It can be seen that the sorption of the more polar CO_2/1,4-dioxane mixture saturated with SA by PA-6 at 140°C and 8 MPa (–◆–) not only goes faster than the sorption of pure CO_2 saturated with SA by PA-6 at 140°C and 8 MPa (–●–), but also the equilibrium amount of fluid absorbed by the polymer is higher, i.e. the equilibrium sorption of CO_2/1,4-dioxane saturated with SA is estimated to be 0.048 g/g PA-6, which is 2.4 times higher than the equilibrium sorption of CO_2 saturated with SA, which is 0.02 g/g PA-6. The D_1 values for the mixtures, calculated for the first 60 min with Eq. (2), are respectively 4×10^{-12} m^2 s^{-1} and 2×10^{-12} m^2 s^{-1}. For longer sorption times, for which the equilibrium sorption M_∞ was approached, Eq. (3) was used, yielding D_2 values of respectively 6×10^{-12} and 5×10^{-12} m^2 s^{-1}, indicating that the sorption rate in CO_2/1,4-dioxane is still higher than the sorption rate in pure CO_2 at longer sorption times. Clearly, the diffusion of CO_2 containing 10 mol% 1,4-dioxane into the PA-6 granules is much faster than the diffusion of pure CO_2 into the PA-6 granules.

Comparing the sorption of scCO$_2$/1,4-dioxane saturated with SA by PA-6 granules at 140°C (–◆–, supercritical conditions) and 50°C (–□–, subcritical conditions), the initial diffusion is faster at 140°C. The D_1 values are respectively 4×10^{-12} m^2 s^{-1} at 140°C and $8\cdot10^{-13}$ m^2 s^{-1} at 50°C. However, after approximately 8 h the amount of fluid absorbed in both measurements is the same within the experimental error of $\pm2\,\mu$g. From that time on the diffusion rate is also approximately the same in both measurements ($D_2 = 6\times10^{-12}$ m^2 s^{-1}).

In Fig. 13.4, the concentration of the amine end groups (determined by potentiometric end-group titrations) after blocking with succinic anhydride in CO_2/10 mol% 1,4-dioxane is given vs time.

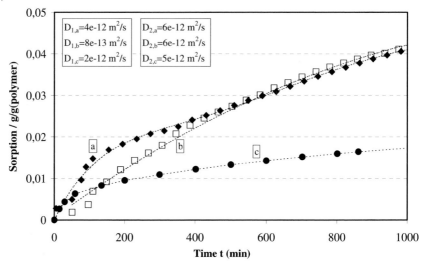

Fig. 13.3 Sorption by PA-6 versus time. $-\bullet-$ sorption of subcritical CO_2 saturated with SA by PA-6 at 140 °C and 8 MPa; $-\blacklozenge-$ sorption of subcritical CO_2/1,4-dioxane saturated with SA by PA-6 at 140 °C and 8 MPa; $-\square-$ sorption of subcritical CO_2/1,4-dioxane saturated with SA by PA-6 at 50 °C and 8 MPa. (Reproduced with permission from Elsevier).

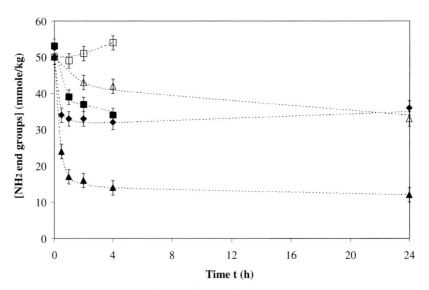

Fig. 13.4 Amine end-group modification of PA-6 with succinic anhydride in supercritical or subcritical CO_2/10 mol% 1,4-dioxane. $-\blacklozenge-$ 50 °C, 8 MPa, subcritical fluid; $-\square-$ ref. exp. without SA, 100 °C, 8 MPa, sc fluid; $-\blacksquare-$ 100 °C, 8 MPa, sc fluid; $-\triangle-$ ref. exp. without SA, 140 °C, 8 MPa, sc fluid; $-\blacktriangle-$ 140 °C, 8 MPa, sc fluid. (Reproduced with permission from Elsevier).

When PA-6 was placed in SA-free CO$_2$/10 mol% 1,4-dioxane at 100°C (Fig. 13.4, –□–), a small increase in amine end-group concentration from 50 to 54 mmol kg^{-1} after 4 h residence time was observed, and the η_{rel} decreased from the initial value of 2.58 (measured in 90% w/w formic acid) to a value of 2.52. These observations imply that the PA-6 did show some thermal degradation when kept in CO$_2$/10 mol% 1,4-dioxane at 100°C for 4 h. The reference experiments (SA free) in CO$_2$/10 mol% 1,4-dioxane at 140°C (Fig. 13.4, –△–) showed a decrease in amine end-group concentration (from 50 to 33 mmol kg^{-1} in Fig. 13.4) after 24 h reaction time. The η_{rel} had increased, which implies that some amine and carboxylic acid end groups had reacted with each other and therefore had resulted in chain extension. From these results we can conclude that there exists equilibrium between degradation and post-condensation of the PA-6 and that post-condensation is favored at higher temperatures.

It is known that the diffusion rate of a supercritical fluid into a polymer is much higher than that of a liquid [56]. The modification of amine end groups with SA in CO$_2$/10 mol% 1,4-dioxane was performed under subcritical conditions, i.e. at 50°C (in a homogeneous liquid mixture) and under supercritical conditions, i.e. at 100 and 140°C, leaving the pressure unchanged (Fig. 13.4, –◆– 50°C, –■– 100°C, –▲– 140°C). Even at 50°C, the blocking of the amine end groups with succinic anhydride proceeded rapidly, but the amine end-group concentration leveled off at a value of 30–35 mmol kg^{-1}. The constant but relatively high value which is reached can possibly be explained in terms of (1) limited mobility of the amorphous phase of the PA-6, because the temperature is below its T_g (55°C), and (2) worse solubility of the SA in the CO$_2$/10 mol% 1,4-dioxane mixture at this low temperature. Furthermore, since the sorption of this CO$_2$/10 mol% 1,4-dioxane saturated with SA is slower at 50°C than at 140°C for the first 4 h but does reach the same value after 8 h (Fig. 13.3), this means that an insufficient amount of SA is impregnated into the PA-6 granules. For the modification in scCO$_2$/10 mol% 1,4-dioxane at 100°C, the amine end-group concentration decreases to a value of 34 mmol kg^{-1} after 4 h, but is expected to decrease further with time, since this temperature is approximately 50°C above the T_g, which means enhanced flexibility and swellability of the amorphous phase. It was expected that the modification of the amine end groups with succinic anhydride in CO$_2$ at 100°C would proceed faster than the modification at 50°C. On the contrary, at 50°C the modification of the amine end groups seems to go faster. The reason for this is that at 100°C degradation of the polymer occurs, resulting in the generation of new amine end groups, which partially compensates the end-group modification (see reference experiment at 100°C in Fig. 13.4). This results in an overall less complete end-group modification. At 50°C, hardly any or no degradation of the polymer occurs. For the modification in scCO$_2$/10 mol% 1,4-dioxane at 140°C, again a large decrease in amine end-group concentration was found (from 50 to 14 mmol kg^{-1}, Fig. 13.4, –▲–).

Now that we have shown that the succinic anhydride does react with the amine end groups under the applied conditions, it is of interest to check whether only the amine end groups close to the surface of the granules or also the amine end groups in the core of the granules are modified. To get a better insight into the diffusion of the SA into the polyamide granules, a simple and straightforward approach is applied: cutting a cross section of the granule and investigate the granule from the surface to the core visually and with FTIR spectroscopy. In doing so, one should realize that imide formation usually implies a yellow to slightly brown discoloration. Observation with the naked eye of a cross section of a colored granule showed that the outer shell of the granule was colored, whereas the core of the granule was less colored or even remained colorless. This observation suggests that the end-group modification took place mainly in the outer shell of the PA-6 granules. To reinforce these observations, FTIR spectroscopy was used. A few modified PA-6 samples were investigated. The results obtained for PA-6 modified with SA in $scCO_2$/1,4-dioxane at 140 °C and 8 MPa for 4 h are given in Fig. 13.5. At 1725 cm^{-1}, an absorption peak is observed for the surface of the modified granule. This absorption is related to the formed acid end group and decreases for spectra obtained for positions in the granule closer to the core (a–c). In the core of the granule, this absorption is completely absent and the spectrum is similar to that of unmodified polyamide 6 (d). If the modification was allowed to continue for 24 h, reaction of SA with the amine end groups occurred even in the core. If the very same modification was performed in subcritical CO_2/1,4-dioxane at 50 °C and 8 MPa for 4 h, again an absorption peak was observed at 1725 cm^{-1} for the surface of the modified granule. However, when approaching the core, the absorption peak drastically diminishes. Already at 40 µm from the surface this absorption peak is

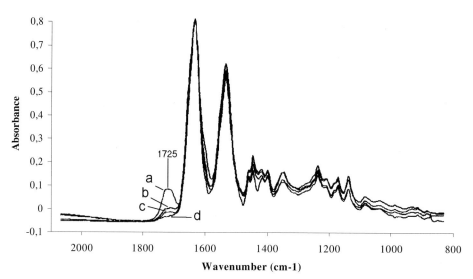

Fig. 13.5 FTIR spectra of a cross section of a PA-6 granule modified with SA in $scCO_2$/10 mol% 1,4-dioxane at 140 °C for 4 h. a = surface, b = 200 µm from the surface, c = core, d = surface of unmodified PA-6. (Reproduced with permission from Elsevier).

hardly visible, and at a distance of 200 μm from the surface it is totally absent and the spectrum is similar to that of the unmodified PA-6. Thus, in 4 h succinic anhydride penetrates much deeper into the polymer particles during the modification under supercritical conditions than during the modification under subcritical conditions. This is in agreement with the sorption measurements (see Fig. 13.3).

From these results it is obvious that the diffusion of CO_2, and therewith the transportation of the blocking agent into the polymer, is insufficient to result in blocking of the end groups in the core of the granules. These findings are in agreement with the sorption of CO_2 in the granules and with the end-group titration results, i.e. the fact that the end-group concentration does not go to zero.

To investigate the possible advantage of the supercritical technique compared with the modification of granules in liquids at atmospheric pressures, modifications were performed with succinic anhydride in decane/10 mol% 1,4-dioxane or in decalin/10 mol% diethylene glycol diethyl ether. These additives were chosen because of their high boiling points (allowing sorption measurements at atmospheric pressure) and a solvent polarity reasonably close to that of 1,4-dioxane to allow a reasonable comparison. Although these mixtures approach the (non-)polarity of the CO_2/1,4-dioxane mixture, it remains difficult to compare these modifications in common liquids with those in the applied supercritical fluids and draw hard conclusions, since not only the phase of the reaction medium changes, but also the reaction medium itself (i.e. molecular size, polarity etc.). This was unavoidable, since CO_2 is a gas at ambient temperatures and pressures. Comparing the blocking at 100 °C and 140 °C in decalin/10 mol% diethylene glycol diethyl ether, amine end-group concentrations respectively decrease from 50 to 39 mmol kg^{-1} and from 50 to 21 mmol kg^{-1} after 4 h reaction time. For a similar modification under supercritical conditions, the amine end groups decrease to 34 mmol kg^{-1} (100 °C) and 14 mmol kg^{-1} (140 °C). Clearly, under supercritical conditions, the end-group modification at 140 °C is much more complete, and a lower concentration of amine end groups is reached compared with modification in organic liquids, although the difference is less pronounced than expected.

13.3.2.2 Melt Stability of Modified and Unmodified PA-6

Since the aim of modifying the amine end groups of PA-6 was to enhance the stability in the melt with respect to chain extension and accordingly to avoid the generation of gel particles in PA films, this was checked by the authors. In order to keep other properties of the polyamide (like flow, crystallization behaviour and tensile properties) the same, the molecular weights ($\bar{M}_w$) of the PA-6 samples should not change upon the supercritical modification, and should even remain at a constant level after a residence time in the melt of 20–30 min, which is typical for PA-6 film extrusion. This was checked by performing SEC (Size Exclusion Chromatography) measurements on starting material and on PA-6 with blocked amine end groups, both before and after a stay in the melt.

The $\bar{M}_w$ of unmodified PA-6 significantly increases after a residence time of 30 min in the melt from 23 to 64 kg mol^{-1}. Clearly, the post-condensation reaction which occurs between the amine and carboxylic acid end groups in the melt at 260 °C cannot be avoided. For PA-6 whose amine end groups have partially been blocked, a comparable treatment in the melt leaves the $\bar{M}_w$ nearly unaffected (a change from 20 to 21 kg mol^{-1}). These results clearly point to an increased melt stability of the partially end-capped PA-6. The fact that the molecular weight of PA-6 slightly decreases upon the modification in the supercritical fluid indicates that more low-molecular-weight material is present as a result of some chain scission.

13.3.2.3 Conclusions

In this section it has been shown that it is possible to block the amine end groups present in PA-6 granules with succinic anhydride in supercritical and subcritical fluids. Sorption measurements showed that the addition of 10% 1,4-dioxane to CO_2 resulted in an improved swelling and sorption of the granules, which is favorable for the amine end-group modifications. The modified PA-6 samples clearly showed improved melt stability compared with the unmodified PA-6.

13.3.3
Amine and Carboxylic Acid End-Group Modification with 1,2-Epoxybutane

In contrast to the results described in Section 13.3.2, this section describes the blocking of both the amine and the carboxylic acid end groups of PA-6 in supercritical and subcritical fluids. This time 1,2-epoxybutane (1,2-EB) was used as the blocking agent, which is reactive toward both carboxylic acid and amine end groups [81, 82] and is a relatively small compound, which is favorable for its impregnation into the swollen PA-6 granules The aim of blocking the end groups of PA-6 is to avoid undesired side reactions, thereby improving the melt stability during processing without changing the molecular weight distribution. As in the previous section, 10 mol% 1,4-dioxane was added to the CO_2.

The chemistry taking place in the swollen PA-6 granules is shown In Fig. 13.6. For the supercritical and subcritical modifications, similar temperatures and pressures to those described in Section 13.3.2 were applied. Attention was focused on the blocking of the carboxylic acid end groups by titrating these groups. Since, under similar conditions, epoxides are more reactive toward primary amines than toward carboxylic acids, it can be expected that the blocking of the amine groups will be at least as efficient as the simultaneous blocking of the carboxylic acid groups; however, this cannot be determined by standard titration experiments.

In Fig. 13.7 the results for the modification of the carboxylic acid end groups in CO_2/10 mol% 1,4-dioxane at 50, 100 and 140 °C are presented. The results show that under subcritical conditions, i.e. at 50 °C, hardly any reaction between

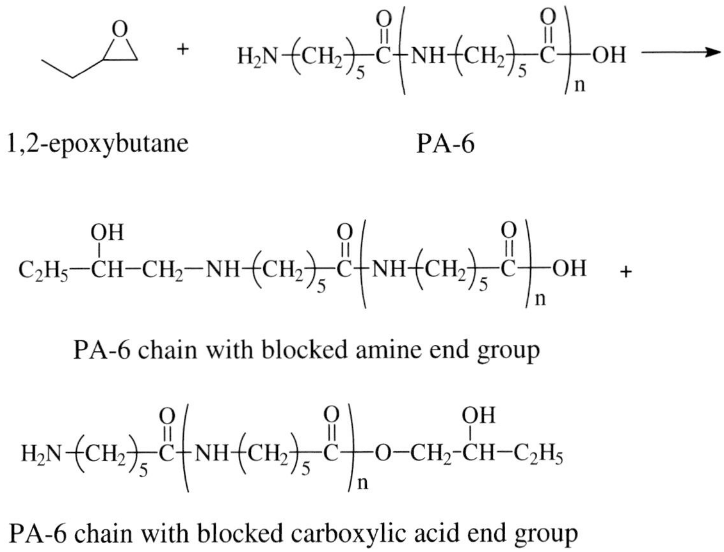

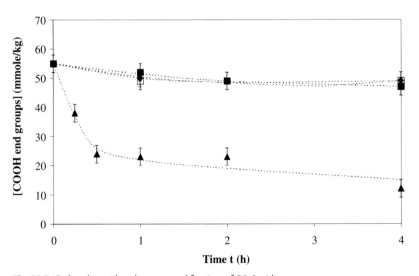

1,2-epoxybutane   PA-6

PA-6 chain with blocked amine end group   +

PA-6 chain with blocked carboxylic acid end group

Fig. 13.6 Schematic representation of the amine and carboxylic acid end-group modification of PA-6 with 1,2-epoxybutane. (Reproduced with permission from Elsevier).

Fig. 13.7 Carboxylic acid end-group modification of PA-6 with 1,2-epoxybutane in supercritical or subcritical CO$_2$/10 mol%1,4-dioxane. –◆– 50 °C, 8 MPa, subcr. fluid; –□– ref. exp., 100 °C, 8 MPa, sc fluid; –■– 100 °C, 8 MPa, sc fluid; –▲– 140 °C, 8 MPa, sc fluid. (Reproduced with permission from Elsevier).

the epoxide and the carboxylic acid end groups had occurred, but we have to keep in mind that the amine end groups might have been end capped by the 1,2-epoxybutane at these temperatures. Increasing the temperature to 100 °C, thereby bringing the $CO_2/10$ mol% 1,4-dioxane mixture into the supercritical state, does not result in a decrease in carboxylic acid end-group concentration. With increasing temperature from 100 °C to 140 °C, the blocking of the carboxylic end groups improves. In $scCO_2/10$ mol% 1,4-dioxane of 140 °C, already after 30 min a decrease from 55 mmol kg^{-1} to 24 mmol kg^{-1} carboxylic acid end groups is observed, implying effective blocking. The better modification is a result of increased reactivity of the 1,2-epoxybutane at 140 °C compared to 100 °C.

A preliminary conclusion from these results is that going from subcritical (a homogeneous liquid) to supercritical conditions (from 50 to 100 °C) is not enough to block the carboxylic acid end groups of PA-6. Not only is a quick sorption of $CO_2/1,4$-dioxane containing 1,2-epoxybutane needed, but so also is a temperature exceeding 100 °C to get the chemistry going.

As reported for the amine end-group modification with the highly reactive succinic anhydride, an FTIR spectroscopic analysis of the granule from the surface to the core showed that the end-group modification with the epoxide in $scCO_2/10$ mol% 1,4-dioxane at 140 °C occurred in a quite homogeneous way, since there was no difference between the spectra for the surface and the core of the modified granule, but there was a clear difference from the spectrum of the unmodified PA-6. This could be concluded by focusing on the absorption peak at 1725 cm^{-1}, which is ascribed to the carbonyl of the ester group. Interestingly, this result differs considerably from the results obtained for the blocking of the amine end groups with succinic anhydride, described in Section 13.3.2. The extremely high reactivity of the SA with the amine end groups is thought to result in a reduction of the free volume close to the edge of the PA-6 granules after partial blocking of the chain ends. The reduction of free volume is thought to slow down the sorption process. The epoxide is less reactive toward the end groups than the SA, resulting in a slower reduction of the free volume in the outer part of the granules, which allows a deeper penetration of the epoxide solution into the PA-6 granules. The result is a more homogeneous modification.

As described in Section 13.3.2, the supercritical modifications shown in Fig. 13.7 are compared with modifications in organic liquids at atmospheric pressure at 100 and 140 °C. The carboxylic acid end-group concentration of PA-6 decreased from 55 to 49 and 40 mmol kg^{-1} after 4 h reaction time at respectively 100 °C and 140 °C in decalin/10 mol% diethylene glycol diethyl ether. Although the end-group modification at 140 °C is somewhat more effective than at 100 °C (because the reaction at 100 °C is too slow), the modification is not as effective as in supercritical fluids at 140 °C, which resulted in a carboxylic acid concentration of 12 mmol kg^{-1}. Comparing the modifications at 100 °C but in different solvent mixtures, the carboxylic acid end-group concentrations are the same within the experimental error. So, as for the amine end-group modifica-

tion with SA, for carboxylic acid end-group modification with 1,2-EB a difference in blocking efficiency in liquids and sc fluids seems to exist in favor of the
sc fluids, provided that the temperature is high enough to let the carboxylic acid
end groups react with the epoxy groups. However, as was found for the amine/
SA system, the advantage of using supercritical fluids for polymer modification
is less pronounced than expected.

13.3.3.1 Melt Stability of Modified and Unmodified PA-6

As reported for the PA-6 samples whose amine end groups had been blocked
with SA, the molecular weights ($\bar{M}_w$) before and after modification with 1,2-EB
proved to be the same within experimental error. Moreover, the melt stability of
these modified PA-6 samples proved to be virtually invariant after a treatment
for 30 min at 260 °C. Consequently, the supercritical modification of PA-6 with
1,2-epoxybutane also results in a polyamide film grade with an enhanced melt
stability.

13.3.3.2 Conclusions

In this section it is shown that it is possible to block the carboxylic acid end
groups, present in PA-6 granules, with 1,2-epoxybutane in supercritical and subcritical fluids. It was found that better blocking of these end groups occurred at
higher temperatures, and an improved melt stability was obtained.

   If we compare this with the end-group modification with succinic anhydride
(SA) described in Section 13.3.2, the modification with 1,2-epoxybutane is to be
preferred, since 1,2-epoxybutane is less reactive than SA, which favors the diffusion of the blocking agent to the core of the granules, resulting in a more uniform modification of the granules.

13.3.4
Amine End-group Modification with Diketene and Diketene Acetone Adduct

In this section, the blocking of the amine end groups of PA-6 with liquid diketene
(the dimer of ketene) and the diketene acetone adduct (Fig. 13.8) in supercritical
CO$_2$ is discussed. Ketene itself is an extremely reactive, unstable, and very toxic
gas. Diketene and the diketene acetone adduct have frequently been used in industry since they are reactive toward a large variety of functional groups such as
amines, alcohols, and carboxylic acids [83–85], but are not reactive toward the
amide groups in the PA-6 chain without a catalyst, whereas ketene is. This makes
them useful for the modification of polymer particles in supercritical CO$_2$ under
very mild reaction conditions, thereby avoiding side reactions.

   In this section, the reactive reagent diketene and the somewhat less reactive
diketene acetone adduct are impregnated into the polyamide granules, using
pure CO$_2$ as a carrier. Similar experimental conditions to those described for
the amine end-group modification with SA (Section 13.3.2) were applied.

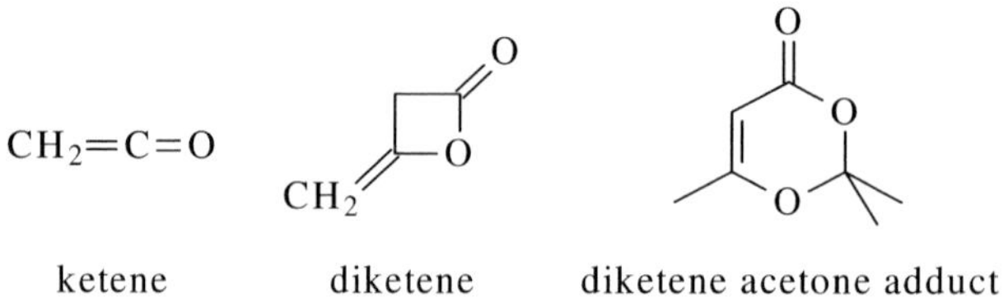

$$CH_2=C=O$$

ketene   diketene   diketene acetone adduct

Fig. 13.8 Structure of ketene, its dimer, and the diketene acetone adduct.

13.3.4.1 Modification of PA-6 Granules with Diketene and Diketene Acetone Adduct in Supercritical and Subcritical CO₂

The amine end groups of PA-6, which are more reactive toward the blocking agents than the carboxylic acid end groups [86, 87], are converted into acetoace-tamide end groups. If the carboxylic acid end groups reacted at all with the blocking agent, this would result in aliphatic anhydrides, which are hydrolyti-cally unstable and therefore would be readily reconverted into the carboxylic acid end groups (Fig. 13.9). Therefore, in this section only the modification of the amine end groups is considered. To prevent the post-condensation between amine and carboxylic acid end groups during melt processing, and accordingly

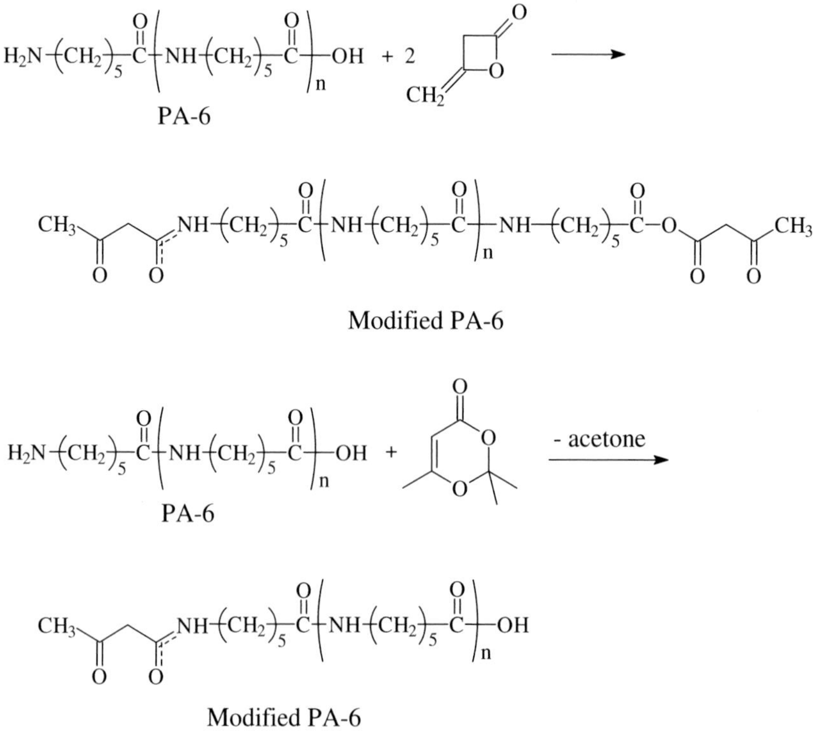

PA-6

Modified PA-6

PA-6   - acetone

Modified PA-6

Fig. 13.9 Acetoacetylation of end groups of PA-6 in scCO₂.

to avoid an undesired raise in melt viscosity, blocking of only one of the end groups is sufficient.

13.3.4.2 Molecular Characterization

Results of the amine end-group modification of PA-6 granules performed under supercritical conditions, namely at 75 °C/30 MPa, 100 °C/30 MPa and 100 °C/10 MPa, as well as under subcritical conditions (75 °C/6 MPa), are described here. The obtained amine end-group concentration found for modified PA-6 granules was compared with the value for unmodified PA-6 granules (i.e. 55 ± 2 mmol kg^{-1}).

Upon raising the temperature from 75 °C to 100 °C at constant pressure (i.e. either 30 or 10 MPa) for modifications with either diketene or diketene acetone adduct, more amine end groups are blocked. Better blocking of the amine end groups at higher temperature is expected: the reactivity of the diketene and the diketene acetone adduct increases with increasing temperature and the sorption of CO$_2$ by PA-6 granules is considered to be better at higher temperature, resulting in an increased penetration into the polymer of diketene and the diketene acetone adduct, which are dissolved in the CO$_2$. The decrease in amine end-group concentration for the diketene acetone adduct is less pronounced than it is for the end-group modification with diketene at similar conditions. This can be explained in terms of reactivity; under similar reaction conditions the diketene is more reactive than the acetone adduct. In addition, the acetone adduct molecule is more bulky, which might hamper its diffusion into the amorphous phase of polyamide granules. An illustrative example is the fact that for the diketene after 24 h at 100 °C/10 MPa an amine end-group concentration of 11 mmol kg^{-1} was obtained, whereas for the less reactive and more bulky acetone adduct under the same reaction conditions a concentration of 34 mmol kg^{-1} was obtained.

In general, when the pressure is increased the polymer swelling increases, resulting in better diffusion of CO$_2$ and hence of blocking agent [3]. However, the results show that, for modifications with either diketene or diketene acetone adduct, upon decreasing the pressure from 300 via 10 to 6 MPa at constant temperature (either 75 or 100 °C) more amine end groups are blocked, even though the lowest pressure of 6 MPa implies subcritical conditions. This is probably caused by the so called "concentration effect". The pressure is raised from 6 to 30 MPa by pumping more CO$_2$ into the reactor at constant temperature, which obviously reduces the concentration of the blocking agent in the supercritical mixture. As a consequence, at 6 MPa the concentration of blocking agent is at its highest.

For this reason, the fact that for the modification with diketene at 75 °C and 6 MPa for 24 h a low concentration of amine end groups is found (i.e. 8 mmol kg^{-1}, which is in the same order of magnitude as the 11 mmol kg^{-1} obtained under supercritical conditions at 100 °C/10 MPa, see earlier) is ascribed to both the very high reactivity of the diketene and the concentration effect. Interestingly, after 4 h the supercritical modification with diketene at 100 °C and 10 MPa is more complete than the subcritical modification at 75 °C and 6 MPa (amine concentrations 27 and 36 mmol kg^{-1}, respectively). Obviously, the supercritical modification is faster, but after 24 h an equilibrium value around 10 mmol kg^{-1} is reached.

As reported for the amine end-group modification with the highly reactive succinic anhydride, an FTIR spectroscopic analysis of a granule from the surface to the core showed that the end-group modification with the small and reactive diketene molecule mainly took place in the outer shell of the PA-6 granules. This could be concluded by focusing on the acetoacetamide end-group absorption at 1720 cm^{-1}. In the core of the granule, the FTIR spectrum was similar to the spectrum of a non-modified PA-6 granule. A similar study on a granule treated under the same conditions with the less reactive and more bulky diketene acetone adduct showed a less extensive but more homogeneous modification throughout the entire granule. This difference in homogeneity is explained in Section 13.3.3.

13.3.4.3 Conclusions
In this section it is shown that diketene, and to a lesser extent the diketene acetone adduct, are useful reagents for the blocking of amine end groups of polyamide-6 granules. A decreased amine end-group concentration was obtained with increasing temperature and decreasing pressure. FTIR spectra showed that most extensive modification of the granules occurred at the surface, implying that the diffusion of CO_2, and hence of the blocking agent, into the core of the granules is hampered.

13.3.5
General Conclusions on Polyamide Modification

Comparing the results of the modification of PA-6 end groups with either succinic anhydride, 1,2-epoxybutane, or diketene and the diketene acetone adduct, it can be concluded that the optimal blocking agent must be small and must exhibit an intermediate reactivity toward the end groups in order to allow extensive penetration into the core of the granules before the blocking reaction starts to hinder further penetration.

13.4
Carboxylic Acid End-group Modification of Poly(Butylene Terephthalate) with 1,2-Epoxybutane in Supercritical CO_2

13.4.1
Background

Poly(butylene terephthalate), PBT, is a material which is widely used in a variety of applications, one of which is for optical fiber tubing. These fiber tubes are buried in the ground, and therefore it is necessary for this application that the polymer should be hydrolytically stable. However, the hydrolytic stability of PBT is not optimal and needs to be improved. The hydrolysis of the ester bonds in the PBT chain is an auto-catalytic process catalyzed by the carboxylic acid end groups [88–90]. By reducing the number of carboxylic acid end groups the hydrolysis can be retarded. Known routes to reducing the carboxylic acid end-group concentration are, for example, changing the polymerization process conditions, adding a polymerization co-catalyst, or using chain-extension reactions in the melt. A new and clean route is the chemical modification of these carboxylic acid end groups in supercritical fluids. Supercritical CO_2 is frequently used as a medium for dyeing polyester fibers [41, 91, 92]. Chemically modifying the carboxylic acid end groups of PBT in supercritical fluids is therefore a logical step, with good commercial prospects.

Supercritical fluids do not dissolve the polyester but do have the possibility to swell the amorphous phase of the polymer. This is possible under relatively mild reaction conditions, which could to a large extent limit undesired side reactions or degradation reactions that could possibly occur during modification in the melt. Since the hydrolysis of ester functionalities of PBT takes place in the amorphous phase, it is important to block most of the carboxylic acid end groups, which are in this amorphous phase as well. The swellability of PBT depends on the interaction of the supercritical fluid with the PBT and on the crystallinity of this polymer. The higher the percentage of crystallinity, the more difficult it is to swell the PBT. The crystallinity of this type of semi-crystalline polymer is approximately 40%. This is relatively high, and therefore the solubility of CO_2 in PBT would be expected to be poor (in the range of 2–5 wt%). However, if low concentrations of carboxylic acid end groups can nevertheless be reached, the supercritical fluid modification technique still proves to be useful.

13.4.2
Chemical Modification of PBT with 1,2-Epoxybutane

In our investigation into the chemical modification of swollen polymer particles in supercritical fluids, PBT with 44 mmol kg^{-1} carboxylic acid end groups was modified with 1,2-epoxybutane in supercritical CO_2 or in a supercritical or subcritical mixture of CO_2 with 10 mol% 1,4-dioxane. The critical pressures (P_c) and temperatures (T_c) of CO_2, and the P_c and T_c values for mixtures of CO_2 with 10 mol% 1,4-dioxane, calculated using Kay's rule [79], have been given in Section 13.3.1.

Temperatures of 120 and 180 °C and pressures of 6, 15 and 30 MPa were applied. The reactions were performed for periods of 1, 2, 4 and 24 h. Modification at a pressure of 6 MPa (subcritical conditions, i.e., both CO_2 and 1,4-dioxane are in the gas phase) was chosen in order to be able to trace possible differences between supercritical and subcritical end-group modification. The reaction temperature needs to be above the T_g of the PBT granules to ensure sufficient chain mobility, but below the T_m to avoid deformation or melting of the granules and, after depressurization, to regain their original form and their free-

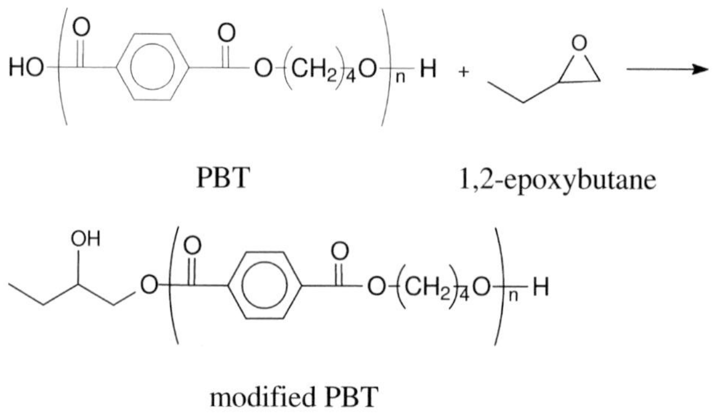

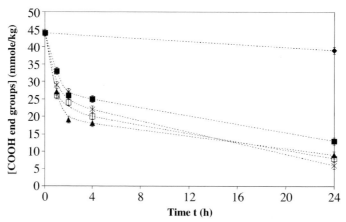

PBT   1,2-epoxybutane

modified PBT

Fig. 13.10 Reaction scheme of the carboxylic acid end-group modification of PBT.

flowing character without sticking together. The chemistry of the modification of the carboxylic acid end groups with 1,2-epoxybutane is given schematically in Fig. 13.10. Although the hydroxyl end groups could also react with 1,2-epoxybutane, the carboxylic acid groups are expected to be much more reactive, namely by a factor 10 to 20 [81].

In Fig. 13.11, the carboxylic acid end-group concentrations of PBT samples, which were modified with 1,2-epoxybutane under several different conditions, are given versus modification time (experimental error is ± 2 mmol kg^{-1}).

Comparing the results of the modification of the carboxylic acid end groups with 1,2-epoxybutane in scCO$_2$ at 120 °C ($-\blacklozenge-$) and at 180 °C ($-\blacksquare-$), it is found that at

Fig. 13.11 Carboxylic acid end-group concentration of PBT granules modified with 1,2-epoxybutane vs time. $-\blacklozenge-$ scCO$_2$ at 120 °C and 30 MPa; $-\blacksquare-$ scCO$_2$ at 180 °C and 30 MPa; $-\blacktriangle-$ scCO$_2$ with 10 mol% 1,4-dioxane at 180 °C and 30 MPa; $-\times-$ scCO$_2$ with 10 mol% 1,4-dioxane at 180 °C and 15 MPa; $-\square-$ subcritical CO$_2$ with 10 mol% 1,4-dioxane at 180 °C and 6 MPa. Error bars are ± 2. (Reproduced with permission from Elsevier).

120 °C hardly any end groups are blocked, whereas at 180 °C blocking of the carboxylic acid end groups occurs rapidly. This is explained in terms of reactivity: at 120 °C, the 1,2-epoxybutane is not reactive enough toward the carboxylic acid end groups in the polymer. Other researchers also observed that the curing of COOH-containing powder coatings with multifunctional epoxides in the presence of a catalyst occurs readily at 180 °C or higher temperatures, and found that almost no curing takes place at lower temperatures [93]. Therefore a temperature of 180 °C was chosen for further modification experiments, with the additional advantages of better swelling of the granules and increased chain mobility in the amorphous phase of the polymer with increasing temperature. At 180 °C and 30 MPa in scCO$_2$, after 2 h a decrease from 44 to 26 mmol kg^{-1} carboxylic acid end groups is observed, and after 24 h a decrease to even 13 mmol kg^{-1} carboxylic acid end groups is obtained, which is close to the desired concentration of ca. 10 mmol kg^{-1}, where the hydrolysis process is drastically diminished [94]. Comparison of the results of the modification of the carboxylic acid end groups with 1,2-epoxybutane in scCO$_2$ (–■–) with those using scCO$_2$/1,4-dioxane (–▲–) at 180 °C and 30 MPa shows that the modification proceeds better in the scCO$_2$/1,4-dioxane mixture, although the difference is relatively small. A better modification in this mixture would be expected, as was shown above in Section 13.3.2. After 24 h modification in scCO$_2$/1,4-dioxane at 180 °C and 30 MPa (–▲–) a concentration of 9 mmol kg^{-1} carboxylic acid end groups is obtained, which is below the target value of 10 mmol kg^{-1}. To investigate the effect of decreasing the pressure to below the critical pressure of the CO$_2$/1,4-dioxane mixture, additional modifications were performed at 15 and 6 MPa, the latter resulting in a subcritical medium. Decreasing the pressure is accomplished by pumping a smaller amount of CO$_2$/1,4-dioxane into the high-pressure cell. The CO$_2$/1,4-dioxane density decreases as well, which results in a 5 times higher concentration of 1,2-epoxybutane in this medium at 6 MPa compared to the medium at 30 MPa. Looking at the modifications in scCO$_2$/1,4-dioxane at 180 °C and 30, 15 and 6 MPa (Fig. 13.11), all modifications resulted after 24 h in approximately the same carboxylic acid end-group concentration. It was expected that the modification at 30 MPa would result in the best blocking of the end groups, because a general observation is that a higher density of the medium results in a better swelling and sorption of the polymer [4, 95]. But on the other hand the above-mentioned 1,2-epoxybutane concentration effect is expected to result in better end-group modification as well. So for long exposure times (24 h) a higher system pressure seems to compensate for a lower 1,2-epoxybutane concentration, resulting in a similar carboxylic acid end-group concentration.

For the modifications of PA-6 with diketene and diketene acetone adduct given in Section 13.3.4, the concentration effect proved to be significant: a higher concentration of the blocking agent resulted in an improved blocking of the end groups. This implies that the diketene and its acetone adduct are more reactive toward the amine end groups of PA-6 than 1,2-epoxybutane is towards the carboxylic acid end groups of PBT.

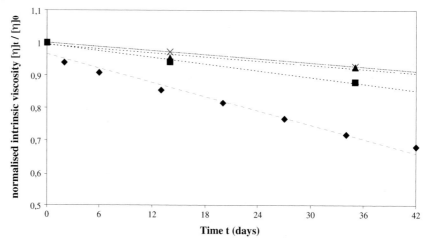

Fig. 13.12 Normalized intrinsic viscosity of PBT granules hydrolyzed vs time.
–◆– unmodified PBT; –■– PBT, modified in scCO₂ at 180 °C and 30 MPa
for 24 h; –▲– PBT, modified in scCO₂ with 10 mol% 1,4-dioxane at 180 °C and
30 MPa for 24 h; –×– PBT, modified in scCO₂ with 10 mol% 1,4-dioxane at
180 °C and 15 MPa for 24 h.

13.4.2.1 Influence of Acid End-Group Concentration on Hydrolytic Stability

A reliable check for determining the hydrolytic stability is to determine the intrinsic viscosity of the modified PBT granules which were kept in water at 90 °C for a certain time, and compare the values with unmodified PBT granules which were also kept in water of 90 °C for the same time. The investigated samples were PBT modified for 24 h in (1) scCO₂ at 180 °C and 30 MPa, (2) scCO₂ with 10 mol% 1,4-dioxane at 180 °C and 30 MPa, and (3) scCO₂ with 10 mol% 1,4-dioxane at 180 °C and 15 MPa. The carboxylic acid end-group concentrations of these modified PBT samples are respectively 13, 9 and 6 mmol kg^{-1}. In Fig. 13.12, the normalized intrinsic viscosity of hydrolyzed PBT granules is given as a function of hydrolysis time.

From this Figure, it is clear that the intrinsic viscosity of unmodified PBT (–◆–) decreased significantly more rapidly than that of the end-capped PBT samples and that the hydrolytic stability increased with decreasing end-group concentrations. Clearly, modifying the carboxylic acid end groups in supercritical fluids has a positive effect on the hydrolytic stability.

13.4.2.2 Determination of Molecular Weights

The hydrolysis experiments showed that the modification of the carboxylic acid end groups of PBT in supercritical fluids resulted in an increased hydrolytic stability compared to the unmodified PBT. The sensitivity of the intrinsic viscosity towards hydrolysis conditions is a measure of the hydrolytic stability. Hydrolysis of the ester functionalities in the polymer chain is also reflected in the molecu-

of the polymer. Therefore, the number and weight average molecular weights ($\bar{M}_n$ and $\bar{M}_w$) of the three modified PBT samples and the unmodified PBT, before and after hydrolysis, were investigated.

For unmodified PBT, the $\bar{M}_n$ and $\bar{M}_w$ are respectively 16 200 and 34 000 g mol^{-1} ($\bar{M}_n$ accurately calculated, based on the end groups) before hydrolysis. For the three modified PBT samples the $\bar{M}_n$ and $\bar{M}_w$ are 15 400 and 30 100 g mol^{-1} (which corresponds to a decrease of 5–11%). This is an indication that some hydrolysis of the polymer chains had occurred during the modifications as a result of traces of water, which is always present even in dried PBT samples.

The hydrolysis of the unmodified PBT samples results in significant decreases in molecular weights. After 5 weeks in water at 90 °C, the $\bar{M}_n$ and $\bar{M}_w$ decreased about 40%.

The hydrolysis of the modified PBT samples results in a much less pronounced decrease in molecular weights. The decrease in $\bar{M}_n$ and $\bar{M}_w$ is only around 16% for modified PBT samples. Clearly, the modification results in a hydrolytically more stable polymer.

13.4.3
General Conclusions Concerning PBT Modification

In this section we show that it is possible to block carboxylic acid end groups of PBT with 1,2-epoxybutane in supercritical and subcritical fluids. The best results were obtained for the modification in scCO$_2$ with 10 mol% 1,4-dioxane at 180 °C. A decrease in the carboxylic acid end-group concentration from 44 mmol kg^{-1} to approximately 10 mmol kg^{-1} was achieved. This low concentration of acid end groups resulted in a significantly enhanced hydrolytic stability, which was demonstrated by measuring the intrinsic viscosity as a function of hydrolysis time and measuring the number and weight average molecular weights ($\bar{M}_n$ and $\bar{M}_w$) before and after modification.

13.5
Concluding Remarks and Outlook

As we come to the end of this chapter, we will attempt to put the results in perspective. We have explored the scope and limitations of modifying a variety of polymers in supercritical and subcritical fluids, while trying to provide a better understanding of the process. It is obvious that the use of supercritical fluids in polymerization and polymer modification has several advantages over more conventional techniques such as melt and solution modification.

In this chapter it has been shown that CO$_2$ can be used as the reaction medium for polymer modification. Depending on a variety of factors, such as the chemistry of the modification, the polarity of the polymer, and the temperature and pressure, which are needed for sufficient modification of the polymer, one can decide which supercritical fluid is most suitable.

The modification of the reactive chain ends of PA-6 with a variety of blocking agents can be successfully performed in CO_2 to which 1,4-dioxane has been added. The addition of 1,4-dioxane enhances the polarity of the supercritical fluid and hence the interaction with the polar polymer (i.e. sorption by and swelling of the polymer). Because of the limited swelling of the polymer, modification of the end groups mainly occurred in the outer layer of the granules, but in some special cases the core of the granules was also reached by the blocking agent.

The reactivity of the blocking agent proved to be an important factor. The blocking agent needs to be reactive enough to block the end groups under relatively mild reaction conditions, but must not be too reactive, otherwise then it will instantaneously react with the end groups located in the shell of the granules, reducing the free volume in the swollen amorphous phase, thereby retarding (or even preventing) further transportation of the supercritical fluid and especially the blocking agent toward the core of the granules. It is obvious that very bulky blocking agents are less suitable for a quick penetration into the polymer granules.

Furthermore, if the blocking agent is too reactive, side reactions such as chain branching and discoloration reactions can occur, which are undesirable from the point of view of utility of the polymer. For this reason, optimal and most homogeneous end-capping results were obtained using relatively small blocking agents with intermediate reactivity. As an example, good results were obtained using 1,2-epoxybutane for blocking the end groups of PA-6 and PBT.

Comparing supercritical polymer modification with (a) polymer granule modification in organic liquids (in which only the blocking agent is dissolved), and (b) more conventional solution and melt modification processes, it was found that the benefits of using supercritical fluids are less pronounced than was hoped for. Nevertheless, it is a necessity to further optimize polymer modification processes in supercritical CO_2, because we are convinced that the importance of CO_2 technology will increase when the environmental legislation on the use of organic solvents becomes more stringent.

Notation

P_c	Critical pressure of a fluid	[Pa]
T_c	Critical temperature of a fluid	[°C]
T_g	Glass transition temperature of a polymer	[°C]
T_m	Melting temperature of a polymer	[°C]
ΔH_m	Melting enthalpy of a polymer	[kJ]
M_t	Mass of absorbed gas or fluid in a polymer particle	[kg]
M_∞	Equilibrium amount of absorbed gas or fluid in a polymer particle	[kg]
D	Diffusion coefficient into a polymer particle	[m^2 s^{-1}]
R	Radius of the initially non-swollen polymer particle	[m]

t	Sorption time of absorbed gas or fluid in a polymer particle	[s]
η_{rel}	Relative viscosity of a polymer	[–]
$[\eta]$	Intrinsic viscosity	[dl g^{-1}]
$\bar{M}_n$	Number average molecular weight of a polymer	[kg mol^{-1}]
$\bar{M}_w$	Weight average molecular weight of a polymer	[kg mol^{-1}]

References

1 Y. T. Shieh, J. H. Su, G. Manivannan, P. H. C. Lee, S. P. Sawan, W. D. Spall, *J. Appl. Polym .Sci.*, **1996**, *59*, 695, 707.

2 W. C. V. Wang, E. J. Kramer, W. H. Sachse, *J. Polym. Sci. Polym. Phys.*, **1982**, *20*, 1371.

3 J. V. Schnitzler, R. Eggers, *High Press. Chem. Eng.*, **1999**, 59.

4 R. G. Wissinger, M. E. Paulaitis, *J. Polym. Sci. Part B, Polym. Phys.*, **1987**, *25*, 2497.

5 N. H. Brantley, S. G. Kazarian, C. A. Eckert, *J. Appl. Polym. Sci.*, **2000**, *77*, 764.

6 S. G. Kazarian, N. H. Brantley, B. L. West, M. F. Vincent, C. A. Eckert, *Appl. Spectroscopy*, **1997**, *51*, 491.

7 A. Venema, H. J. F. M. Van de Ven, F. David, P. Sandra, *J. High Resolut. Chromatogr.*, **1993**, *16*, 522.

8 S. R. Porter, L. T. Taylor, *J. Chromatogr. A*, **1999**, *855*, 715.

9 A. M. Pinto, L. T. Taylor, *J. Chromatogr. A*, **1998**, *811*, 163.

10 S. M. Lambert, M. E. Paulaitis, *J.Supercrit.Fluids*, **1991**, *4*, 15.

11 E. Beckman, R. S. Porter, *J. Polym .Sci., Part B: Polym. Phys.*, **1987**, *25*, 1511.

12 Z. Zhong, S. Zheng, Y. Mi, *Polymer*, **1999**, *40*, 3829.

13 J. S. Chiou, J. W. Barlow, D. R. Paul, *J. Appl. Polym. Sci.*, **1985**, *30(9)*, 3911.

14 S. M. Gross, M. D. Goodner, G. W. Roberts, D. J. Kiserow, J. M. DeSimone, *Book of Abstracts*, 218th ACS National Meeting, New Orleans, Aug. **1999**, 22.

15 S. Küppers, *Chromatographia*, **1992**, *33*, 434.

16 C. Lutermann, W. Dott, J. Hollender, *J. Chromatogr., A*, **1998**, *811 (1+2)*, 151.

17 J. K. Sekinger, G. N. Ghebremeskel, L. H. Concienne, *Rubber Chem. Technol.*, **1996**, *69, (5)*, 851.

18 X. Chaudot, A. Tambute, M. Caude, *Analysis*, **1997**, *25 (4)*, 81.

19 U. Petersson, K. E. Markides, *J. Chromatogr., A*, **1996**, *734 (2)*, 311.

20 X. Lou, H. G. Janssen, C. A. Cramers, *J. Chromatogr. Sci.*, **1996**, *34*, 282.

21 T. A. Berger, *J Chromatogr. A*, **1997**, *785*, 3.

22 G. O. Cantrell, J. A. Blackwell, *J. Chromatogr. A*, **1997**, *782*, 237.

23 L. T. Taylor, *ACS Symp. Ser.*, **1997**, 134.

24 S. M. Gross, G. W. Roberts, D. J. Kiserow, J. M. DeSimone, *Macromolecules*, **2000**, *33*, 40.

25 J. M. DeSimone, R. Givens, Y. Ni, *WO Pat. 98/34975 A1* **1998**, pp. 26.

26 E. Kung, A. J. Lesser, T. J. McCarthy, *Macromolecules*, **1997**, *31*, 4160.

27 M. L. O'Neill, M. Z. Yates, K. P. Johnston, C. D. Smith, S. P. Wilkinson, *Macromolecules*, **1998**, *31*, 2848.

28 D. A. Canelas, J. M. DeSimone, *Adv. Polym. Sci.*, **1997**, *133*, 103.

29 S. Lee, S. Kwak, F. Azzam, *US Patent 5663237*, **1997**.

30 F. M. Kerton, G. A. Lawless, S. P. Armes, *J. Mater. Chem.*, **1997**, *7 (10)*, 1965.

31 C. Lepilleur, E. J. Beckman, *Macromolecules*, **1997**, *30*, 745.

32 M. Super, E. J. Beckman, *Macromol. Symp.*, **1998**, *127*, 89.

33 T. J. Romack, E. E. Maury, J. M. DeSimone, *Macromolecules*, **1995**, *28 (4)*, 912.

34 E. Kiran, *Supercritical Fluids*, **1994**, *273*, 541.

35 A. I. Cooper, W. P. Hems, A. B. Holmes, *Macromol. Rapid Commun.*, **1998**, *19*, 353.

36 T. J. De Vries, R. Duchateau, M. A. G. Vorstman, J. T. F. Keurentjes, *Chem. Commun.*, **2000**, 263.

37 J. Jagur-Grodinski, *Heterogeneous Modification of Polymers. Matrix and Surface Reactions*, 1st edition. Wiley, 1997.

38 M. Yalpani, *Polymer*, **1993**, *34*, 1102.

39 P. Rajagopalan, T. J. McCarthy, *Macromolecules*, **1998**, *31*, 4791.

40 E. Bach, E. Cleve, E. Schollmeyer, *Proceedings of the 7th meeting on Supercritical fluids, Part 1, Antibes/Juan-Les-Pins*, **2000**, 385.

41 M. R. De Giorgi, E. Cadoni, D. Maricca, A. Piras, *Dyes and Pigments*, **2000**, *45*, 75.

42 V. Phuvanartnuruks, T. J. McCarthy, *Macromolecules*, **1998**, *31*, 1906.

43 X. Ma, D. L. Tomasko, *Ind. Eng. Chem. Res.*, **1997**, *36*, 1586.

44 J. J. Watkins, T. J. McCarthy, *Macromolecules*, **1995**, *28 (12)*, 4067.

45 S. M. Howdle, J. M. Ramsay, A. I. Cooper, *J. Polym. Sci. Part B, Polym. Phys.*, **1994**, *32*, 541.

46 A. R. Berens, G. S. Huvard, R. W. Korsmeyer, F. W. Kunig, *J. Appl. Polym. Sci.*, **1992**, *46*, 231.

47 M. Jobling, S. M. Howdle, M. Poliakoff, *J. Chem. Soc., Chem. Commun.*, **1990**, *24*, 1762.

48 P. G. Jessop, W. Leitner, *Chemical synthesis using supercritical fluids*, 1st edn., Wiley-VCH, Weinheim, 1999.

49 R. S. Oakes, A. A. Clifford, C. M. Rayner, *J. Chem. Soc., Perkin Trans 1*, **2001**, 917.

50 M. E. Paulaitis, G. C. Alexander, *Pure Appl. Chem.*, **1987**, *59(1)*, 61.

51 T. W. Randolph, H. W. Blanch, J. M. Prausnitz, C. R. Wilke, *Biotechnol. Lett.*, **1985**, *7(5)*, 325.

52 D. A. Hammond, M. Karel, A. M. Klibanov, V. J. Krukonis, *Appl. Biochem. Biotechnol.*, **1985**, *11(5)*, 393.

53 K. Nakamura, Y. M. Chi, Y. Yamada, T. Yano, *Chem. Eng. Commun.*, **1985**, *45*, 207.

54 A. J. Mesiano, E. J. Beckman, A. J. Russell, *Chem. Rev.*, **1999**, *99*, 623.

55 P. G. Jessop, T. Ikariya, R. Noyori, *Science*, **1995**, *269*, 1065.

56 M. McHugh, V. Krukonis, *Supercritical Fluid Extraction, Principles and Practice*, 1986.

57 C. A. Eckert, B. L. Knutson, P. G. Debenedetti, *Nature*, **1996**, *383*, 313.

58 M. Lora, I. Kikic, *Sep. Purif. Methods*, **1999**, *28(2)*, 179.

59 E. J. Beckman, *Proceedings of the 7th meeting on Supercritical fluids, Part 1, Antibes/Juan-Les-Pins*, **2000**, 215.

60 S. G. Kazarian, B. J. Briscoe, C. J. Lawrence, D. Coombs, G. Poulter, *Proceedings of the 6th meeting on Supercritical fluids, Nottingham*, **1999**, 11.

61 K. A. Arora, A. J. Lesser, T. J. McCarthy, *Macromolecules*, **1998**, *31*, 4614.

62 J. J. Watkins, T. J. McCarthy, *Macromolecules*, **1994**, *27*, 4845.

63 E. Kung, K. Arora, H. Hayes, P. Rajagopalan, T. J. McCarthy, *Book of Abstracts, 216th ACS National Meeting*, **1998**, PMSE-063.

64 O. Muth, T. Hirth, H. Vogel, *J. Supercrit. Fluids*, **2000**, *17*, 65.

65 G. Friedmann, Y. Guilbert, J. M. Catala, *Eur. Polym. J.*, **2000**, *36*, 13.

66 G. Filardo, S. Gambino, G. Silvestri, E. Calderaro, G. Spadaro, *Radiation Phys. Chem.*, **1994**, *44*, 597.

67 C. Dispenza, G. Filardo, G. Silvestri, G. Spadaro, *Coll. Polym. Sci.*, **1997**, *275*, 390.

68 G. Filardo, C. Dispenza, G. Silvestri, G. Spadaro, *J. Supercrit. Fluids*, **1998**, *12*, 177.

69 G. Spadaro, R. De Gregorio, A. Galia, A. Valenza, G. Filardo, *Polymer*, **2000**, *41*, 3491.

70 A. Galia, R. De Gregorio, G. Spadaro, O. Scialdone, G. Filardo, *Macromolecules*, **2004**, *37*, 4580.

71 J. M. De Gooijer, M. Scheltus, H. W. Lösch, R. Staudt, J. Meuldijk, C. E. Koning, *J. Supercrit. Fluids*, **2004**, *29*, 129.

72 J. M. De Gooijer, M. Scheltus, C. E. Koning, *J. Supercrit. Fluids*, **2004**, *29*, 153.

73 J. M. De Gooijer, J. Ellmann, M. Möller, C. E. Koning, *J. Supercrit. Fluids*, **2004**, *31*, 75.

74 J. M. De Gooijer, M. Scheltus, M. A. G. Jansen, C. E. Koning, *Polymer*, **2003**, *44*, 2201.

75 H. W. Lösch, R. Kleinrahm, W. Wagner, *Chem. Ing. Tech.*, **1994**, *66*, 1055.

76 J. V. Schnitzler, R. Eggers *J. Supercrit. Fluids*, **1999**, *16*, 81.

77 J. Crank, G. S. Park, *Diffusion in Polymers*, Academic Press, London, 1968.

78 O. Muth, Th. Hirth, H. Vogel, *J. Supercrit. Fluids*, **2001**, *19*, 299.

79 W. B. Kay, *Ind. Eng. Chem.*, **1936**, *28*, 1014.

80 M. Van Duin, M. Aussems, R. J. M. Borg-greve, *J. Polym. Sci.: Part A: Pol. Chem.*, **1998**, *36*, 179.

81 Y. J. Sun, G. H. Hu, M. Lambla, H. K. Kotlar, *Polymer*, **1996**, *37(18)*, 4119.

82 J. March, *Advanced Organic Chemistry; Reactions, mechanisms and structure*, 3rd. edition, Wiley Interscience, 1985, 368.

83 R. Miller, C. Abaecherli, A. Said, *Encyclo-pedia of Industrial Chemistry*, Ullmann's, 1985 Edn.

84 R. J. Clemens, *Kodak Lab. Chem. Bull.*, **1984**, *55 (3)*, 1.

85 R. L. Rosas, J. M. McCormick, *US Pat. 6099635*, **2000**, 5pp.

86 S. S. Sabri, M. M. El-Abadelah, M. F. Za'ater, *J. Chem. Soc. Perkin Trans. 1*, **1977**, *11*, 1356.

87 F. Filira, C. Di Bello, A. C. Veronese, F. D'Angeli, *J. Org. Chem.*, **1972**, *37*, 3265.

88 H. Zimmermann, N. T. Kim, *Polym. Eng. Sci.*, **1980**, *20*, 680.

89 S. Sawada, K. Kamiyama, S. Ohgushi, K. Yabuki, *J. Appl. Polym. Sci.*, **1991**, *42*, 1041.

90 L. H. Buxbaum, *Angew. Chem., Int. Ed. Engl.*, **1968**, *7*, 182.

91 W. H. Th. Veugelers, H. Gooijer, J. W. Gerritsen, G. F. Woerlee, *Eur. Pat. Appl.*, 1126072, **2001**, pp. 7

92 E. Bach, E. Cleve, E. Schollmeyer, *Process Technol. Proc.*, **1996**, *12*, 581.

93 Personal communication with T. Loont-jens, DSM Research, Geleen, The Neth-erlands.

94 Personal communication with L. Lee-mans, DSM Research, Geleen, The Netherlands.

95 J. Ellmann, Ph. D. Thesis, *"Polymeriza-tion and reactive coating in supercritical CO_2"*, University of Ulm, Germany, **2001**.

14

Reduction of Residual Monomer in Latex Products Using High-Pressure Carbon Dioxide *

Maartje F. Kemmere, Marcus van Schilt, Marc Jacobs, and Jos Keurentjes

14.1
Introduction

The concentration of residual monomer that remains in the product after the polymerization process is one of the main issues in the production of polymeric materials. Large-scale industrial polymerization processes are preferably operated at relatively low temperatures in order to control the rate of heat production and to prevent thermal runaway, as polymerizations can suffer from considerable heat transfer limitations [1, 2]. A major disadvantage of this kind of operation is the difficulty to reach complete monomer conversion at processing temperatures typically below the glass transition temperature (T_g) of the polymer. This phenomenon is primarily explained by the fact that the diffusion coefficients in the polymer depend upon the weight fraction of polymer. As a consequence, the diffusion-controlled propagation rate coefficient decreases at higher conversion [3]. Therefore, at polymerization temperatures below the T_g and at high conversion, the reaction rate becomes extremely low before monomer conversion is complete. These difficulties often result in the presence of substantial amounts of residual monomer in the final polymer product. This is undesirable for many applications, both from an environmental as well as from a health point of view. For these reasons, one of the most important incentives for the polymer industry today is to reduce the residual monomer content. Moreover, the legislation for the level of these monomers in polymer products is becoming substantially stricter. To illustrate the possibilities of high-pressure CO_2 technology for the reduction of residual monomer, this chapter focuses on the development of a post-emulsion polymerization process in which the amount of monomer in a product latex is significantly decreased (Fig. 14.1). Latex consists of submicron polymer particles dispersed in an aqueous medium. In particular, the reduction of methyl methacrylate (MMA) in PMMA latexes using high-pres-

---

* The symbols used in this chapter are listed at the end of the text, under "Notation".

Supercritical Carbon Dioxide: in Polymer Reaction Engineering
Edited by Maartje F. Kemmere and Thierry Meyer
Copyright © 2005 WILEY-VCH Verlag GmbH & Co. KGaA, Weinheim
ISBN: 3-527-31092-4

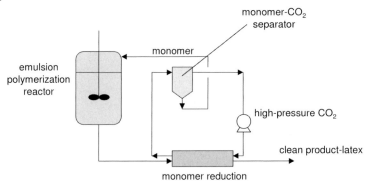

Fig. 14.1 Process concept for the reduction of residual monomer from latex products using high-pressure CO_2 in a countercurrent apparatus.

sure CO_2 has been studied. This system has been taken as a representative case study, since it involves a monomer which is extensively used in industrial polymerization processes and for which many data are available in literature. The effect of high-pressure CO_2 has been investigated in the liquid as well as the supercritical regime.

In the following section, an overview of the existing techniques for the reduction of residual monomer in latex products is given, including the advantages and drawbacks of each technology. Subsequently, the potential of using high-pressure carbon dioxide as an alternative and more sustainable method to reduce the amount of residual monomer is explained. Both the enhanced polymerization of monomer and the extraction of the residual monomer in high-pressure CO_2 have been studied. Based on the results obtained, a process design is presented for which the technical and economical feasibility of the process are discussed. Finally, a short assessment of the application possibility of high-pressure CO_2 for reduction of residual monomer in other polymer products is presented.

14.2
Overview of Techniques for Reduction of Residual Monomer

For health, safety, and environmental reasons, significant efforts have been made to reduce the residual monomer content in commercial latexes. Some of the techniques are already being applied industrially, whereas others are still being developed. In general, the available methods for the reduction of residual monomer are based on two different concepts. The first of these aims at further conversion of the monomer by increasing the diffusion-controlled propagation rate of the polymerization reaction, while the second involves removal of the residual monomer.

14.2.1
Conversion of Residual Monomer

The slow rate of diffusion-controlled propagation can be accelerated by increasing the temperature, thus lowering the particle viscosity and enhancing the diffusivity of the monomer. Starting the polymerization at a relatively low temperature ensures a controlled conversion of the greater part of the monomer. Subsequently, the reaction rate decreases, and the temperature can safely be increased to complete the conversion of monomer [4]. Although this technique is relatively simple and no additional chemicals are required, the need for an additional large post-reaction tank and the high energy consumption hampers its economic viability, especially when future regulations on residual monomer become more severe.

In addition to thermally enhanced monomer diffusion, swelling the polymer particles with an inert diluent, thus reducing the weight fraction of polymer, has been shown to increase monomer diffusivity as well [5]. A plasticizing effect causes mass transport within the polymer particle to be significantly enhanced. Consequently, the diffusion-controlled propagation rate does not decrease as much as it normally does at high conversions. However, this implies the introduction of an extra component that later has to be removed from the final product, and this requires an additional post-processing step.

Biological degradation of the monomer by means of adding a peroxide-generating enzyme has been shown to reduce the amount of monomer in a latex, as the latter reacts with the radicals originating from the peroxides. This process runs at ambient temperatures, but generally takes days to weeks to reduce the amount of monomer to an acceptable level [6].

14.2.2
Removal of Residual Monomer

Stripping removes the residual monomer indirectly by sweeping the aqueous phase with steam, gases, or mixtures of gases [7]. Steam stripping is the most widely used commercial process for the removal of residual monomer and reaction by-products. Typically, the process is carried out at about 353 K under reduced N_2 atmosphere, the volume of steam distillate being equal to that of the latex being collected [8–11]. Although steam stripping is an efficient process for the removal of residual monomer, the foaming tendency and high viscosity of the latex are known to complicate the reactor design. These parameters limit the rate at which the aqueous and gas phases can be contacted, while a high contact area is of vital importance to the viability of this technique. Moreover, the use of high temperatures may cause thermal degradation of the polymer, as well as problems with the colloidal stability of the latex and the polymer morphology. Furthermore, the recovery of the vaporized monomers from an inert gas can be difficult.

Taylor has described a method for the removal of residual monomer by a combination of chemical and steam stripping [12]. An added amount of fresh free-radical initiator, which decomposes at the temperature of treatment, induces the chemical depletion, while steam extracts the residual amount of monomer from the polymer particles.

This method appears to be twice as efficient in reducing the residual monomer as conventional steam stripping. Kelly [13] has reported a similar method using steam stripping under vacuum instead of at atmospheric pressure. The chemical depletion of monomer with initiator is done prior to the vacuum steam stripping.

One of the major disadvantages of steam stripping is the high energy consumption. Olivares et al. [14] have reported a chemical depletion method based on a redox system consisting of an oxidizing agent generator of free radicals and a reducing agent, which makes any additional steam stripping step redundant.

A promising technique for residual monomer removal is pervaporation, as no additional chemicals are needed for this membrane process and the energy costs are typically low. It has been shown that pervaporation can remove a considerable amount of acrylic monomer from polymethylmethacrylate (PMMA) latexes [15]. Apparently, the limiting factor for mass transfer does not occur in the polymer particles, mainly because of the high specific area of the polymer-water interface as compared to the membrane area. Although the high initial costs, as well as fouling of the membrane surface with the polymer particles, are potential drawbacks, pervaporation may thus be expected to provide a viable alternative.

Table 14.1 Overview of various methods for reduction of residual monomer from latex products.

Technique	Advantages	Disadvantages
Temperature increase	Simple to operate No toxic solvents required	High energy costs Time consuming
Addition of an inert diluent	Simple to operate	Removal of inert diluent required Large reactor volume
Biological degradation	Ambient temperature	Time consuming Large bath volumes required Specific chemicals required
Stripping	No toxic solvents required	Tendency to foam Possible thermal degradation of polymer Colloidal instability of the latex High throughputs of the stripping medium required
Pervaporation	No extra separation step No toxic chemicals needed No extra energy required	High membrane resistance High investment costs Low fluxes

Serum exchange [16], film evaporation, hollow fiber dialysis, and micro- and ultrafiltration [17] can also be effectively used to remove residual monomer on the laboratory scale. However, for larger scale applications these techniques are less suitable.

Table 14.1 summarizes several techniques for the reduction of residual monomer. At present, the industrial practice of emulsion polymerization mainly involves thermally enhanced diffusion and a combination of steam and chemical stripping. As mentioned above, all these techniques come with their specific advantages and disadvantages. Since legislation on levels of residual monomer in consumer products is expected to become more severe in the near future, there is considerable room for improvement of the existing techniques.

14.2.3
Alternative Technology: High-Pressure Carbon Dioxide

High-pressure carbon dioxide potentially forms an interesting alternative for the reduction of residual monomer, because it combines both approaches mentioned above, i.e. simultaneous conversion and removal of the monomer. The plasticizing effect is expected to enhance diffusivity in the polymer particles, as they swell in high-pressure CO_2 [18]. This would first of all increase the diffusion-controlled propagation rate of the polymerization and thus further complete the conversion of monomer within an acceptable time of operation. Secondly, the enhanced diffusivity within the polymer particles would be expected to facilitate the transport of monomer through the polymer particles to the aqueous phase. Moreover, high-pressure CO_2 is an excellent extraction medium for a variety of monomers, as little water and virtually no polymer dissolves in it, while most monomers have a relatively high solubility in pressurized carbon dioxide. In addition, relatively small changes in pressure and temperature can tune the solubility in supercritical carbon dioxide (scCO_2), as well as the diffusion and the viscosity. Sections 14.3 and 14.4 discuss the enhanced polymerization and extraction of residual monomer in the presence high-pressure CO_2.

14.3
Enhanced Polymerization in High-Pressure Carbon Dioxide

To investigate the effect of high-pressure CO_2 on the polymerization reaction as well as to determine the amount of monomer inside the polymer particles, electron beam experiments have been performed [19]. Pulsed electron beam polymerization involves the generation of radicals in the aqueous phase, this being activated by an electron beam. These radicals initiate the polymerization of the residual monomer inside the latex particles. Based on the molecular weight of the newly formed polymer chains, the local monomer concentration in the polymer particles can be calculated. The growth time of a polymer chain is directly determined by the time between two pulses (t_p). The chain length (L_i) of the produced polymer in this time t_p is given by:

$$L_i = i \cdot k_p \cdot [M] \cdot t_p \tag{1}$$

where k_p and [M] are the propagation rate constant and monomer concentration in the polymer particles, respectively. Higher-order peaks in molecular weight $(i=2,3,...)$ may occur when growing chains survive termination by one or more subsequent pulses. In this case, only the primary peak $(i=1)$ has been observed. Between two pulses, bimolecular termination or transfer can occur, resulting in the so-called background polymer.

14.3.1
Procedure for Pulsed Electron Beam Experiments

In this study, electron beam experiments were performed using a LINAC SL75-5 electron accelerator (M.E.L., Sussex, England). During the electron beam experiments the energy of the electrons was adjusted to 5 MeV, resulting in pulse energies of 0.8 J. A high-pressure view cell of 2.9 mL was used for irradiating the sample, a crosslinked (0.6 wt% ethylene glycol dimethacrylate, EGDM) PMMA latex that was swollen to its maximum with MMA [20]. The cell was equipped with a stainless steel window for the electron beam according to Wishart and van Eldik [21] and one sapphire window. The electron beam experiments were performed at 348 K with a pulse frequency of 10 Hz. The number of pulses varied from ca. 300 to ca. 600. Before the view cell was filled and electron beam experiments were performed, the latex-scCO$_2$ mixture was stabilized for one hour at 16.0 MPa in a high-pressure 80 mL vessel (40 vol% CO$_2$). As a reference, an electron beam experiment was performed at atmospheric conditions without carbon dioxide. In all cases, the molecular-weight distribution (MWD) of the sol fraction of the crosslinked polymer was determined using gel permeation chromatography (GPC). The inflection point was taken as a measure of the local monomer concentration experienced by the newly formed polymer chains [22]. Taking the propagation rate coefficient based on the Arrhenius parameters for MMA emulsion polymerization as recommended by IUPAC $(k_p = 1.179 \times 10^3$ L mol^{-1} s^{-1}, the intrinsic value), the time between two electron pulses t_p, and $i=1$, the local monomer concentration in the polymer particle can be calculated according to Eq. (1).

14.3.2
Results and Discussion

In Fig. 14.2, the results of the electron beam experiments performed at ambient-pressure and high-pressure CO$_2$ conditions are presented. At ambient pressure, Fig. 14.2a clearly shows the additional peak of the newly formed polymer of M_w 79 800 Da produced by the radicals in the aqueous phase induced by the electron accelerator. According to Eq. (1), this results in a monomer concentration inside the polymer particles of 6.8 mol L^{-1}, which is in agreement with the maximum concentration of monomer in the PMMA particles at 348 K [23]. These results illustrate that pulsed electron beam experiments indeed can be used to determine the local monomer concentration in this system. It should be noted that the position of the MWD of the sol fraction is somewhat shifted to higher molecular weight after irradiation as compared to the original MWD of the sol fraction of the latex, since the interaction of the polymer chains with the applied electron beam can result in some crosslinking as well as scission.

It is expected that in the presence of high-pressure CO_2 a change in monomer concentration will be reflected in a shift of the molecular weight of the additional peak. However, experiments at $scCO_2$ conditions (Fig. 14.2 b) do not

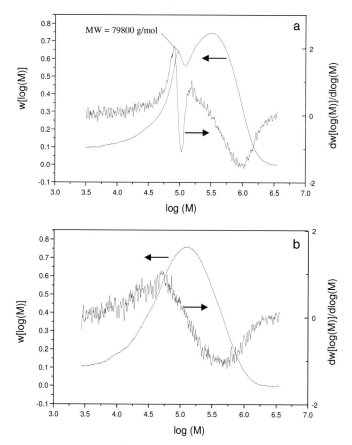

Fig. 14.2 Molecular weight distributions of a PMMA latex, swollen to its maximum with MMA, irradiated with electron beam, a at atmospheric conditions without CO_2 present, and b in the presence of high-pressure CO_2 at 16.0 MPa.

show any newly formed polymer produced by the pulsed electron beam. Considering the results of the electron beam experiments with respect to the relatively high partition coefficient of MMA between the CO_2 and water phase (see Section 14.4), it is very likely that the CO_2 phase has extracted the monomer during the electron beam experiments to such an extent that there was hardly any residual monomer left in the polymer particles to polymerize. As a consequence, parameters such as contact time and volume ratio of the water and the CO_2 phase determine the amount of monomer left in the polymer particles in this kind of experiment.

The effect of CO_2 on the emulsion polymerization of MMA has been investigated previously (at 348 K and 1–35 MPa) [24], when a decrease in average molecular weight and a much steeper log (number distribution) at higher pressure was observed. According to the authors, the reason for this effect is the swelling of polymer by CO_2, which becomes more significant at higher CO_2 pressures. This swelling results in a reduction of the internal viscosity and a delay of the gel effect. However, to draw these conclusions, the effect of $scCO_2$ on the polymerization rate and the monomer concentration in the polymer particles should be known. The interpretation of these experiments is complicated, because the observed position of the MWD is mainly influenced by experimental factors such as contact time and interfacial area at non-equilibrium conditions. This determines the extent of extraction and does not give information on the plasticizing effect. However, it can be expected from diffusion coefficient measurements of monomer in polymer particles [25, 26] that it will be possible to perform electron beam experiments under conditions where monomer is still present in the particles at $scCO_2$ conditions. Nevertheless, extraction will be the dominating mechanism for reduction of MMA in PMMA latexes based on CO_2 technology.

14.4
Extraction Capacity of Carbon Dioxide

Monomer extraction from latex products involves mass transfer in a heterogeneous system. Three phases have to be considered: the polymer particles, the aqueous phase, and the CO_2 phase (see Fig. 14.3). In this work, the film model has been used to describe the diffusion of MMA in this three-phase system, for which the following assumptions have been made:

- No mass transport of monomer directly from the polymer to the CO_2 phase occurs.
- The zones for mass transfer resistances can be represented by two hypothetical layers, in which the transfer is entirely determined by molecular diffusion and the concentration gradient is linear in each of these layers.
- There is equilibrium at the interfaces, so that the interface concentrations are determined by equilibrium conditions between the phases.

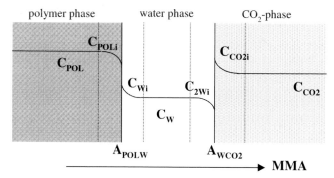

Fig. 14.3 Schematic view of mass transfer of monomer from the polymer particles, through the aqueous phase, into the CO_2 phase, based on the film model.

The mass transport flux of MMA in each phase is defined as:

$$N = A_{POLW} \cdot k_{POL} \cdot (C_{POL} - C_{POLi}) \tag{2}$$

$$N = A_{POLW} \cdot k_w \cdot (C_{Wi} - C_W) \tag{3}$$

$$N = A_{WCO_2} \cdot k_w \cdot (C_W - C_{2Wi}) \tag{4}$$

$$N = A_{WCO_2} \cdot k_{CO_2} \cdot (C_{CO_{2i}} - C_{CO_2}) \tag{5}$$

The equilibrium equation of MMA between two phases is defined by a partition coefficient:

$$m_1 = \frac{C_{Wi}}{C_{POLi}} \tag{6}$$

$$m_2 = \frac{C_{CO_{2i}}}{C_{2Wi}} \tag{7}$$

To quantify the extraction capacity of carbon dioxide, the partition coefficient of MMA between the water and the CO_2-phase (m_2) is a prerequisite. Since this partition coefficient is not reported in literature, m_2 has been measured in a laboratory scale extraction unit at different temperatures and CO_2 pressures and has also been predicted using the Peng-Robinson equation of state.

14.4.1
Modeling Phase Behavior with the Peng-Robinson Equation of State

The Peng-Robinson equation of state (eos) [27] is one of the most frequently used semi-empirical cubic equations of state, and is able to describe high-pressure and supercritical fluid systems. The Peng-Robinson eos expresses pressure as a summation of a repulsive term and a attractive term and is given by

$$P = P_R + P_A = \frac{RT}{v-b} - \frac{a(R,\omega)}{v(v+b)+b(v-b)} \tag{8}$$

where a reflects the temperature-dependent intermolecular attraction and b reflects the Van der Waals hard sphere volume. For pure components, a and b are given by

$$a = 0.45724 \, \frac{(RT_c)^2}{P_c} \left(1+(0.37464+1.54226\,\omega - 0.26992\,\omega^2)\left(1-\sqrt{T_r}\right)\right)^2 \tag{9}$$

$$b = 0.07780 \, \frac{RT_c}{P_c} \tag{10}$$

where T_r is the reduced temperature and ω the acentric factor.

For the description of multi-component systems, the parameters a and b can be obtained from mixing rules:

$$a_m = \sum_i \sum_j x_i x_j a_{ij} \qquad b_m = \sum_i \sum_j x_i x_j b_{ij} \tag{11}$$

Various expressions have been proposed for the cross coefficients a_{ij} and b_{ij}, which are shown in Table 14.2.

The Quadratic mixing rule is only applicable to mixtures of similar components showing no specific interactions. The Panagiotopoulos-Reid [28] and Stryjek-Vera [29] mixing rules introduce a second parameter for a, thereby allowing non-symmetrical behavior to be described. Both these mixing rules suffer from a so-called dilution problem and the Michelsen-Kistenmacher syndrome [30]. Melhem et al. [31] proposed a mixing rule that solved the dilution problem. The Mathias-Klotz-Prausnitz mixing rule [32] is an extension of the Panagiotopoulos-

Table 14.2 Cross-coefficients a_{ij} and b_{ij} for various mixing rules.

Mixing rule	Parameter a		Parameter b
Quadratic	$a_{ij} = \sqrt{a_i a_j}\,(1-k_{ij})$	$k_{ij} = k_{ji}$	
Panagiotopoulos-Reid	$a_{ij} = \sqrt{a_i a_j}\,(1-k_{ij}+\lambda_{ij}\,x_i)$	$\lambda_{ij} = k_{ij} - k_{ji}$	
Stryjek-Vera	$a_{ij} = \sqrt{a_i a_j}\left(1-\dfrac{k_{ij}k_{ji}}{x_i\,k_{ij}+x_j k_{ji}}\right)$		$b_{ij} = \dfrac{b_i+b_j}{2}(1-\ell_{ij})$
Melhem	$a_{ij} = \sqrt{a_i a_j}\left(1-k_{ij}+\lambda_{ij}\dfrac{x_i}{x_i+x_j}\right)$		$\ell_{ij} = \ell_{ji}$
Mathias-Klotz-Prausnitz	$a_m = \sum_i \sum_j x_i x_j \sqrt{a_i a_j}\,(1-k_{ij}) +$	$k_{ij} = k_{ji}$	
	$\sum_i x_i \left(\sum_j x_j \left(\sqrt{a_i a_j}\,\lambda_{ij}\right)^{1/3}\right)^3$	$\lambda_{ij} = -\lambda_{ji}$	

Reid mixing rule without the dilution problem and the Michelsen-Kistenmacher syndrome. Moreover, when applied to binary mixtures, it becomes identical to the Panagiotopoulos-Reid mixing rule. As shown in Table 14.2, b_{ij} is identical for all mixing rules, and generally ℓ_{ij} is set to zero. There are almost no thermodynamic restraints on mixing rules, except for the fact that in the low-density limit a second virial coefficient is required, which is a quadratic function of composition. Wong and Sandler [33] proposed a mixing rule to solve this issue. It includes an excess Helmholtz free energy at infinite pressure, which allows the mixing rule to consist of interaction parameters, equation-of-state coefficients and mole fractions only.

The computer program Phase Equilibria [34], which has been developed by Brunner and co-workers at the Technical University of Hamburg, includes the Peng-Robinson eos and the above-described mixing rules, thereby allowing a complete description of the phase equilibria of the ternary system and its binary subsystems. In this work, the Phase Equilibria program has been used to model the MMA-CO_2-water system, applying a Simplex-Nelder-Mead algorithm to optimize the binary interaction parameters.

14.4.2
Procedure for Measuring Monomer Partition Coefficients

A high-pressure unit with on-line sampling of the aqueous phase has been developed to measure monomer partition coefficients [19]. Fig. 14.4 shows a schematic set-up of the equipment and a description of the relevant components and features. With this equipment, the partition coefficient m_2 given by Eq. (7) can be measured as a function of pressure and temperature.

The experimental procedure was started by adding a carefully weighed amount of aqueous MMA solution to the equilibrium cell, filling it approximately 50% full. The cell was immediately sealed and heated to the desired temperature. The liquid phase was stirred vigorously, creating a deep vortex. At the desired temperature, CO_2 was introduced and was pressurized by the high-pressure syringe pump. Equilibrium was assumed if pressure and temperature did not vary more than 0.02 MPa and 0.2 K, respectively, within a period of 30 min. Then, a sample of the aqueous phase was taken by turning the HPLC valve, the sample loop was depressurized, and the sample was led through the first tube of chilled methanol (MeOH) containing butanol as an internal standard, thus removing the MMA from the sample. The second tube was only used to check whether no measurable amounts of MMA had passed through the first tube. The contents of the first tube were analyzed by GC/FID in order to determine the concentration of MMA.

In the proposed method, the presence of only two phases was assumed. Moreover, any significant change of volume of the phases was neglected. From the measured MMA concentration in the aqueous phase at a certain pressure and temperature, the partition coefficient can be calculated using Eq. (12). Here $C_{W,0}$ stands for the initial MMA concentration in the aqueous phase.

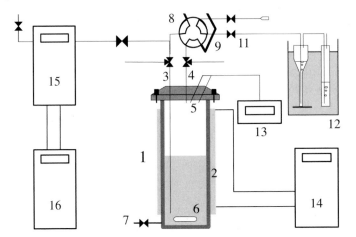

Fig. 14.4 Schematic view of the high-pressure laboratory scale extraction equipment used for the determination of the partition coefficient m_2.

1. Stainless steel high-pressure equilibrium cell (64.33 mL)
2. Stainless steel heating jacket
3. Sample line for extracting fluid phase samples and venting valve
4. Sample line for extracting light phase samples and venting valve
5. PT 100 temperature sensor and miniaturized high-pressure transducer
6. Magnetic stirrer for fluid phase agitation

7. Equilibrium cell outlet
8. 6-port HPLC sample valve (sample valve in sampling position)
9. Calibrated 500 μL sample loop
10. Open connection for syringe
11. Needle valve
12. Chilled beaker containing glass vials for sample collection
13. Display of temperature and pressure
14. Thermostatic water heater
15. Pulse-free 150 mL high-pressure syringe pump
16. Cryostat for cooling the syringe pump

$$m_2 = \left(\frac{C_{W,0}}{C_W} - 1\right) \cdot \frac{V_W}{V_{CO_2}} \tag{12}$$

Note that the initial MMA concentration is limited by the solubility of MMA in water, whereas the experimental set-up restricts the allowable V_{CO_2}/V_W ratio.

In addition to Eq. (7), a partition coefficient based on MMA mole fractions in the aqueous (x_{MMA}) and in the CO_2-rich phase (y_{MMA}) can be calculated using a modified Benedict-Webb-Rubin equation of state [35] to estimate the molar densities of the phases assuming pure CO_2 and water density, respectively.

$$m_2^* = \frac{y_{MMA}}{x_{MMA}} = m_2 \cdot \frac{\rho_W}{\rho_{CO_2}} \tag{13}$$

14.4.3
Validation of the Experimental Determination of Partition Coefficients

The reproducibility of the described method for determination of partition coefficients has been assessed by three experiments at 313 K and 8.5 MPa. The results are given in Table 14.3 and clearly show that these experiments produced data that are very similar.

For the evaluation of the reliability of this method, a means of comparison with other experimental data is essential. Since a direct validation of the data is impossible because of a lack of specific experimental data on the ternary system H_2O-CO_2-MMA, a system displaying qualitatively similar behavior and for which reliable experimental data are available has been used instead. For this purpose, measurements on the four-phase region of the ternary system CO_2-H_2O-1-BuOH are compared with experimental data published by Panagiotopoulos and Reid [36] as well as by Winkler and Stephan [37]. The results of the validation experiments are given in Table 14.4.

The experiments were carried out following the outlined procedure and yielded a 1-BuOH concentration in water (g/mL). By using the reported density of 1.00 g/mL for the aqueous phase, the weight fraction of 1-BuOH has been calculated. According to Table 14.4 the results of these experiments agree with literature data. The overall composition of the system has been carefully chosen to lie close to the four-phase region of the system in such a way that the volumetric phase ratio is approximately 5:1:1:3 in order of decreasing density. This means that approximately 50% of the system consists of a water-rich phase and

Table 14.3 Reproducibility measurements of MMA partition coefficients obtained with the high-pressure extraction equipment.

Sample	Pressure (MPa)	C_W (mg mL^{-1})	m_2 (–)
A	8.56	1.44	6.79
B	8.58	1.70	6.19
C	8.51	1.69	6.14

Table 14.4 Validation experiments in the high-pressure extraction equipment. Measured weight fraction of 1-BuOH in the aqueous phase in the four-phase equilibrium CO_2-H_2O-1-BuOH.

T (K)	P (MPa)	$w_{1\text{-BuOH}}$ (–)	Reference
313.1	8.25	0.043	Panagiotopoulos and Reid [36]
313	8.26	0.031	Winkler and Stephan [37]
313.2	8.22	0.040	This work
313.7	8.20	0.049	This work

about 30% consists of a CO_2-rich phase. Small variations in the relative quantity of the four phases enable the system to compensate for deviations in the overall composition. Moreover, the composition of the aqueous phase in the three-phase regions at pressures just below or above the four-phase region does not significantly differ from the composition of the aqueous phase in the four-phase region, so that we do not need to be concerned with small deviations from the four-phase point. Note that the concentrations of this system are typically one order of magnitude lower than the MMA concentrations used in this study. In general, the outcome of both the reproducibility and validation experiments strongly suggests that the developed method is reliable as well as accurate.

14.4.4
Measured Partition Coefficients of MMA over Water and CO_2

First, high-pressure view-cell experiments were performed to study the phase behavior of the CO_2-H_2O-MMA system at 298 K. Below 5.8 MPa a two-phase system [CO_2(g), H_2O(l)] was observed, whereas between 5.8 and 6.3 MPa a three-phase region [CO_2(g), CO_2(l), H_2O(l)] exists. This region was omitted during the extraction experiments. At 6.3 MPa the beginning of the liquid-liquid phase system [CO_2(l), H_2O(l)] was observed. Similar results have been described in detail by Adrian et al. [38]. The experimental conditions for the determination of the partition coefficients are given in Table 14.5. In Table 14.6 the results are presented per isothermal series.

Considering the results of the extraction experiments performed at 298 K shown in Fig. 14.5 a, the partition coefficient increases significantly above 6.3 MPa. The increasing density of the CO_2-rich phase at this point explains the sudden rise of the partition coefficient. Fig. 14.5 b–c presents the measured partition coefficients at 313, 323 and 333 K, respectively. View-cell experiments at 313 K show that three phases resembling a boiling system can be distinguished close to 7.3–7.4 MPa. This effect has not been observed at 323 or 333 K. In general, the system exhibits the same behavior at temperatures above the critical temperature of CO_2. Around 5 MPa all partition coefficients are close to unity, and they increase with pressure. The rate of this increase gradually rises with pressure and decreases with temperature. Similarly to the results at lower tem-

Table 14.5 Experimental conditions for the determination of m_2 using the high-pressure extraction equipment.

T (K)	Pressure range (MPa)	$C_{w,0}$ (g g water^{-1})	V_w (mL)
298	5.2–9.5	0.01207	31.7
313	5.1–8.6	0.01207	32.3
323	5.0–10.0	0.01207	32.1
333	4.9–8.3	0.01207	31.4

Table 14.6 Predicted partition coefficients with the Peng-Robinson eos and various mixing rules as compared to the experimentally obtained values in the high-pressure extraction unit.

T (K)	P (MPa)	Exp. data m_2 (–)	m_2^* (–)	PR m_2 (–)	m_2^* (–)	M m_2 (–)	m_2^* (–)	SV m_2 (–)	m_2^* (–)	MKP m_2 (–)	m_2^* (–)
298	4.84	1.71	32.6	0.13	2.48	0.12	2.27	0.021	0.39	0.19	3.45
298	5.26	2.19	36.1	0.19	2.98	0.17	2.72	0.030	0.48	0.26	4.20
298	6.27	3.76	41.2	0.60	6.29	0.54	5.71	0.096	1.03	0.87	9.13
298	6.36	8.59	89.2	0.70	7.00	0.63	6.36	0.15	1.51	1.02	10.2
298	6.50	12.4	119.8	0.99	8.92	0.90	8.10	0.25	2.27	1.44	13.0
298	7.98	48.8	151.8	58.9	191.2	53.4	173.3	1.09	8.15	87.3	283.1
313	5.07	0.87	18.3	0.22	4.41	0.20	4.12	0.024	0.50	0.27	5.61
313	5.96	1.30	21.4	0.36	5.74	0.34	5.36	0.044	0.69	0.48	7.54
313	7.13	2.69	32.0	0.85	9.67	0.81	9.13	0.11	1.29	1.17	13.2
313	7.54	2.95	30.9	1.26	12.5	1.20	11.9	0.18	1.81	1.76	17.4
313	8.06	4.79	41.5	2.37	19.3	2.30	18.7	0.38	3.09	3.39	27.6
313	8.56	6.79	46.1	5.80	36.6	5.05	31.9	1.05	6.64	8.67	54.5
323	4.97	1.02	23.6	0.076	1.72	0.070	1.58	0.027	0.62	0.31	6.95
323	5.98	1.19	21.4	0.12	2.09	0.11	1.91	0.044	0.76	0.51	8.86
323	7.15	1.95	26.6	0.22	2.93	0.20	2.66	0.084	1.11	1.00	13.0
323	8.11	2.25	24.5	0.42	4.32	0.37	3.89	0.15	1.59	1.93	19.9
323	9.10	2.71	22.9	0.94	7.45	0.85	6.69	0.33	2.59	4.55	35.9
323	10.01	3.89	25.7	2.55	15.2	2.27	13.5	0.86	5.02	12.9	76.1
333	4.99	0.84	20.7	0.31	7.51	0.29	7.06	0.031	0.74	0.36	8.63
333	6.00	0.87	16.8	0.48	8.87	0.44	8.27	0.045	0.85	0.56	10.4
333	7.12	1.55	23.5	0.79	11.5	0.73	10.6	0.075	1.10	0.94	13.7
333	7.50	1.70	23.8	0.94	12.7	0.87	11.8	0.091	1.22	1.13	15.3
333	8.29	2.16	25.8	1.35	15.4	1.22	13.9	0.13	1.53	1.73	19.7

peratures, the partition coefficient at 333 K is therefore expected to increase more significantly beyond the measured pressure range. Generally, these results can be explained by the density of the CO_2-rich phase increasing continuously with pressure above 7.4 MPa. This effect is most pronounced at temperatures just above the critical temperature of CO_2.

In Fig. 14.6, the partition coefficient m_2^* is plotted against pressure. As m_2^* is expressed as a ratio of mole fractions, the changes in density of the CO_2-rich phase are accounted for. Fig. 14.6 shows that the increase of the partition coefficients is only significant for the experiment performed at 313 K, and in this case it only triples over the entire pressure range. It is therefore assumed that the density of the CO_2-rich phase is the decisive factor in the behavior of the partition coefficient as a function of temperature and pressure.

It should be noted that the reported extraction results comprise the partitioning behavior of MMA between water and CO_2 without polymer particles present.

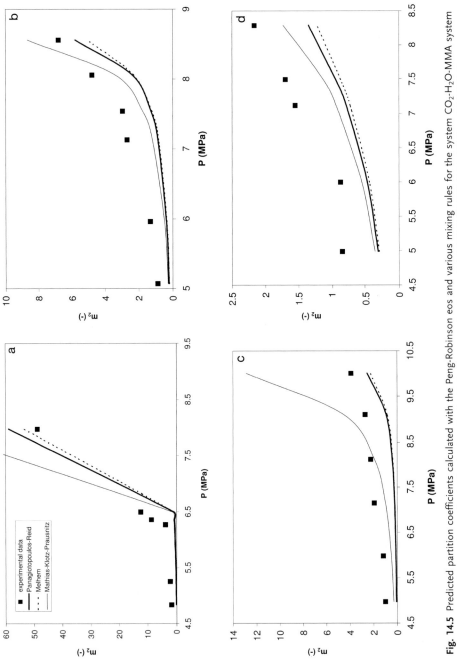

Fig. 14.5 Predicted partition coefficients calculated with the Peng-Robinson eos and various mixing rules for the system CO_2-H_2O-MMA system as compared to the experimentally obtained values in the high-pressure extraction unit (a) 298 K (b) 313 K (c) 323 K and (d) 333 K.

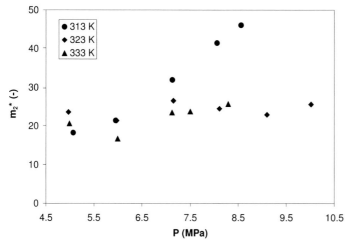

Fig. 14.6 Partition coefficient m_2^* based on mole fractions calculated from measured phase equilibrium compositions at 313, 323, and 333 K, respectively.

However, it is expected that the latex-CO_2 system will also exhibit a strong extraction rate toward the CO_2-phase as the rate-limiting step, for mass transfer of MMA appears to be situated in the water phase at the CO_2 side, while the mass transfer inside the polymer particles appears to be very fast (see Section 14.5.1). This is mainly because of a relatively small water-CO_2 surface area as compared to the overall polymer-water interfacial area [39].

14.4.5
Prediction of Partition Coefficients of MMA Over Water and CO_2

In addition to the experimental data, the partitioning behavior of MMA between water and CO_2 has been modeled. The Peng-Robinson equation of state combined with various mixing rules as described in Section 14.4.1 has been assessed on the ability to correlate phase equilibrium data from literature of the binary subsystems CO_2-H_2O, MMA-CO_2 and MMA-H_2O. Subsequently, the model has been used to predict the phase equilibrium behavior of the ternary system CO_2-H_2O-MMA. Partition coefficients were calculated at four different temperatures at pressures ranging from 5 to 10 MPa. In order to provide a means for comparison, the experimentally determined partition coefficients obtained in the high-pressure extraction unit were used to evaluate the results of the predictive model for phase equilibrium behavior.

14.4.5.1 Modeling the Two-Component Systems CO_2-H_2O, MMA-CO_2 and MMA-H_2O

In the following, a brief overview of modeling the binary subsystems is given; a more detailed description is given by Jacobs [40]. Since the subsystem CO_2-H_2O contributes over 99% to the ternary system, it is considered the key system for describing the phase equilibrium composition. In this work, the experimental data reported by Wiebe and Gaddy [41, 42] and King et al. [43] have been used. As expected, the Adachi-Sugi (AS), Melhem (M) and Mathias-Klotz-Prausnitz (MKP) mixing rules reveal similar results to those obtained from the calculations using the Panagiotopoulos-Reid (PR) mixing rule, since all these mixing rules become identical when applied to binary systems. For this reason, only the PR mixing rule is further discussed. Both the results of the calculations with PR and SV show an increase in the deviations near the critical temperature of CO_2, which results from the fact that the Peng-Robinson eos gives a poor representation of the equilibrium composition near the critical point. Concerning the PR results, a linear dependency on temperature can be observed in k_{ij} in contrast to λ_{ij}. With respect to the results of calculations based on the SV mixing rule, a poor correlation with the experimental equilibrium data has been obtained, especially in the vapor or supercritical phase.

Concerning the MMA-CO_2 system [44], the Peng-Robinson eos and PR mixing rule cannot accurately describe the entire phase envelope. Forcing it to reproduce the mixture's critical point results in a poor correlation for the liquid branch of the envelope. The interaction parameter has been chosen optimally to reproduce the equilibrium composition of the liquid phase.

The main problem with modeling the MMA-H_2O system is the absence of reliable phase equilibrium data. The only data available in literature on the MMA-water system consisted of mutual solubility data of water in MMA at standard pressure and different temperatures [45]. In this study, the interaction parameters are fitted to isobaric data. Because of this restriction, the parameters are fitted to one single set of data points, i.e. the mutual solubility of water and MMA at a certain temperature. This procedure provides interaction parameters that are temperature and pressure dependent. However, the effect of pressure on the mixing of MMA has been neglected, as the compressibility of this liquid-liquid system is generally assumed to be negligible. A similar temperature dependency can be observed for the Stryjek-Vera parameters.

14.4.5.2 Modeling the Three-Component System CO_2-H_2O-MMA

Based on the interaction parameters for the mixing rules obtained from the binary subsystems, the ternary phase behavior has been modeled by inter- and extrapolation of the interaction parameters. Because of lack of sufficient experimental data, extrapolation is considered to be the only option in some cases, although it obviously will introduce errors. Table 14.7 shows the interaction parameters of the various mixing rules used for the prediction of the ternary phase behavior, whereas the resulting partition coefficients per isothermal series

Table 14.7 Binary interaction parameters for various mixing rules at different temperatures used in the prediction of the phase behavior of the three-component system CO_2-H_2O-MMA.

T (K)	H_2O–CO_2		MMA–CO_2		H_2O–MMA	
	k_{ij}	λ_{ij}	k_{ij}	λ_{ij}	k_{ij}	λ_{ij}
Panagiotopoulos-Reid/Melhem						
298	0.170	0.304	−0.0556	0.280	−0.0459	0.254
313	0.173	0.286	−0.0669	0.217	−0.0293	0.255
323	0.184	0.286	−0.0744	0.175	−0.0174	0.258
333	0.201	0.286	−0.0819	0.133	−0.00485	0.261
Stryjek-Vera						
298	0.168	0.274	−0.0615	0.268	−0.0708	0.233
313	0.166	0.260	−0.0765	0.207	−0.0610	0.228
323	0.171	0.254	−0.0865	0.166	−0.0545	0.225
333	0.182	0.254	−0.0965	0.125	−0.0483	0.223
Mathias-Klotz-Prausnitz						
298	0.0183	0.304	−0.196	0.280	−0.173	0.254
313	0.0298	0.286	−0.176	0.217	−0.157	0.255
323	0.0407	0.286	−0.162	0.175	−0.146	0.258
333	0.0582	0.286	−0.149	0.133	−0.135	0.261

have been included in Table 14.6 and Fig. 14.5. The results obtained with the Stryjek-Vera mixing rule deviate significantly from the experimental data. Therefore, this mixing rule is omitted in further discussions. The Panagiotopoulos-Reid, Melhem, and Mathias-Klotz-Prausnitz mixing rules will be discussed.

According to Fig. 14.5 a, similar trends can be observed in both the experimental and the calculated partition coefficients at 298 K, although the calculated values of m_2 are initially lower than the experimentally determined values. At approximately 6.3 MPa a phase transition can be observed in the experimental data described in Section 14.4.3.2. The system changes from vapor-liquid to liquid-liquid, with an intermediate region in between. Calculations with the Peng-Robinson eos suggest that only one liquid phase and a vapor phase are present at 6.5 MPa. Adrian et al. [38], however, describe the presence of two liquid phases at that pressure. This illustrates again that the Peng-Robinson eos is not accurate near phase transitions. At higher temperatures the differences of the predicted m_2-values from the experimental partition coefficients decreases. This implies that the accuracy of predictions with the model increases when the CO_2-rich phase is above critical conditions.

Table 14.6 and Fig. 14.5 b–d show that above the critical point of the mixture the calculations are fairly predictive at 313, 323, and 333 K. The predicted values for the partition coefficient are slightly lower than the experimental m_2 values as observed at 298 K. Although calculations with the MKP mixing rule result in

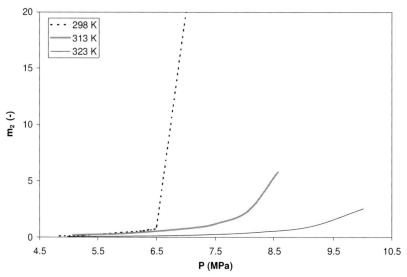

Fig. 14.7 Partitioning behavior for the system CO_2-H_2O-MMA calculated with the Peng-Robinson eos and Panagiotopoulos-Reid mixing rule at different temperatures.

values for m_2 which are closest to the experimental data, at all temperatures above T_c of the CO_2-rich phase the observed trend with this mixing rule is somewhat different from the trend of both the experimental data and the other mixing rules. Especially at higher pressures the MKP-mixing rule progressively overestimates the values for m_2. The difference between the partition coefficients calculated from the Panagiotopoulos-Reid and the Melhem mixing rules is very small, although the predictions with the Panagiotopoulos-Reid mixing rule are slightly better.

The large increase in the partition coefficient observed at the critical pressure and temperatures above T_c shows that MMA is far better soluble in the supercritical phase than in the vapor phase. As discussed earlier, the increased density of the CO_2-rich phase causes this increase in partition coefficient. The steep increase in partition coefficient m_2 at higher pressure is illustrated in Fig. 14.7 for the PR-mixing rule at different temperatures. Since the transition from the vapor phase to the supercritical fluid phase occurs at higher pressures, this increase moves to higher pressures at increased temperatures. Concerning the dilution problem as described in Section 14.4.1, this appears not to occur in this system even though MMA concentrations are typically low. Since the system contains only three components and all three components are chemically different, the Michelsen-Kistenmacher syndrome does not cause any problem either. In addition to the regular m_2, the calculated partition coefficient m_2^* based on MMA mole fractions is in reasonable agreement with the experimentally determined m_2^* values.

In summary, from an extensive comparison of the calculations with the experimentally obtained results [40], it can be concluded that the phase behavior of the ternary system MMA-H$_2$O-CO$_2$ can be qualitatively predicted by the Peng-Robinson equation of state in combination with the Panagiotopoulos-Reid mixing rule. In general, more experimental data are required concerning the binary systems for a better optimization of the interaction parameters. To obtain a more quantitative description of the phase behavior, especially close to the critical point, a more rigorous approach such as the SAFT-VRX model [46, 47] can be used.

14.5
Process Design for the Removal of MMA from a PMMA Latex Using CO$_2$

The value of the monomer partition coefficient between the CO$_2$ and the water phase indirectly determines the ratio between the effect of enhanced polymerization and the effect of extraction on the reduction of residual monomer. Depending on the process conditions, i.e. temperature, pressure, and the phase behavior of the system involved, this ratio between enhanced polymerization and extraction may vary for different latex systems. With respect to the PMMA latex, the high partition coefficient m_2 as shown in Section 14.4, causes extraction to be the predominant effect as compared to conversion of the monomer. Therefore, a preliminary process design has been developed based on CO$_2$-extraction. For this purpose, a mass transfer model has been set up to determine the rate-limiting step in the extraction process. In addition, a process flow diagram, including equipment sizing has been developed. Finally, an economic evaluation has been performed to study the viability of this technique for the removal of residual monomer from latex-products.

14.5.1
Extraction Model

In order to design an extraction unit, the rate-limiting step in the extraction process has to be determined. As explained in Section 14.4 and Fig. 14.3, the film model can be used to describe extraction in the three-phase system. The overall mass transfer flux of monomer in this extraction process is given by Eq. (14)

$$N = K_{OV, POL} \cdot A_{POLW} \cdot (C_{POL} - C^*_{CO_2}) = K_{OV, POL} \cdot \left(C_{POL} - \frac{C_{CO_2}}{m_1 \cdot m_2} \right) \tag{14}$$

in which $C^*_{CO_2}$ is the concentration of MMA in the CO$_2$ phase that is in equilibrium with its concentration in the polymer phase, given by $C_{CO2}/(m_1 m_2)$. The overall mass transfer coefficient $K_{OV,POL}$ defined on the polymer phase, is a summation of the resistances in the three phases and can be obtained by rearranging Eqs. (2) to (7) and (14).

$$\frac{1}{A_{POLW} \cdot K_{OV, POL}} = \frac{1}{A_{POLW} \cdot k_{POL}} + \frac{1}{A_{POLW} \cdot k_W \cdot m_1} + \cdots$$

$$\cdots + \frac{1}{A_{WCO_2} \cdot k_{W \cdot m_1}} + \frac{1}{A_{WCO_2} \cdot k_{CO_2} \cdot m_1 \cdot m_2} \quad (15)$$

By calculating the time constant of each resistance, the largest resistance for mass transfer can be determined. For this purpose, the partition coefficients m_1 and m_2, the mass transfer coefficients k_{POL}, k_W and k_{CO_2}, as well as the interfacial surface areas A_{POLW} and A_{WCO2} have been determined in various ways as discussed below. For a more detailed description we refer to Cleven [48].

14.5.1.1 Diffusion and Mass Transfer Coefficients

In order to calculate the mass transfer coefficient as the diffusion coefficient D divided by the diffusion length Δz, the diffusion coefficients of MMA in each phase have to be known. The diffusion coefficient of MMA in the polymer particles is determined using the Vrentas-Duda model, which is based on the free-volume theory [49]. In general, the diffusion coefficient of MMA in PMMA decreases with increasing weight fraction of polymer in the particles [23]. Because of the plasticizing effect of CO_2, however, the diffusion coefficient inside the polymer particles will be significantly higher in the presence of pressurized CO_2 than at atmospheric conditions. Few examples of the enhancement in diffusivity of components in polymers in the presence of CO_2 have been reported. According to Chapman et al. [26], the diffusion coefficient of solutes in a polymer can increase by at least 6 orders of magnitude in the presence of CO_2 at subcritical pressures. In the extraction of ethylbenzene from polystyrene using scCO₂, Alsoy and Duda [25] have reported an increased diffusivity of 2–3 orders of magnitude as the weight fraction of CO_2 increases from 0 to 10%. Assuming that the diffusion coefficient of MMA in PMMA in the presence of CO_2 is three orders of magnitude larger than at atmospheric conditions, a value of approximately 10^{-9} m² s⁻¹ is obtained.

The diffusion coefficient of MMA in water has been estimated using the Wilke-Chang equation [50]:

$$D_W = \frac{1.173 \cdot 10^{-16} \cdot \Phi_w^{0.5} \cdot M_w^{0.5} \cdot T}{\mu \cdot V_{MMA}^{0.6}} \quad (16)$$

Using Eq. (16) a value of $1.9 \cdot 10^{-9}$ m² s⁻¹ at 333 K is obtained, for which the viscosity of the water is assumed to be unaffected by the presence of CO_2.

In the literature some diffusion coefficients of naphthol, naphthalene, and phenanthrene in scCO₂ are reported in the range of $2 \cdot 10^{-8}$ to $4 \cdot 10^{-8}$ m² s⁻¹ [51, 52]. The same order of magnitude is calculated for MMA in scCO₂ ($6.3 \cdot 10^{-8}$ m² s⁻¹ at 333 K) using the empirical equation proposed by Fuller and coworkers [53] based on the corresponding states principle:

$$D_{CO_2}(1 \ bar, \ 333 \ K) = \frac{0.00143 \cdot T^{1.75} \cdot 0.5 \cdot \left(\frac{1}{M_{MMA}} + 1\frac{1}{M_{CO_2}}\right)^{0.5}}{P \cdot \left[\left(\sum_v\right)_{CO_2}^{1/3} + \left(\sum_v\right)_{MMA}^{1/3}\right]^2} \quad (17)$$

First, the diffusion coefficient is determined at 0.1 MPa. Subsequently, the diffusion coefficient at higher pressures can be calculated using a function of the reduced pressure and temperature.

The diffusion length for mass transport in the polymer particles can be estimated from a worst case assumption in which the diffusion length is half the diameter of these particles (Sherwood = 2, d_p = 80 nm). The diffusion lengths in the water and CO_2 phase are estimated to be 10 μm, the characteristic diffusion length in a laminar fluid. The resulting mass transfer coefficients are given in Table 14.8.

Table 14.8 Overview of the calculated mass transfer resistances in the CO_2 extraction process of MMA from PMMA latex particles.

Phase	Resistance	Parameters			Time constant [s]
Polymer particles	$res_{POL} = \frac{1}{A_{POLW} \cdot k_{POL}}$	D_{POL} Δz_{POL} k_{POL} A_{POLW}	(m^2/s) (m) (m/s) (m^2/m^3)	10^{-9} $4 \cdot 10^{-8}$ 0.025 $1.06 \cdot 10^7$	$3.77 \cdot 10^{-6}$
Water at polymer side	$res_{W1} = \frac{1}{A_{POLW} \cdot k_W \cdot m_1}$	D_W Δz_W k_{W1} A_{POLW} m_1	(m^2/s) (m) (m/s) (m^2/m^3) $(-)$	$1.9 \cdot 10^{-9}$ $1 \cdot 10^{-5}$ $1.9 \cdot 10^{-4}$ $1.06 \cdot 10^7$ 0.067	$7.43 \cdot 10^{-3}$
Water at CO_2 side	$res_{W2} = \frac{1}{A_{WCO_2} \cdot k_W \cdot m_1}$	D_w Δz_W k_{W2} A_{WCO_2} m_1	(m^2/s) (m) (m/s) (m^2/m^3) $(-)$	$1.9 \cdot 10^{-9}$ $1 \cdot 10^{-5}$ $1.9 \cdot 10^{-4}$ 2000 0.067	39.4
CO_2	$res_{CO_2} = \frac{1}{A_{WCO_2} \cdot k_{CO_2} \cdot m_1 \cdot m_2}$	D_{CO_2} Δz_{CO_2} k_{CO_2} A_{WCO_2} m_1 m_2	(m^2/s) (m) (m/s) (m^2/m^3) $(-)$ $(-)$	$6.3 \cdot 10^{-8}$ $1 \cdot 10^{-5}$ $6.3 \cdot 10^{-3}$ 2000 0.067 5	0.240

14.5.1.2 **Partition Coefficients**

The partition coefficient m_1, which is defined as the concentration MMA in the water phase divided by the corresponding concentration in the polymer phase (Eq. 6), is given by Ballard et al. [54]. With respect to m_2, a value of 5 has been taken as representative of the experimentally determined partition coefficients shown in Section 14.4.3.2.

14.5.1.3 **Interfacial Surface Areas**

The specific surface area of the polymer particles (A_{POL}) is a function of the radius of the polymer particles (r_p), the volume of latex relative to the volume of the total extraction volume, and the weight fraction of polymer in the latex (ε). This results in a specific surface area of $1.06 \cdot 10^7 \, m^2/m^3$. Considering the specific surface area between the water and CO_2 phase (A_{WCO_2}), an industrially reasonable mass transfer area of 2000 m^2/m^3 has been taken.

  With these data, the resistance to mass transport in each phase can be calculated. According to Table 14.8, the largest resistance to mass transport appears to be situated in the water phase at the CO_2 side.

14.5.2
Process Flow Diagram, Equipment Selection, and Equipment Sizing

Fig. 14.8 shows the process design for the reduction of residual monomer from 9000 to 100 ppm with CO_2, calculated for a latex production of 30 000 tonnes per year [48]. As will be obvious from the previous section, an important issue in the process design is the selection of the extractor, for which a high interfacial mass transfer area between the water and CO_2 phase is required. Mixer-settlers are ruled out as they contain less than one single mass transfer unit. Towers with rotating stirrers cannot be used because of potential fouling, and a pulsed column has relatively high investment and operating costs and a limitedthroughput. In the current process design, a packed-bed extractor is chosen, as it provides a high mass transfer area, lacks rotating parts, and has the possibility of a higher number of mass transfer units, depending on column height.

  The latex from the polymerization reactor (333 K) is fed to a packed-bed extractor, in which it is contacted with high-pressure CO_2 in counter-current flow. The start-up pump pressurizes the CO_2 from the supply tank (5.5 MPa) to 8.0 MPa. The scCO$_2$ is conditioned by a heat exchanger (HEX1). The residual monomer is further polymerized and extracted from the polymer particles in the extractor at a pressure of 8.0 MPa. A controlled pressure relief will minimize foaming problems of the cleaned latex stream. The CO_2 and the residual monomer are separated in a gas/liquid separator by reducing the pressure from 8.0 to 3.0 MPa. As a consequence of the adiabatic CO_2 expansion the temperature will decrease also. The extracted monomer can in principle be re-used in the polymerization process. The gaseous CO_2 flow from the separation vessel is

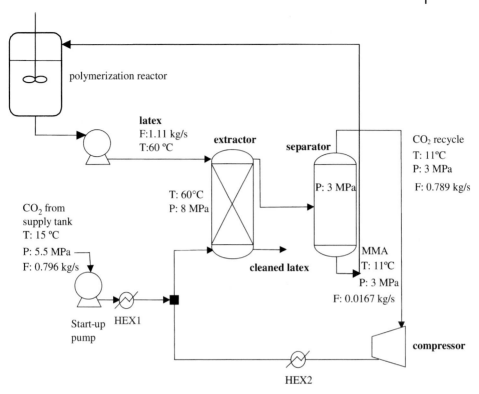

Fig. 14.8 Process flow diagram for the extraction of residual monomer from latex products using pressurized CO$_2$.

recycled to the extractor after pressurizing by the recycle compressor and cooling to 333 K by a heat exchanger (HEX2).

The different process units are sized according to general engineering rules [55] as well as by using simulations in the commercial process design package AspenPlus. It should be noted that the calculated sizes of the extractor and separator shown in Table 14.9 are relatively small as compared to the typical size of an industrial emulsion polymerization reactor (50 m^3).

In contrast to most conventional techniques for removal of residual monomer, such as temperature rise, there is considerable flexibility in the CO$_2$-based process. For example, doubling the processing capacity would only require twice the extractor volume [48]. Fig. 14.9 shows that if the permitted amount of residual monomer in the product latex is limited to 10 ppm, a less than twofold increase in the extractor volume will be needed. In addition, an increase in the partition coefficient, m_2, by a factor of 10 results in a decrease in the extractor volume of approximately 20%.

Table 14.9 Overview of purchased and installed costs per unit for the
high-pressure CO_2 extraction process of residual monomer [44].

Latex pump (stainless steel)		Extractor (vertical vessel)	
Flow (m^3/hr)	4.0	Height (m)	4.25
Feet head	2592.0	Diameter (m)	0.60
Power (kW)	12.5	Wall thickness (m)	0.024
Purchased price 2000 (USD)	10,500	Purchased price 2000 (USD)	30,000
Installed costs 2000 (USD)	21,000	Installed costs 2000 (USD)	51,000

Start up pump (centrifugal, cast iron)		Separator (vertical vessel)	
Flow (m^3/hr)	3.45	Height (m)	3.38
Feet head	995.7	Diameter (m)	1.12
Power (kW)	3.44	Wall thickness (m)	0.045
Purchased price 2000 (USD)	19,100	Purchased price 2000 (USD)	61,200
Installed costs 2000 (USD)	38,200	Installed costs 2000 (USD)	104,040

Heat exchanger HEX1 (shellCS/tubeSS)		Heat exchanger HEX2 (shellCS/tubeSS)	
Heating duty (kW)	80.5	Heating duty (kW)	(–) 51.57
U (W/m^{2}K)	850.0	U (W/m^{2}K)	850.0
T_m (K)	69.12	T_m (K)	57.00
Required area	1.4	Required area	4.52
Purchased price 2000 (USD)	3,360	Purchased price 2000 (USD)	4,150
Installed costs 2000 (USD)	5,712	Installed costs 2000 (USD)	7,055

Compressor (screw)	
Total power (kW)	53.66
Purchased price 2000 (USD)	37,400
Installed costs 2000 (USD)	56,100

14.5.3
Economic Evaluation

Based on the described process design, an economic feasibility study has been
performed [48]. The total processing costs are calculated from the costs of the
process equipment, the chemicals, and the utilities required. In addition, the
start-up costs (5% of total fixed capital), maintenance (5% of inside battery limits, a standard term denoting the part of the plant considered), plant overheads
(80% of labor costs and 20% maintenance), and capital charges (10 year depre-
ciation) are included in the calculations. As the infrastructure is already present
at the current polymerization plant, no additional costs are assumed to be in-
curred outside battery limits. The operators on duty are assumed to handle the
extraction process: therefore no additional labor costs are made. The purchased
costs of the process equipment shown in Table 14.9 were determined using the
cost estimation program Chemical Process Equipment Individual Equipment
Costing [55].

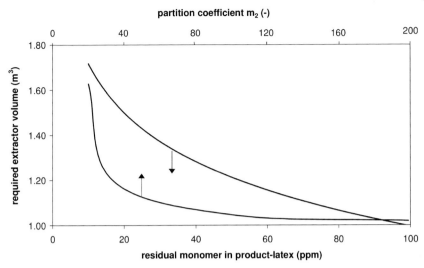

Fig. 14.9 Sensitivity of the required extractor volume to the ppm level of monomer allowed in the product latex and to the partition coefficient m_2.

The calculated total capital required is approximately 300 000 USD. The calculated processing cost per tonne latex is approximately 4.7 USD. The gas/liquid separator appears to have the largest fixed cost. Optimizing the process conditions in the separator results in a processing cost of approximately 4.5 USD/ton latex. Increasing the processing capacity to 63 000 tonnes latex/year and limiting the permitted amount of residual monomer in the product latex to 60 ppm decreases the processing cost to 3.1 USD/tonne latex. Taking into consideration the current production price of latex (approximately 800 USD/tonne), the additional processing cost for the new process is only 0.6% of the production costs. Comparison with the processing cost using conventional techniques for reducing residual monomer is rather difficult, because no data about the economics of the conventional techniques are available in the open literature. Araujo and coworkers give an extensive review of techniques for reducing residual monomer content in polymers from an industrial point of view [56]. The technique that can be employed is influenced by the polymer application and required polymer quality. Moreover, the costs for the removal of residual monomer depend significantly on the properties of the latex produced. Taking into consideration the fact that the conventional techniques are hardly able to meet more stringent requirements (e.g., 10 ppm residual monomer), the novel technique based on pressurized CO_2 is expected to become a viable alternative in the near future.

14.6
Conclusion and Future Outlook

In this chapter, the potential of a novel process for the reduction of residual monomer from latex particles based on high-pressure CO_2 has been evaluated. Typically, the method comprises a counterflow process in which part of the residual monomer is converted by the increased diffusion inside the polymer particles due to the swelling by CO_2. Moreover, the amount of residual monomer is further reduced by the extraction capacity of pressurized CO_2. Pulsed electron beam experiments have been performed to study the effect of CO_2 on the polymerization reaction. In addition, the extraction capacity of CO_2 has been measured in a laboratory scale high-pressure extraction unit and modeled using the Peng-Robinson equation of state. The results have shown that the CO_2 extraction of MMA is the predominant effect as compared to the enhanced polymerization due to plasticization. From a mass transfer model, the greatest resistance to mass transfer in the extraction process appears to be situated in the water phase near the CO_2 interface because of a relatively small interfacial area. Finally, a viability study based on a preliminary process design, including equipment sizing and economic evaluation, has shown that the removal of residual monomer from latex products using high-pressure CO_2 is in principle both technically and economically feasible. Moreover, there appears to be significant flexibility in the CO_2-based process as compared to the existing methods in terms of processing capacity, extraction capacity, and the amount of residual monomer left in the polymer. In principle it is possible to extend the CO_2-extraction to bulk products such as polymer films and pellets. In contrast to the conventional techniques, monomer reduction using CO_2 technology is capable of meeting future, more stringent requirements for residual monomer levels in polymer products.

Acknowledgments

The authors thank the technical staff of the Department of Chemical Engineering and Chemistry, especially Chris Luyk, for the construction of the high-pressure extraction unit. The students Rob Wering and Mascha Cleven are acknowledged for their contribution to this chapter. Part of this work was supported by a grant from the Foundation of Emulsion Polymerization (SEP).

Notation

λ_{ij}	Interaction parameter	[–]
ρ_{CO_2}	Molar density of CO_2	[kg m^{-3} mol^{-1}]
ρ_W	Molar density of water	[kg m^{-3} mol^{-1}]
ω	Acentric factor	[–]

Δz_{CO_2}	Characteristic diffusion length in CO_2 phase	[m]
Δz_{POL}	Characteristic diffusion length in polymer phase	[m]
Δz_W	Characteristic diffusion length in water phase	[m]
[M]	Monomer concentration in polymer particles	[mol m^{-3}]
a	Temperature-dependent attractive parameter	[Nm^4 mol^{-2}]
A_{POLW}	Specific surface area between polymer and water phase	[m^2/m^3]
A_{WCO_2}	Specific surface area between water and CO_2 phase	[m^2/m^3]
b	Van der Waals hard sphere volume parameter	[m^3 mol^{-1}]
C_{2Wi}	Monomer concentration in water phase at CO_2 interface	[kg m^{-3}]
C_{CO_2}	Monomer concentration in bulk CO_2 phase	[kg m^{-3}]
C_{CO_2i}	Monomer concentration in CO_2 phase at water interface	[kg m^{-3}]
C_{POL}	Monomer concentration in bulk polymer phase	[kg m^{-3}]
C_{POLi}	Monomer concentration in polymer phase at water interface	[kg m^{-3}]
C_W	Monomer concentration in bulk water phase	[kg m^{-3}]
$C_{W,0}$	Initial monomer concentration in water phase	[kg m^{-3}]
C_{Wi}	Monomer concentration in water phase at polymer interface	[kg m^{-3}]
D_{CO_2}	Monomer diffusion coefficient in CO_2 phase	[m^2 s^{-1}]
D_{POL}	Monomer diffusion coefficient in polymer phase	[m^2 s^{-1}]
D_W	Monomer diffusion coefficient in water phase	[m^2 s^{-1}]
k_{CO_2}	Mass transfer coefficient in CO_2 phase	[m s^{-1}]
k_{ij}	Interaction parameter	[–]
$K_{OV,POL}$	Overall mass transfer coefficient	[m s^{-1}]
k_p	Propagation rate constant	[m^3 mol^{-1} s^{-1}]
k_{POL}	Mass transfer coefficient in polymer phase	[m s^{-1}]
k_W	Mass transfer coefficient in water phase	[m s^{-1}]
L_i	Chain length	[–]
m_1	Monomer partition coefficient between polymer and water phase	[–]
m_2	Monomer partition coefficient between CO_2 and water phase	[–]
m_2^*	Monomer partition coefficient between CO_2 and water phase, based on mole fractions	[–]
M_W	Weight average molecular weight of a polymer	[g mol^{-1}]
N	Monomer flux	[kg m^{-3} s^{-1}]
P	Pressure	[MPa]
P_A	Attractive pressure	[MPa]
P_c	Critical pressure	[MPa]
P_R	Repulsive pressure	[MPa]
T	Temperature	[K]
T_c	Critical temperature	[K]
t_p	Time between two electron beam pulses	[s]
V_{CO_2}	Volume of CO_2 phase	[m^3]
V_W	Volume of water phase	[m^3]

References

1 B. W. Brooks, *Ind. Eng. Chem. Res.*, **1997**, *36*, 1158.

2 G. W. Poehlein, *Reaction engineering for emulsion polymerization*, in *Polymer Dispersions: Principles and Applications*, J. M. Asua (ed.), Kluwer Academic Publishers, **1997**.

3 J. A. Maxwell, E. M. F. J. Verdurmen, A. L. German, *Makromolecular Chemistry*, **1992**, *193*, 2677.

4 M. J. J . Mayer, J. Meuldijk, D. Thoenes, *Chem. Eng. Sci.*, **1995**, *50*, 3329.

5 S. Chen, S. Lee, *Macromolecules*, **1992**, *25*, 1530.

6 G. C. Overbeek, Y. W. Smak, *EP 521620/ US Patent 5292660*, **1993**.

7 S. M. Englund, *Chem. Eng. Progr.*, **1981**, 55.

8 D. H. Everett, M. E. Gultepe, M. C. Wilkinson, *J. Colloid Inter. Sci.*, **1979**, *71*, 336.

9 M. A. Barrett, D. H. Everett, M. E. Gultepe, *Polymer Colloids II*, Plenum Press, New York, **1980**, 313.

10 D. H. Everett, M. E. Gultepe, *Preprints of Polymer Colloids*, NATO advanced Study Institute, **1975**.

11 J. Hearn, M. C. Wilkinson, A. R. Goodall, P. Cope, *Polymer Colloids II*, Plenum Press, New York, **1980**, 379.

12 M. A. Taylor, *EP 0158523/US Patent 4529753*, **1985**.

13 P. Kelly, *US Patent 5430127*, **1995**.

14 M. A. Olivares, F. J. Archundia, F. L. Ramos, *US Patent 5886140*, **1999**.

15 J. T. F. Keurentjes, E. Mills, F. J. Kruizenga, *Engineering Foundation, New directions in separation technology*, Noordwijkerhout, **1993**.

16 S. M. Ahmed, M. S. El-Aasser, G. H. Powli, G. W. Poehlein, J. W. Vanderhof, *J. Colloid Interface Sci.*, **1980**, *73*, 388.

17 M. C. Wilkinson, J. Hearn, P. A. Steward, *Adv. Colloid Interface Sci.*, **1999**, *81*, 77.

18 J. L. Kendall, D. A. Canelas, J. L. Young, J. M. DeSimone, *Chem. Rev.*, **1999**, *99*, 543.

19 M. F. Kemmere, M. A. van Schilt, M. H. W. Cleven, A. M. van Herk, J. T. F. Keurentjes, *Ind. Eng. Chem. Res.*, **2002**, *41* (11), 2617.

20 A. M. van Herk, H. de Brouwer, B. G. Manders, L. H. Luthjens, M. L. Hom, A. Hummel, *Macromolecules*, **1996**, *29*, 1027.

21 J. F. Wishart, R. Vaneldik, *Rev. Scientific Instruments*, 1992, *63*, 3224.

22 O. F. Olaj, I. Bitai, F. Hinkelmann, *Makromol. Chem.*, **1987**, *188*, 1689.

23 R. G. Gilbert, *Emulsion Polymerization, A Mechanistic Approach*, Academic Press, **1995**.

24 M. A. Quadir, R. Snook, R. G. Gilbert, J. M. DeSimone, *Macromolecules*, **1997**, *30*, 6015.

25 S. Alsoy, J. L. Duda, *AIChE J.*, **1998**, *44*, 582.

26 B. R. Chapman, C. R. Gochanour, M. E. Paulaitis, *Macromolecules*, **1996**, *29*, 5635.

27 D. Peng, D. B. Robinson, *Ind. Eng. Chem. Fund.* **1976**, *15*, 59.

28 A. Z. Panagiotopolous, R. C. Reid, *ACS Div. Chem., Prepr.*, **1985**, *30*, 46.

29 R. Stryjek, J. H. Vera, *Can. J. Chem. Eng.* **1986**, *64*, 820.

30 M. L. Michelsen, H. Kistenmacher, *Fluid Phase Equilibria* **1990**, *58*, 229.

31 G. A. Melhem, R. Saini, C. F. Leibovici, *Proceedings of the 2nd International Symposium on Supercritical Fluids*, Boston, **1991**, 475.

32 P. M. Mathias, H. C. Klotz, J. M. Prausnitz, *Fluid Phase Equilibria*, **1991**, *67*, 31.

33 D. S. H. Wong, S. I. Sandler, *AIChE Journal*, **1992**, *38*, 671.

34 O. Pfohl, S. Petkov, G. Brunner, *Phase Equilibria for Windows*, V2.085 Release December 2000, Hamburg, **2000**.

35 D. Bush, Equation of State for Windows 95, version 1.01.14, Georgia Institute of Technology, Atlanta, **1994**.

36 A. Z. Panagiotopoulos, R. C. Reid, *Fluid Phase Equilibria*, **1986**, *29*, 525.

37 S. Winkler, K. Stephan, *Fluid Phase Equilibria*, **1997**, *137*, 247.

38 T. Adrian, M. Wendland, H. Hasse, G. Maurer, *J. Supercritical Fluids*, **1998**, *12*, 185.

39 M. F. Kemmere, M. H. W. Cleven, M. A. van Schilt, J. T. F. Keurentjes, *Chem. Eng. Sci.*, **2002**, *57*, 3929.

40 M. A. Jacobs, *Measurement and modeling of thermodynamic properties for processing of polymers in supercritical fluids*, PhD thesis, Technische Universiteit Eindhoven, ISBN 90-386-2596-0, Eindhoven, **2004**.

41 R. Wiebe, V. L. Gaddy, *J. Am. Chem. Soc.*, **1940**, *62*, 815.

42 R. Wiebe, V. L. Gaddy, *J. Am. Chem. Soc.*, **1941**, *63*, 475.

43 M. B. King, A. Mubarak, J. D. Kim, T. R. Bott, *J. Supercritical Fluids*, **1992**, *5*, 296.

44 M. Lora, M. A. McHugh, *Fluid Phase Equilibria*, **1999**, 157, 285.

45 C. T. Kauter, *Ullmann's Enzyklopädie der Technische Chemie*, VCH, Weinheim, **1960**, *12*, 392.

46 S. B. Kiselev, *Fluid Phase Equilibria*, **1998**, *147*, 2839.

47 C. McCabe, S. B. Kiselev, *Ind. Eng. Chem. Res.*, **2004**, *43*, 2839.

48 M. H. W. Cleven, *Removal of residual monomer from latex products using supercritical carbon dioxide*, Graduate Report in Process and Product Design, Eindhoven University of Technology, **2001**, ISBN 90-444-0083-5.

49 A. Faldi, M. Tirrell, *Macromolecules*, **1994**, *27*, 4184.

50 C. R. Wilke, P. Chang, *AIChE J.*, **1955**, *1*, 264.

51 K. Abaroudi, F. Trabels, B. Calloud-Gabriel, F. Recasens, *Ind. Eng. Chem. Res.*, **1999**, *38*, 3505.

52 G. Madras, C. Thibaud, C. Erkey, A. Akgerman, *AIChE J.*, **1994**, *40*, 777.

53 E. N. Fuller, P. D. Schettler, J. C. Giddings, *Ind. Eng. Chem. Res.*, **1966**, *58(5)*, 18.

54 M. J. Ballard, D. H. Napper, R. G. Gilbert, *J. Pol. Sci., Polymer Chemistry Edition*, **1984**, *22*, 3225.

55 S. M. Walas, *Chemical Process Equipment, Selection and Design*, Butterworths-Heinemann, **1990**.

56 P. H. H. Araujo, C. Sayer, R. Giudici, *Polym. Eng. Sci.*, **2002**, *42(7)*, 1442.

Subject Index

a

accumulation term 93–94
acoustic
– compression 45
– method 96
– time 45
activation parameter 72
activation volume 65, 75
amine 278
amphiphilic 199
analytical technique 82
azo-bis-isobutyronitrile 61

b

β-scission 192
back leakage flow 257
binary interaction parameter 161
biological degradation 305
biomaterials 230
blocking agent 286
boundary layer 45
branch-on-branch structure 180
branching 22
Brookhart catalyst 163

c

calibration 48
– probe 49, 93, 96
calorimetric method 90
calorimetric signal 96
carboxylation 277
carboxylic acid 278
– end group 293
catalyst
– decay 176
– ethylene complex 183
– methyltrioxorhenium (MTO) 166
– palladium based 165
– solubility 163
catalytic CTA 73

chain mobility 295
chain walking 177, 179, 181
chain transfer 73, 141
chemical fluid deposition 252
chemical shift 87
chemically initiated polymerization 70
chlorofluorocarbon (CFC) 189
closed microcellular 244
cloud-point measurement 158
cloud-point pressure 158
co-solvent 21
coating 139
coefficient 324
– decomposition rate 57
– free radical polymerization 55
– heat transfer 50
– over all heat transfer 94
– partition 315 f., 319, 326
– propagation rate 62
– termination rate 69
commercial latexes 304
commercial polyolefin production 184
comprehensive model 108
compressed fluid sedimentation
   polymerization 240
computational fluid dynamics (CFD) 39
conduction 44
conformal metal film 247
continuous process 184
continuous reactor 129
continuous precipitation polymerization
   of fluoropolymer 240
convection 44
conventional chain transfer agent
   (CTA) 73
conversion of residual monomer 305
copolymerization 67
– of ethylene and methyl acrylate 180
– of norbornene and ethylene 165

Supercritical Carbon Dioxide: in Polymer Reaction Engineering
Edited by Maartje F. Kemmere and Thierry Meyer
Copyright © 2005 WILEY-VCH Verlag GmbH & Co. KGaA, Weinheim
ISBN: 3-527-31092-4

copolymer   20
cost estimation   328
CO_2-phile:hydrophile balance   152
CO_2-philic catalyst   164
CO_2-philicity   143
crosslinking   277
crystallinity   177
crystallization   208, 210

d
D_1 values   281
D_2 values   281
decomposition rate coefficient   57
density relaxation   45
determination of reaction rate   171
development trajectory   11
diacyl peroxide   58
dichloromethane   169
diffusion   213, 215, 324
– coefficient   279
– limitation   112
– time   262
diffusivity   307
1,1-dihydroperfluorooctyl acrylate
   (FOA)   197
diketene   289
– acetone adduct   289
Dimroth-Reichardt empirical parameter
   61
dipole moment   8
discoloration   284
dispersion   149
– polymerization   97, 106, 240
divergence   50
dosing term   93
"dry" chemical mechanical planarization
   251
dyeing   229
dynamic viscosity   37

e
economic evaluation   328
economically feasible   11
emulsion polymerization   139
end group   277, 281
energy balance   91
enhanced polymerization   307
enthalpy of polymerization   98
environmental   275
1,2-epoxybutane   286, 293
equations of state   27
equipment
– selection   326

– sizing   326
etching   250
ethylene-propylene-diene monomer
   copolymer (EPDM)   157
external film   50
extraction   316
– capacity   310
– model   323
extruder   255 f.
extrusion   220, 255

f
film model   310
flexibility   147
fluorinated polymer   108
fluorinated surfactant   143
fluoroacrylate   153
fluoroalkyl acrylate   197
fluoropolymer   189
flux   311
foaming   215 f.
focal plane array   226
Fourier transform infrared spectroscopy
   (FTIR)   82
free-volume   16, 147, 288
– theory   113
– imaging   226
fully filled zone   257

g
gas chromatography   99
gel-effect   75
generally regarded as safe (GRAS)   12
glass transition temperature   207
grafting   276
green chemistry   5

h
heat and mass transfer   53
heat balance   92
heat capacity   37, 95
heat flow   48
– calorimetry   91
heat generation   96
– rate   99
heat of polymerization   93
heat transfer   39, 45
– coefficient   50
– over all   94
– resistance   48
heterogeneous polymerization   105, 139
high solid content   101
high-pressure cell   87

high-pressure probe 87
homogeneous modification 292
homogeneous phase 55
homogeneous polymerization 195
hybrid 202
– solvents 6
hydrodynamic behavior 39
hydrolytic stability 293, 296

i
implicit penultimate unit effect 69
impregnation 228, 273
in-line FT-NIR spectroscopy 56, 70
incorporation of MA 181
initiation 57
interfacial surface areas 326
interfacial tension 211–212, 263
intermediate reactivity 298
internal film 49
interphase transport 116
intrinsic viscosity 296
inverse emulsion polymerization 139
ionomer 195
IR spectroscopy 206
isothermal compressibility 39

l
lower critical solution temperature 16
laser-Doppler velocimetry 39
late transition metal catalyst 163
Lee-Kessler-Plöcker 171
legislation on levels of residual monomer 307
Lewis acid 146
Lewis base 146
life-cycle assessment 5
limitations 297
lithography 249
local concentration 68
– monomer 65–66
loop reactor 184
low solubility 170
low viscosity 218

m
macromonomer 97
mass transfer coefficient 323
material
– mesoporous 244
– nanoporous 244
– porous 243
measuring monomer partition coefficient 313

melt seal 258
melt stability 286, 289
methyl methacrylate (MMA) 114
– dispersion polymerization 115
methyltrioxorhenium (MTO) catalyst 166
microelectronics 249
mixing 39, 53
– rule 313
modeling 27
modification 275
molecular weight 296
– distribution 107
monitor reaction 97
– rate 175
monitoring 81
– technique 82
monomer 68
MP2 (second order Møller-Plesset perturbation theory) calculations 144
multidentate interactions 147
multiphase systems 89

n
Nafion® 192
nanocomposite 247
nanoparticle 247
nanoscale material 247
nanowire 248
neutral catalyst 165
NMR spectroscopy 85, 87
non-fluorous CO_2-phile 144
non-linear optical 230
Nusselt 52
– number 48

o
on-line measurement 100
open nanoporous filament 244
optimal blocking agent 292
ordered mesoporous material 244
organic solvent replacement 3
overall heat transfer coefficient 94

p
PA-6 granules 281
palladium-based catalyst 165
Panagiotopoulos-Reid (PR) mixing rule 320
partition coefficient 315 f., 326
partitioning 215
penetration 288
Peng-Robinson 171
– equation of state (eos) 311, 320

perfluorodecaline 168
peroxyesters 58
perturbation theory 28–29
pervaporation 306
pharmaceutical material 230
phase
– behavior 15, 17, 158, 206, 311
– boundary 89
– separation 224–225
photooxidation 200
photoresist drying 250
piston effect 45
plasticization 10, 207
polar compound 279
polarity 19
polarizability 8
polyamide 277
poly(butylene terephthalate) 292
polydispersity 16
polymer
– blending 262
– blend 222 f., 264, 276
– solubility 19
polymerization 149, 276
– chemically initiated 70
– compressed fluid sedimentation 240
– continuous precipitation 240
– dispersion 97, 106, 240
– emulsion 139
– enhanced 307
– enthalpy of 98
– free-radical 55
– heterogeneous 105, 139
– homogeneous 195
– inverse emulsion 139
– MMA dispersion 115
– of 1-hexene 168
– precipitation 107, 149
– pulsed electro-beam 307
– pulsed laser initiated 62
– rate 148, 176
– ring-opening 166
– single phase 56
– VDF precipitation 124
polytetrafluorethylene (PTFE) 202
porous material 243
– by chemical syntheses 245
– by supercritical fluid (SCF)
  processing 243
power compensation
– calorimeter 90
– calorimetry 90
power consumption 43

power number 43
Prandtl 48
precipitation polymerization 107, 149
prediction of partition coefficient 319
process control 90
process design 323
process flow diagram 326
process intensification 4
process optimization 90
processing cost 329
propagation rate coefficient 62
propagation reaction 94
pulsed electron beam polymerization 307
pulsed-laser-initiated polymerization 62
pump number 41

q
quadrupole dipole 144
quadrupole moment 8, 160

r
radiation 44
Raman 84
– spectroscopy 208
rate coefficients for free-radical
  polymerization 55
rate of polymerization 148, 176
reaction
– calorimeter 46
– calorimetry 91
– loci 115
– rate 94
reactivity 295
– ratios 67
reflectometry 89
reproducibility 315
residual monomer 303
– reduction 304
– removal 305
resist deposition 249
reversed pitch element 259
review 275
Reynolds 48
rheology 218
ring-opening metathesis polymerization
  of norbornene 166

s
safe process 81
Sanchez-Lacombe 109
scope 297
screw profile 258
sealing 258–259

segmental diffusion   71
selective blocking   278
self-interactions   147
short chain branching   177
silicone surfactants   152
single-phase polymerization   56
single pulse pulsed laser polymerization
   (SP-PLP) technique   69
size exclusion chromatography   99
slow solubilization   170
small blocking agent   298
solubility   206 f., 307
solvent recovery   1, 157
solvent-free process   2
sorption   10, 279
– of CO_2 256
sound waves   89
spectroscopic analyses   288
spectroscopic method   82
spectroscopy   226
spin coating   249
stabilizer   86, 240
static mixer   259, 262
Statistical Associating Fluid Theory
   (SAFT) equation of state (eos)   161,
   171
status   12
steam stripping   305
subcritical   280
succinic anhydride   279, 283
supercritical   15, 189, 280
– chromatography   6
– fluid chromatography   88
– fluid extraction   6
– modification   275
– water   3
supported catalyst   165
surface modification   243
surfactant   142, 240
sustainability   4
swelling   10, 273
syndiotactic polystyrene   210

t
Teflon® AF   193
telomerization   196
templating   246
termination   69
– rate coefficient   69
ternary cloud point   162
terpolymerization   56
tetrafluoroethylene   191
thermal conductivity   37
thermal monomer conversion   99
topology of synthesized polyethylene   177
total capital required   329
translational diffusion process   71
transport property   37
transportation of the blocking agent   285
turbidimetry   86
turbidity   86
turbulent kinetic energy   42
turbulent regime   41

u
upper critical solution temperature   16
ultrasonic sensor   90, 100
ultraviolet-visible (UV-Vis)   85

v
validation   315
vinylidene fluoride (VDF)   115
– precipitation polymerization   124
velocity   41
vibrational spectroscopy   85
view cell   83
viscometer   219
viscosity   218–220
– reduction   255
volatile organic compound   1

w
Weber number   260
Wilson plot   49
– analyses   48